evening primrose 113

mallows 114–115

rockroses 116

Mignonette and Weld 117

cabbages and cresses 118–140

cabbages and cresses 118–140

Bastard Toadflax 141

Moschatel 141

sea-lavenders 142–143

Thrift 144

Sea Heath 144

sorrels, bistorts 145–149

docks 150–152

sundews 153

campions, stitchworts, chickweeds, mouse-ears, pearlworts, spurries etc. 154–171

Fat Hen 172–173

oraches 174–175

glassworts 176

Sea Beet 177

balsams 178–179

Jacob's Ladder 179

Springbeauty, Blinks 180

Chickweed Wintergreen 181

Cyclamen 181

primulas 182–183

Brookweed 184

pimpernels 185–186

yellow loosestrifes 187

D

centauries 195 and 198

For Anne,

'the flower lady'

First published in 2013

Copyright © 2013 text by Simon Harrap
Copyright © 2013 photographs by Simon Harrap (except Cornish Heath on p. 399 Deborah Ward)
Copyright © 2013 maps by Botanical Society of Great Britain and Ireland

The right of Simon Harrap to be identified as the author of this work has been asserted by him in
accordance with the Copyright, Designs and Patents Act 1988.

Bloomsbury Publishing Plc, 50 Bedford Square, London WC1B 3DP
www.bloomsbury.com

Bloomsbury Publishing, London, New Delhi, New York and Sydney

A CIP catalogue record for this book is available from the British Library

Commissioning editor: Nigel Redman
Design by Simon Harrap

ISBN (print) 978-1-4081-1360-8
ISBN (ePub) 978-1-4081-8987-0

Printed in China by C&C Offset Printing Co Ltd.

10 9 8 7 6 5 4 3 2 1

HARRAP'S
WILD
FLOWERS

A Field Guide to the Wild Flowers of Britain & Ireland

SIMON HARRAP

B L O O M S B U R Y

LONDON · NEW DELHI · NEW YORK · SYDNEY

CONTENTS

ACKNOWLEDGEMENTS

I have had much help in the acquisition of the photographs for this guide, notably from Gillian Beckett, Alec Bull, Francis Farrow, Jeremy Halls, Bob Leaney, Tony Leech, Stephen Martin, Richard Porter, Craig Robson, Frances Schumann, Darrell Stevens and Robin Stevenson in Norfolk. Further afield, Roy Frost, Peter Garner, Geoffrey Halliday, David Hawker, John Knowler, Martin Rand, Martin Sandford, Trevor Taylor, Stephen Westerberg and Bryan Yorke were most helpful, and Alex Lockton facilitated access to the BSBI database. Above all, I would like to thank Bob Ellis, the BSBI recorder for East Norfolk, for his help and advice, and for the answers to all too many questions that began, 'Where can I find...'. For allowing access to their land, I thank Nick and Sue Emmett (Norfolk) and Tony Grey (Herefordshire). A big thanks to Richard and Julia Barnes at Tehidy Holiday Park, Cornwall, for their hospitality and enthusiasm, and to Charlie and Helen Thurston and Sarah and Teucer Wilson for their patience on 'holidays'. Nigel Redman at Bloomsbury was as supportive (and patient) as ever, and Julie Dando of Fluke Art advised on design matters.

In preparing the text I have made extensive use of the literature, but three sources of information stand out as invaluable: the *New Atlas of the British & Irish Flora* for data on habitat and distribution, *The Vegetative Key to the British Flora*, which is full of new insights into plant identification, and the *New Flora of the British Isles* by Clive Stace: perceptive, scholarly and meticulously accurate, I can only begin to appreciate just how much work the 1,232 pages of the *New Flora* represent. For permission to use the maps I thank the Botanical Society of Great Britain and Ireland (BSBI), and especially Tom Humphrey for his expertise and hard work in actually outputting the maps.

INTRODUCTION

Aim of this book

Harrap's Wild Flowers is designed to be a portable, easy-to-use field guide to the wild flowers of the British Isles. The text has, as far as possible, been kept free of jargon and abbreviations, while at the same time providing the information necessary to accurately name the plant. The photographs have been selected to give a good impression of plants in their natural habitat, and to illustrate their identification features. It is also hoped that the photographs are attractive, highlighting the intricacy and beauty of wild flowers.

Coverage

To include every species of flower to be found growing wild in Britain, be they native, ancient introductions or more recent garden escapes, and to cover them in sufficient detail to allow confident identification, would have resulted in an unmanageably large book. I have therefore concentrated on the plants that are likely to be most frequently encountered; this naturally results in a bias towards the more densely populated regions of Britain and the places and habitats that we tend to visit on days out or holidays. I have excluded a large proportion of the uncommon upland plants that are only found at higher altitudes on a few of the mountains of N Wales, N England and Scotland, as well as most of the species confined within the British Isles to the Isles of Scilly, Northern Isles and W Ireland, and all those confined to the Channel Islands. I have also excluded most rarities, many of which are extremely unlikely to be seen without a hand-held GPS, 8-figure grid reference and detailed directions – those rarities that I have included are conspicuous plants found on much-visited reserves, or iconic flowers that regularly receive a mention in the natural history and conservation literature.

Grasses, sedges, rushes, ferns, horsetails and clubmosses are also excluded. These are fascinating and beautiful plants, but they often require detailed study in order to arrive at a correct identification and are best tackled once you are familiar with the easier groups and the botanical jargon. They have their own specialist literature. Also excluded are submerged aquatics – plants that grow and flower underwater; again, these are often hard to identify correctly without specialist knowledge and experience, and have to be specifically looked for.

Sequence of species, and English and scientific names

The sequence of species mostly follows the *New Flora of the British Isles* (3rd edition, 2010). To make identification easier, however, all the trees and shrubs are grouped together, as well as most of the water plants (those species that are always or almost always found in water, either free-floating or rooted to the bottom). For reasons of space, and to facilitate easy comparisons, some species and genera have been moved around within each family, and some families that are represented by a single species have had to be slotted in where space allowed. English and scientific names follow the *New Flora*, although I have reduced the use of hyphens.

How to identify wild flowers

Identifying flowers requires no special skills, merely the ability to look carefully at the plant, make a note of its features, and then match these more or less exactly to the text and illustrations in the book. I have reduced the number of specialist botanical terms to a minimum, but some cannot be avoided, so the first step is to study the **Glossary** (pp. 9–14) and familiarise yourself with the jargon. It is useful to study a variety of blooms from your garden or some common flowers from your nearest patch of countryside: pull them apart and really get to know the various parts of a flower.

Botanists do face particular challenges. Birds, butterflies and other insects do not vary much in their dimensions – every Blue Tit that you see is pretty much the same size and shape. Accordingly, size and shape are very useful clues in the identification process for birdwatchers and entomologists. By contrast, you will never see two oak trees that are exactly the same shape, and as an oak grows it changes its size and shape all the time, from a tiny seedling a few centimetres tall to a mature tree of perhaps 35 metres. In another source of variation, plants may differ greatly in stature depending on the conditions: rich or poor soils, wet or dry weather, sunny or shady sites, and they may differ again if they re-grow after being cut or grazed off. In short, size and shape are of limited value in identifying plants. Instead, botanists tend to use finer details: the shape of the leaves and stem, the number of petals and sepals, the number of stigmas or stamens, and the presence or absence of hairs. To observe these fine details a 10× hand lens is an invaluable tool. A hand lens is a small portable magnifying glass that swivels in and out of a protective cover, and good quality lenses are available for around £15–£30.

In general, it is best to carry the field guide with you – take the book to the plant rather than the plant to the book. With the field guide to hand, you will know what features of the plant are important for identification and can check them on the spot. If you take bits of plant home to look at later they will usually have wilted or shrivelled, and inevitably you will not have the right part of the plant to confirm identification. Photographs also suffer from the same limitations; there is a good chance that you will not have photographed the details that you need for identification.

There is no short cut to flower identification – the ability to recognise several hundred species of wild flower only comes with experience. It helps greatly, however, to learn to recognise the major families – the pea family, mint family, carrot family etc. – and use the **Illustrated Index** on the inside covers. Beginners will inevitably spend a lot of time flicking through the book, trying to match their plant to the photographs, and it may be frustrating at times, but practice is the key to success. Practise naming the plants you see – compile a list of wild plants for your garden, your local park, dog-walking route or any other area that you visit regularly. Keep the list up to date; annually, monthly or even weekly. Mentally name each species as you pass it. When you have time, re-check the plants against the book and note again the salient features. Remind yourself to which family each plant belongs. In this way you will become very familiar with a range of plants and soon recognise many of the families; you will then know just where to look when you encounter a new species.

Most botanical textbooks use keys as an aid to identification. A key is a series of questions, each having two possible answers that in turn lead to further questions. By working through a key all other possible species are eliminated until you get to the correct identification. Keys are especially useful in the larger groups, where there are a lot of similar species, and a well-constructed key highlights the diagnostic features one by one. On the other hand, keys tend to use a lot of technical terms and often require fine judgements to be made in order to answer the questions correctly, which in turn demands a great deal of experience: it has been said that keys work best when you *already know the answer!* It should also be remembered that keys were devised in an era when colour printing was either non-existent or expensive, and thus colour illustrations of plants were few and far between. I have chosen not to use keys, and hope that the many photographs in this guide will do the job as well, if not better.

Wild plants and the law

It is not illegal, despite popular belief to the contrary, to pick wild flowers. Under English Common Law (and specifically noted in the Theft Act) there is a right to gather the 'Four Fs' – flowers, fruit, foliage and fungi – as long as they are from wild plants, you have a right to be on the land (e.g. it is a road, a footpath or some other area with public access, but see below), and it is not for commercial gain. This right is modified, however, by a variety of confusing legislation. Most importantly, for some species (and for some sites) special protection is required.

Under the Wildlife and Countryside Act (1981) it is illegal to uproot any wild plant without the landowner's (or the occupier's) permission. Furthermore, some of the rarest and most vulnerable species are listed in 'Schedule 8' and it is illegal to take *any part of them* (the few Schedule 8 species included in this book are indicated by 'Sch 8'). On SSSIs (Sites of Special Scientific Interest) and some nature reserves some or all of the plants present may be specially protected and there may even be a catch-all ban on causing any damage to any wild plant, although in practice you can still, for example, walk on the grass! On land where access is made possible under the Countryside and Rights of Way Act (2000), there is no right to gather anything, and specific byelaws forbidding the picking of wild flowers may be in operation in some places (e.g. National Trust properties and some nature reserves).

Faced with this confusing situation, first establish that you are not dealing with a specially protected species or a protected site. I would then advocate a common sense approach. If a species is obviously abundant, there is no harm in picking a flower or leaf for closer examination. Indeed, this is a good thing to do, allowing intimate contact with our flora and a real appreciation of the finer details, of the feel and the smell of the plant. Children, in particular, should be encouraged to really get to know common flowers in this way. But, the scarcer the plant is in any one place, or if it is known it to be uncommon or even rare, the more reluctant I am to pick it, and if there are only a few, or even just one, I would always leave it untouched. Be pragmatic. Will you, for example, cause more damage to a plant and its environment by picking a single bloom from a low-growing species for examination, or by lying flat on the ground in order to examine it?

In the past, many people made collections of pressed wild flowers, and the official botanical recorders for each county may still ask for a 'voucher' specimen in order to confirm new and interesting records, especially if a species is hard to identify. However, · pressed flowers lose much of the attractiveness of the living plant and some species (e.g. orchids) are extremely difficult to press successfully. Photographs are now a much more popular way of recording what has been seen, and digital cameras capture colours remarkably well. Careless photography can cause much more damage than picking flowers, however, and some prize flowers at popular sites may be surrounded by heavily trampled ground where admirers have knelt (or lain) to take pictures. This not only looks unattractive, but may also result in the next generation of seedlings being destroyed.

Conservation and recording

Wild flowers can be found everywhere, from pavement cracks to the tops of the highest mountains. Our wild flowers are under threat, however, with changes to their habitats due to development, modern farming practices, or neglect taking their toll, as well as more insidious changes due to 'nutrient overload' as agricultural fertilisers and nitrates from car exhausts build up the fertility of the soil, allowing the few species that thrive in such conditions, such as nettles and docks, to flourish at the expense of the majority of wild plants that are adapted to poor soils and harsh conditions. As a result of all these changes, much of the countryside is far poorer for wild flowers than it was only a few years ago.

You can contribute to plant conservation directly, with cash or by volunteering for your local wildlife trust. Plant conservation must be based upon facts, however, and there is a dearth of competent botanists able to accurately identify wild plants. As your knowledge grows, please consider submitting plant records to the network of county recorders organised by the BSBI (see below) which covers the whole of Britain and Ireland. Many county recorders also run 'flora groups' to promote surveys and recording for national and regional projects – a great way to learn your plants.

Societies to join

Botanical Society of Great Britain and Ireland (BSBI) www.bsbi.org.uk. The learned society for professional and amateur botanists. Publishes the *New Journal of Botany* as well as the less formal (and very readable) *BSBI News*, organises a network of county plant recorders, runs field meetings and has a panel of referees to advise on identification. Notably, the BSBI website is a gold mine of useful information and resources. In short, the BSBI is *the* society for anyone with a keen interest in wild plants.

Plantlife International www.plantlife.org.uk. The wild plant conservation charity. Has a small number of reserves, runs 'back from the brink' projects for many declining species and promotes various surveys to raise awareness of wild plants and their conservation. Publishes a quarterly magazine, *Plantlife*.

Wild Flower Society www.thewildflowersociety.com. Established in 1886 for amateur botanists and wild flower lovers in the UK. Organises meetings to see and photograph British wild plants in their natural habitats.

Further reading

Wild Flowers of Britain & Ireland by Richard & Alastair Fitter with Marjorie Blamey. 2003. Very comprehensive, but the text is consequently brief; no keys.

Orchids of Britain & Ireland: A field and site guide by Anne and Simon Harrap. 2nd edition, 2009. Everything you ever wanted to know about British wild orchids.

The Wild Flower Key: How to identify wild flowers, trees and shrubs in Britain and Ireland by Francis Rose (revised by Clare O'Reilly). 2006. Comprehensive, making extensive use of keys.

New Atlas of the British & Irish Flora by C.D. Preston, D.A. Pearman & T.D. Dines. 2002. Detailed distribution maps with interpretive text for all British wild plants.

New Flora of the British Isles by Clive Stace. 3rd edition, 2010. A selection of line drawings, but no colour illustrations. Technical, and making extensive use of keys. The standard reference work for the identification of plants in the wild in Britain, covering every native, naturalised or recurrent casual species and hybrids.

Collins Flower Guide by David Streeter, C. Hart-Davies, A Hardcastle, F Cole & L. Harper. 2009. Very comprehensive, with many illustrations and a brief text.

GLOSSARY

agg.: aggregate, a group of closely similar species that are often identified and recorded as the group.

alternate: first on one side, then on the other, alternately along a stem (re leaves, lobes).

annual: a plant that germinates, flowers, sets seed and dies within a year.

ancient introduction: introduced to Britain by people, either deliberately or by accident, before AD 1500.

ancient woodland: woodland on a site thought to have been wooded since at least AD 1600 (AD 1750 in Scotland).

anther: a small sac, carried at the tip of a *filament*, that contains *pollen*; anther and filament together make up the *stamen*.

apomictic: a plant the reproduces by *apomixis*.

apomixis: reproduction by seed that originates from unfertilised egg cells (e.g. Bramble, p. 366). The resultant offspring are clones – genetically identical to the parent plant.

auricle: a lobe or projection, usually ear-shaped. See fig. 1.

axil: the angle between a leaf-stalk or *bract* and the stem from which it grows. New shoots or flowers often arise from the axils.

Common Field Speedwell

back-cross: cross between a hybrid and one of its parent species.

base-rich: soil or water with a high concentration of alkaline (basic) salts, especially calcium and magnesium.

beak: a slender projection, e.g. on the seed heads of geraniums (p. 100), the tip of the seed pods of crucifers (p. 118) or the seeds of some hawksbeards (p. 266).

berry: a fleshy fruit containing a seed (used here to include drupes, which have an additional hard case around the seed, e.g. cherries and plums).

biennial: a plant that flowers in the second year

after germination and then usually dies.

bisexual: a flower that has both male and female reproductive parts, cf *dioecious, monoecious*.

blade: the flat portion of a leaf. See fig. 1.

bog: plant community on wet, acidic peat, often dominated by *Sphagnum* mosses.

boss: a swelling.

bract: a structure at the base of a flower stalk (where it joins the main stem), varying in size and shape, but often leaf-like; also the small leaves at the base of an *umbellifer's* primary ring of flower stalks. See fig. 2.

Pink Water Speedwell

bracteole: a small bract, especially the tiny leaves at the base of an *umbellifer's* secondary ring of flower stalks. See fig. 2.

BSBI: Botanical Society of Great Britain and Ireland.

bulbil: miniature bulb-like structures developing along the edge of a leaf or in leaf axils that drop off and are capable of developing into a new plant.

calcareous: rich in calcium carbonate, e.g. chalk, limestone or sea shells.

calyx: a collective term for the sepals, used especially when they are partly or entirely fused together to form a tube or cup at the base of the flower. See fig. 2.

Coralroot

calyx tube: see above.

capsule: a dry fruit that opens into 2 or more sections (*valves*) to shed the seeds, or releases them via holes, pores or a lid.

carpel: the basic female reproductive unit; carpels can be separate, as in buttercups, or fused together into an *ovary*; each carpel produces a *style* at its tip.

Red Campion

carpels

Goldilocks Buttercup

Figure 1: LEAVES

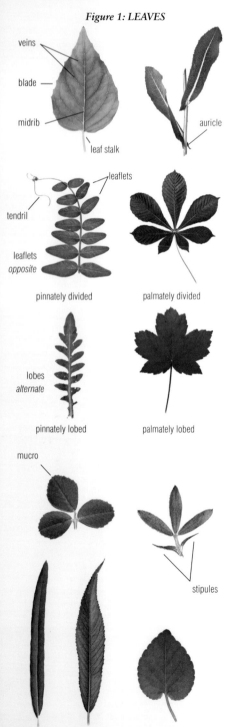

veins
blade
midrib
leaf stalk
auricle

tendril
leaflets
leaflets
opposite

pinnately divided

leaflets

palmately divided

lobes
alternate

pinnately lobed

palmately lobed

mucro

stipules

strap-shaped   lanceolate   heart-shaped

carr: wet woodland, typically with Alder and willows, developing on fens or bogs or alongside lakes and loughs.

casual: an introduction (arising from garden rubbish or bird-sown etc.) that fails to establish itself and thus only persists for as long as the original plants survive.

catkin: a crowded spike of tiny flowers, usually either all male or all female.

chlorophyll: green pigment, important in photosynthesis, found within the cells of plants, especially in the leaves.

clasping (leaf): a leaf, usually stalkless, where the basal lobes appear to clasp the stem.

Almond Willow

clone: a plant with an identical genetic make-up to one or more other individuals, the result of asexual, vegetative reproduction (e.g. from suckers, fragments of shoot or root) or from *apomixis*.

coppice: trees that are cut to near ground level on a regular rotation (typically 8–15 years), which then re-grow, usually with multiple shoots, to provide a sustainable harvest. Coppice stools (i.e. the rootstock and very base, which remain uncut) may be hundreds of years old.

corolla: a collective term for the petals, used especially when they are partly or entirely fused together to form a tube or cup. See fig. 2.

corolla tube: see above. See fig. 2.

Critically Endangered: facing an extremely high risk of extinction in the wild in Britain in the immediate future.

crucifer: member of the cabbage family (Brassicaceae), with 4 petals arranged in a cross (see p. 118).

downs, downland: short, grazed grassland on shallow soils over chalk.

deciduous: a plant that loses its leaves during the winter.

dioecious: a plant that is either male or female, and has flowers with the reproductive parts of one sex only (if the other sex is present, the structures are vestigial, e.g. a 'female' plant may have very small, poorly-developed stamens).

disc-floret: central tube-like florets in a compound flower of the daisy family (see fig. 2, also p. 253).

Endangered: considered to be facing a very high risk of extinction in the wild in Britain.

endemic: confined to a specific area.

epicalyx: a ring of *bracts* below the true *calyx*.

Maiden Pink

calyx
epicalyx

evergreen: a plant that has leaves all year and retains the same basic size and structure.

family: a group of *genera* that are all more closely related to each other than they are to other genera.

fen: plant community on wet soils that are alkaline, neutral or very slightly acidic.

filament: the slender stalk that supports the anther on the *stamen*. See *anther*.

floret: a small flower; used when the 'flower' is actually made up of numerous small flowers, as in the daisy family (see fig. 2, also p. 253).

flush: a waterlogged area where the ground-water is moving through the soil.

fruit: a case surrounding the seed(s) that is formed by the ovary wall; fruits may be dry and papery, or fleshy.

garden escape: cultivated plants that disperse into the wild, usually over short distances (e.g. seeds may be carried by birds or the wind).

garden throw-out: cultivated plants that are introduced to the wild via dumped garden soil or rubbish, usually found in places that are accessible by car or truck.

genus (plural *genera):* a group of species, all more closely related to each other than they are to other species (i.e. a plant's 'immediate family').

gland: a tiny, semi-translucent capsule filled with oils or resins that may be aromatic and/or sticky; glands may be stalkless and sit on the surface of a leaf or stem, or stalked, forming a *glandular hair.* See also spurges, p. 84.

gland

glandular hair: a hair tipped with a *gland*.

grike: a crevice in *limestone pavement.*

heath, heathland: a plant community on poor, acid soils in the lowlands, usually on sands and gravels and typically well drained, dominated by heathers, gorse, a variety of grasses and Bracken. See also *moor.*

hybrid: a plant originating from the fertilisation of one species by another.

indigenous: native to an area or region.

inferior (ovary): placed below the junction of the sepals and petals. See fig. 2.

introduction: a plant brought to a country or region by humans, either intentionally or accidentally.

keel: the lower petal of a pea flower (see fig. 2, also p. 43).

lanceolate: narrowly oval, tapering to a more or less pointed tip. See fig. 1.

leaf stalk: the structure that attaches a leaf to the stem; technically the 'petiole'; leaves may be long-stalked or stalkless.

leaflets: the separate leaf-blades of a compound leaf. See fig. 1.

limestone pavement: a habitat formed by the action of ice sheets and water in which solid, level sheets of limestone are crisscrossed by deep crevices (*grikes*).

lip: a lower petal that differs in shape, size and/or coloration from the rest, e.g. orchids, labiates. See fig. 2.

lowland: up to 310m above sea level.

machair: sandy, lime-rich soil on the landward side of sand dunes supporting a species-rich community of short grasses and herbs. Machair is confined to the W coasts of Ireland and Scotland.

meadow: grassy field from which grazing animals are excluded for at least part of the year so that it can be cut for hay.

microspecies: closely-related species that are very similar, differing in minor characters and only usually identifiable after prolonged study, as in Brambles (p. 366) and Dandelions (p. 265).

midrib: the central vein of a leaf, an elongation of the leaf stalk, that runs all the way to the tip; the midrib may be very obvious, or indistinct. See fig. 1.

monoecious: with distinct male and female flowers, but with the flowers of both sexes on the same plant (cf *dioecious*).

moor, moorland: a plant community on poor, acid soils, typically in the uplands, often poorly drained (or receiving high rainfall), and dominated by heathers, a variety of grasses and Bracken. Often grades into *bog.* See also *heath.*

mucro: a bristle-like projection. See fig. 1.

mucronate: with a *mucro.*

Nationally Rare: recorded in 15 or fewer 10km squares in the UK.

Nationally Scarce: recorded in 100 or fewer 10km squares in the UK.

native: growing in an area where it was not introduced, either accidentally or deliberately, by humans.

naturalised: an introduced plant that has developed self-sustaining populations, reproducing via seed or vegetatively.

Near Threatened: species that does not qualify for *Critically Endangered, Endangered* or *Vulnerable* now, but is close to qualifying for it, or likely to qualify for one of these categories in the near future. Includes species which occur in 15 or fewer 10km squares but which do not qualify as *Critically Endangered, Endangered* or *Vulnerable.*

nectary: a structure that produces nectar.

net-veined: marked with a network of veins (technically 'reticulated').

node: point on a stem from which leaves, flowers or side shoots grow.

nutlet: a very small nut.

ochrea (plural *ochreae*): translucent, papery *stipules* found at the junction of the leaf stalk and the main stem in some members of the dock family (see p. 146).

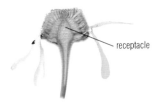

oil gland: see *gland.*

opposite (leaves): growing in pairs, one on either side of the stem.

ovary: female reproductive organ that contains the *ovules.* See fig. 2.

ovule: organ inside the *ovary* that contains the embryo sac, which in turn contains the egg.

palmate: spreading outwards from a central point, as in palmate leaflets, palmate veins (cf *pinnate*). See fig. 1.

papilla (plural *papillae*): a minute, blunt, nipple-like projection.

pappus: fine hairs or bristles, sometimes feathery, that help to catch the wind and disperse the seed, found in members of the daisy family (p. 253) and the valerians (p. 292).

parasite: organism that lives on or at the expense of other organisms.

pasture: grassland that is grazed for some or all of the year but not cut.

perennial: a plant that lives for more than two years; perennials are usually stouter than annuals.

petal: part of the inner ring of a flower; they are often brightly coloured, but in some species may be absent altogether. See fig. 2.

photosynthesis: production of food by green plants. In the presence of *chlorophyll* and light energy from the sun, carbon dioxide and water are converted into carbohydrates with the release of oxygen.

pinnate: a leaf that is cut into opposite pairs of leaflets. See fig. 1.

plume: see *pappus.*

pollard: a tree that is cut at regular intervals at about head height (or higher – out of the reach of grazing animals), in order to produce a crop of timber or poles.

pollen: single-celled spores containing the male gametes (sperm), shed by the *anther.*

prostrate: growing flat on the ground.

ray-florets: the outer, strap-shaped florets in a compound flower of the daisy family (see fig. 2, also p. 253).

recent introduction: introduced to Britain, either deliberately or by accident, by humans after AD 1500.

receptacle: the tip of the flower stalk, which may be a flat disc, domed or cupped, to which all the parts of the flower are attached.

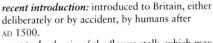

rhizome: creeping underground stem, lasting for more than one year, from which roots and growth buds emerge; also a horizontal stem, either growing along the surface or underground.

root-hemiparasite: partial parasites that have green leaves and photosynthesise, but also extract nutrients from the roots of other plants (e.g. Yellow Rattle, see p. 242).

runner: horizontal stem growing along or just below the surface of the soil that can root at the tip to form a new plant (e.g. Creeping Buttercup, p. 23).

saltmarsh: a plant community that is periodically covered by sea water; the lower saltmarsh may be covered twice a day, while the uppermost parts may only be covered once or twice a year.

Sch 8: a specially protected plant listed on Schedule 8 of the Wildlife and Countryside Act (1981) – it is illegal to remove any part of the plant.

self-pollination: pollination of a flower by pollen from the same plant; usually the pollen comes from the same flower, but sometimes from another flower on the same plant.

sepal: part of the outer ring of a flower; they form the protective covering of the bud and are often green, but in many plants are as brightly coloured as the petals. See fig. 2.

sp.: an abbreviation for species (spp. in the plural).

spadix: a pencil-like stem with closely-packed flowers, often extended beyond the flowers as a fleshy appendix (see *Arum* p. 325).

spathe: a large, leaf-like bract forming a cowl around a *spadix* (see *Arum* p. 325).

species: the basic division of the plant (and animal) kingdom; all members of a species resemble one another and can interbreed, but do not generally cross with members of another species.

speculum: pattern on the lip, often with a metallic lustre or shine, in the bee and spider orchids, genus *Ophrys* (see pp. 323-324).

Figure 2: FLOWERS

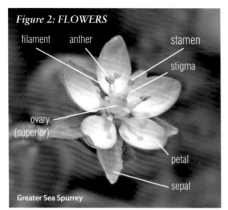

Greater Sea Spurrey

filament, anther, stamen, stigma, ovary (superior), petal, sepal

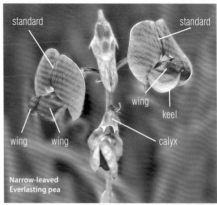

Narrow-leaved Everlasting pea

standard, standard, wing, keel, wing, wing, calyx

Milk Parsley

ray, bracts, bracteoles

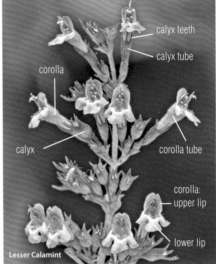

Lesser Calamint

calyx teeth, calyx tube, corolla, calyx, corolla tube, corolla: upper lip, lower lip

Tuberous Comfrey

calyx, calyx teeth, corolla, stigma

Early Purple Orchid

bud, bract, spur, spur, lip, lip

Corn Marigold

disc-florets, ray-floret

spike: a group of flowers all growing from a central stem (technically, a spike is unbranched and the flowers lack stalks, but the term is used here in a more general sense).

spikelet: a small spike.

spur: sac-like extension of the base of an orchid's *lip;* in some orchids the spur contains nectar. See fig. 2.

stamen: male reproductive organ of a flowering plant, comprising a *filament* (stalk) and an *anther.* See fig. 2.

staminode: a sterile *stamen,* sometimes adapted to perform the function of a *petal* or *nectary* (e.g. Grass of Parnassus, p. 82).

standard: the upper petal of a flower of the pea family (see fig. 2, also p. 43).

stellate (hairs): with radiating parts and thus star-shaped.

sterile: incapable of producing fertile *pollen* or *ovules*: the *anthers* will have a small number of poorly-developed pollen grains (or none at all) and the *ovary* will not produce properly developed seed.

stigma: the tip of the *style,* usually sticky, which receives the pollen grains; the style may have one or several stigmas.

stipule: a structure at the junction of a leaf stalk and the main stem, often two together; stipules may be green and leaf-like, or scale-like. See fig. 1.

stolon: a horizontal stem, either growing along the surface or underground, not necessarily forming a new plant at the tip.

strand-line: the line of debris left along the upper part of a beach by the highest tides or winter storms.

strap-shaped: more or less parallel-sided. See fig. 1.

style: the stalk-like structure that forms the upper part of the female organ of a flower, that bears at its tip the *stigma* (or stigmas). See fig. 2.

subspecies (abbreviated ssp.): a geographical or ecological subdivision of a *species,* differing in a consistent way from other subspecies.

sucker: shoots arising from the roots of a tree or shrub. Suckers can develop into new, fully formed trees or shrubs, and may or may not remain attached to the parent (via the roots). All the plants produced by suckers are genetically identical *clones.*

suckering: forming new plants via suckers, often resulting in a thicket.

superior (ovary): placed above the junction of the sepals and petals. See fig. 2.

tall herb fen: a marsh dominated by Meadowsweet, Wild Angelica, Great Willowherb and other tall herbaceous plants.

tendril: slender extension of the *midrib* of a leaf or leaflet, usually coiling around other plants for support. See fig. 1.

tepal: the name used when the sepals and petals are the same colour, size and shape and are indistinguishable.

tuber: swollen underground root or stem, functioning as a storage organ.

tufted: with several stems arising more or less from the same point and clustered together.

umbel: a cluster of flowers, often more or less flat-topped, with the branches or rays all arising from one point, like the spokes of an umbrella. See fig. 2, also p. 296.

umbellifer: a member of the carrot family (Apiaceae), with the flowers arranged into umbels (see p. 296).

variety (abbreviated var.): subdivision of a *species* that differs in a more or less consistent way from the predominant form (or from other varieties). Technically, a *variety* is a lower rank than a *subspecies,* and may not have the geographical or ecological consistency, but subspecies and variety are sometimes used in a confusing and almost interchangeable way by plant scientists.

valve: the part of a fruit that opens or falls off to reveal the seeds, used especially in the seed pods of crucifers (see p. 118).

vein: the thickened, thread- or tube-like portions in a leaf or *petal.* See fig. 1.

Vulnerable: considered to be facing a high risk of extinction in the wild in Britain.

wart: a small, rounded boss or lump (e.g. on the *tepals* of docks).

wart

Marestail

whorl: a group of three or more leaves or flowers growing in a ring from the same point on the stem.

wing: the side (lateral) petals of a flower in the pea family (see fig. 2, also p. 43).

winged (stem, seed pod, etc): with one or more raised flanges (wings) running along part or all of its length.

winter-annual: germinating during the autumn or winter and flowering in the early spring, then dying off and vanishing for the summer.

winter-green: dying down in the winter, but retaining a leafy rosette.

Nette-leaved Bellflower

stigmas   style

Broad-leaved Dock

GUIDE TO THE SPECIES ACCOUNTS

RANGE MAP
The plant's distribution based on data collected by the BSBI, mostly in the period 1987–1999. Each tiny green square indicates its presence in a 10 × 10km square. The maps include both the native range and where recorded as an introduction.

DETAIL
Even for the most 'obvious' species it is worth checking this against the living plant. The more important features are highlighted in italics.

GROWTH
Annual, biennial or perennial.

HEIGHT
The typical range of heights (extremes may be given in parentheses); if sprawling or prostrate, the length of the shoots is given.

FLOWERS
The normal flowering period (extremes may be given in parentheses).

STATUS
Whether a native, a recent introduction (with source and date of introduction), or an ancient introduction (i.e. brought to Britain before AD 1500).

INTRODUCTORY TEXT
Habitat and abundance, followed (after the ✿ symbol) by an outline of the identification features, with the more important features highlighted in italics. *Conservation designations, where appropriate, are also given (for definitions see Glossary).*

ENGLISH AND SCIENTIFIC NAMES

Lesser Celandine *Ficaria verna*

Nationally Scarce. Abundant in woods, meadows and churchyards, and on verges and hedgebanks. ✿ One of the earliest spring flowers, often found in large numbers. **DETAIL** Flowers solitary on long stalks, 10–30mm across, usually with 3 oval sepals and 7–12 narrow-oval petals, 10–15mm long. Seeds finely hairy.
SIMILAR SPECIES Goldilocks Buttercup *R. auricomus* is also found in woodland, but has imperfect flowers. **SEE ALSO** Greater Celandine (p. 314).

SIMILAR SPECIES
Brief details of scarcer species not otherwise covered by this guide.

SEE ALSO *Cross reference to confusion species when these are not on the same spread.*

GROWTH Perennial, growing from small bulbs.
HEIGHT Stems to 25cm, often rooting near base.
FLOWERS Feb–May.
STATUS Native.
ALTITUDE 0–750m.

There are two subspecies:
F. v. fertilis Relatively large flowers (petals 10–20mm long). Fertile, reproducing via seed; found throughout the range.
F. v. verna Small flowers (petals 6–11mm long). Largely sterile, reproducing via bulbils produced in the leaf axils; found throughout the range.

ALTITUDE
The normal altitudinal range in metres above sea level (extremes may be given in parentheses). 'Lowland' indicates below 310m.

ADDITIONAL INFORMATION
on variation, history or uses may be given in pale blue boxes, while further notes on conservation are found in pink boxes.

LEAVES/SEEDS/FRUITS
Illustrated where useful for identification. Note that these are NOT TO SCALE.

GENUS *PAPAVER*: POPPIES

Five species of red poppy occur in Britain. All are ancient introductions that arrived as contaminants of seed-corn with the first Neolithic farmers around 4,000 BC. All are annuals and usually require disturbed soil in order to germinate; they are typically found in arable fields. Poppies are sensitive to herbicides, but can be abundant in unsprayed crops and arable margins. Poppy seed can be very long-lived, however, and following disturbance long-buried seed may germinate. The 5 species differ subtly in flower size and colour and in their leaves, but the easiest and most reliable identification feature is the seed capsule.

Commmon   Rough   Prickly   Long-headed

Common Poppy
Papaver rhoeas

By far the commonest red poppy. ✿ Variable in stature, from tiny plants with 1 flower to robust individuals with dozens. Flowers bright red (mauve, pink or white forms occur rarely, especially around habitation). DETAIL Whole plant bristly-hairy. *Seed capsule hairless, globular.* Flowers mostly 7–10cm across, but frequently dwarfed, and sometimes tiny, petals often with a dark blotch at the base; anthers blue-black. The petals are often dropped by the afternoon.

GROWTH Annual.
HEIGHT 20–60cm.
FLOWERS Jun–Aug.
STATUS Ancient introduction.
ALTITUDE To 410m.

Long-headed Poppy
Papaver dubium

✿ Very like Common Poppy and found in similar habitats (especially on dry, sandy or gravelly soils), and in coastal grassland. Flowers slightly paler, *more orange-red.* DETAIL *Seed capsule hairless, up to 25mm long – at least twice as long as wide.* Anthers brown to blue-black, petals may have a dark blotch at the base. When freshly cut, the stems and leaves bleed *white sap.*

HEIGHT To 60cm.
FLOWERS Late May–Aug.
ALTITUDE To 425m.

Yellow-juiced Poppy
Papaver lecoqii

✿ Very like Long-headed Poppy, including the seed capsule, but typically found on heavier soils, especially chalky boulder clay. DETAIL *Fresh sap custard-yellow* (or quickly turning yellow). In addition, the anthers may be yellow. Flowers orange-red, probably identical to Long-headed Poppy.

HEIGHT To 60cm.
FLOWERS May–Jul.
ALTITUDE Lowland.

Rough Poppy *Papaver hybridum*

The scarcest and most localised of the red poppies, usually found on light, chalky soils. ✿ Flowers relatively small, a unique shade of *pinkish-red.* DETAIL Leaves finely cut. *Seed capsule less than 1.5cm long, globular, with numerous erect, stiff bristles.* Flowers 2–5cm across, anthers blue, petals always with black at the base.

HEIGHT To 50cm.
FLOWERS Jun–Jul.
ALTITUDE To 320m.

Prickly Poppy *Papaver argemone*

Uncommon and declining; listed as Vulnerable. Usually found on light sandy, gravelly or chalky soils, also grows on old walls. ✿ Compared to Common Poppy, *flowers a subtly deeper, blood- or pillar-box red,* also averaging smaller, with the petals well separated. DETAIL Leaves finely cut. *Seed capsule club-shaped,* more than 1.5cm long *(some at least twice as long as wide),* with *erect bristles.* Flowers up to 5cm across, anthers blue, petals often with a dark blotch at the base.

HEIGHT To 45cm.
FLOWERS Jun–Jul.
ALTITUDE Lowland.

Opium Poppy *Papaver somniferum*

GROWTH Annual.
HEIGHT 30–120cm.
FLOWERS Jun–Aug.
STATUS Ancient introduction.
ALTITUDE To 410m.

Opium and its derivatives are made from the sap of the Opium Poppy, but in the British climate little of the active chemical is produced. The seeds are also used in cooking, and the plant has probably been cultivated in Britain since the Bronze Age.

Fairly common around towns and villages and frequently appearing on tipped soil on building sites and waste ground. ❁ *Whole plant conspicuously greyish with a waxy bloom. Flowers large*, usually lilac, but may be white, red or variegated. **DETAIL** More or less *hairless*. Leaves oblong, coarsely-toothed or lobed, *clasping the stem*. Flowers 10–18cm across, petals with a dark blotch at the base; anthers yellow or dark purple. Seed capsule hairless, globular.

Greater Celandine *Chelidonium majus*

GROWTH Perennial.
HEIGHT To 90cm.
FLOWERS May–Aug.
STATUS Introduced in Roman times, and once cultivated as a medicinal plant.
ALTITUDE Lowland.

Fairly common, especially around towns and villages, often growing at the base of walls. ❁ Relatively small, yellow, 4-petalled flowers, growing in loose clusters, and long seed capsules, recall a member of the cabbage family and do not immediately bring to mind a poppy, but the grey-green leaves, with well-rounded lobes, are distinctive. **DETAIL** Sparsely hairy, bleeds orange sap when cut. Flowers 15–25mm across, remaining closed in dull weather, the petals soon falling. Seed capsule cylindrical, 30–50mm long, splitting open from the bottom upwards. **NOTE:** Not closely related to Lesser Celandine (p. 26).

Yellow Horned Poppy *Glaucium flavum*

Locally common on coastal shingle that is not subject to excessive trampling, occasionally also bare cliffs of sand, clay or chalk, always by the sea.
❀ *Large yellow flowers and extraordinarily long, sickle-shaped seed pods ('horns') distinctive.*
DETAIL Leaves lobed, the lower leaves deeply so, grey-green, slightly fleshy but downy; over-winters as a rosette. Stem hairless, multibranched and spreading; when broken, exudes yellow sap. Flowers 6–9cm across. Seed capsule 15–30cm long, pencil-thin, distinctly curved.

GROWTH Biennial to perennial.
HEIGHT To 90cm.
FLOWERS Jun–Sep.
STATUS Native.
ALTITUDE 0–20m.

Welsh Poppy *Meconopsis cambrica*

A Nationally Scarce native of shaded rocky places in the hills of Wales, SW England and Ireland, but more commonly naturalised in similar habitats elsewhere (especially N England and Scotland), as well as around towns and villages and on waste ground. ❀ *Large 4-petalled yellow flowers distinctive.* DETAIL Very sparsely hairy. Leaves deeply-lobed, the lower long-stalked; when cut, bleeds watery white sap. Flowers 5–8cm across, solitary, on long slender stalks, anthers yellow, seed capsule 2–4cm long with the style protruding from the tip.

GROWTH Perennial.
HEIGHT 30–60cm.
FLOWERS May–Aug.
STATUS Native & naturalised.
ALTITUDE To 680m.

GENUS *FUMARIA:* FUMITORIES

Ten closely similar species, arbitrarily divided into 'fumitories' and slightly larger 'ramping fumitories'. Low-growing, the masses of finely cut leaves are the origin of the English name (*fumus terrae*, 'smoke of the Earth'), while the distinctive small flowers are produced in spike-like clusters. The fruits are small and roughly spherical. Identification depends on careful examination of the flowers and fruits, and can be very tricky.

Common Fumitory *Fumaria officinalis*

Much the commonest fumitory in most of Britain, a locally abundant weed of gardens and arable fields. ✿ Sprawling or scrambling, with greyish-green leaves and small pinkish flowers. DETAIL *Flowers 7–8mm long (6–9mm)*, sepals irregularly toothed, 1.5–3.5mm × 1–1.5mm; viewed from below the lower petal is *spatula-shaped*, broadening towards the tip. SIMILAR SPECIES Three other fumitories, with either larger sepals or smaller flowers, are scarce weeds of arable fields on chalk.

GROWTH Annual.
HEIGHT Stems to 100cm.
FLOWERS May–Oct.
STATUS Ancient introduction.
ALTITUDE To 540m.

Common Ramping Fumitory

Fumaria muralis

The commonest fumitory in much of Wales, W England and Ireland. Arable fields, waste ground and hedgebanks. ✿ Rather like Common Fumitory but *flowers larger*, *9–12mm long (8–13mm)* long. DETAIL Sepals toothed towards the base, 3–5mm × 1.5–3mm. Viewed from below the lower petal is *parallel-sided*. SIMILAR SPECIES Five other ramping fumitories, all rather hard to separate from each other, are scarcer and often much more local; most are found in the W.

GROWTH Annual.
HEIGHT Stems to c. 100cm
FLOWERS Apr–Oct.
STATUS Native.
ALTITUDE Lowland.

Climbing Corydalis
Ceratocapnos claviculata

Locally common. Woodland, heathland, scrub, under Bracken and gorse and rocky places, on dry acid soils; can be abundant in recently clear-felled conifer plantations. ❀ Delicate, low-growing and scrambling, climbs by means of tendrils, producing distinctive tangled masses of foliage punctuated by small clusters of creamy flowers. **Detail** Leaves finely cut, terminating as they mature in a branched tendril. Flowers 4–6mm long, fruit *c.* 6mm long, a shiny black pod.

GROWTH Annual.
HEIGHT Stems to 75cm.
FLOWERS May–Dec.
STATUS Native.
ALTITUDE To 430m.

Yellow Corydalis *Pseudofumaria lutea*

Widely and commonly naturalised on walls and rocky waste ground. ❀ The spikes of tubular, rich yellow flowers are distinctive. **Detail** Leaves divided into more or less 3-lobed leaflets, but no tendrils. Flowers 12–18mm long.

GROWTH Perennial.
HEIGHT Stems to 30cm.
FLOWERS May–Aug.
STATUS Introduced to cultivation from the Alps by 1596 and first recorded in the wild in 1796.
ALTITUDE To 570m.

Meadow Buttercup
Ranunculus acris

Locally abundant. Damp, unimproved grassland, including meadows, pastures and, locally, road verges. ☀ Averages taller than other buttercups, with *finely cut leaves*. DETAIL Hairy. Stems branched, but *no runners*. Leaves cut into rather narrow, pointed lobes. *Flower stalks smooth*, flowers 15–25mm across, *sepals spreading*. Seeds in a globular cluster, smooth, hairless, with a short, hooked beak.

GROWTH Perennial.
HEIGHT 20–75cm.
FLOWERS Apr–Aug.
STATUS Native.
ALTITUDE 0–1220m.

seed head

Goldilocks Buttercup *Ranunculus auricomus*

Fairly common in deciduous woodland on lime-rich soils, sometimes also hedgebanks and churchyards. ☀ Spring-flowering (even the leaves dying off by midsummer), the *flowers are usually (but not always) deformed* or have *some or all of the petals missing*. DETAIL *Whole plant more or less hairless*. Basal leaves long-stalked, *kidney-shaped to circular*, very variably 3–5-lobed, upper stem leaves deeply cut. Flowers few, 15–25mm across, with *smooth stalks* and *spreading sepals*. Seeds in a globular cluster, smooth, *finely downy* (but hard to see the hairs - look against the light or a dark background), with a short curved beak.

GROWTH Perennial.
HEIGHT To 40cm.
FLOWERS Apr–Jun.
STATUS Native.
ALTITUDE To 1090m.

basal leaf

Creeping Buttercup
Ranunculus repens

seed head

GROWTH Perennial.
HEIGHT 10–60cm.
FLOWERS Late Apr–Oct.
STATUS Native.
ALTITUDE To 1035m.

Abundant in a wide variety of grassy habitats, from fens to woodland rides, and a common garden weed, annoyingly hard to eradicate. ❀ Puts out *creeping runners that root at the nodes to form new plants.* DETAIL Hairy. Leaves divided into 3 lobes, each lobed in turn, with the central lobe often long-stalked. *Flower stalks grooved*, flowers 20–30mm across, sepals *spreading*. Seeds in a globular cluster, hairless, with a short curved beak.

Bulbous Buttercup *Ranunculus bulbosus*

Locally abundant, *early-flowering* buttercup. Unimproved dry grassland, especially on lime-rich soils, often now restricted to road verges and churchyards. ❀ Very like Creeping Buttercup and easily overlooked, but *lacks runners* and *sepals turned sharply downwards.* DETAIL Hairy. Base of stem expanded into a tuber (hence the name). Leaves 3-lobed, with the central lobe often long-stalked. *Flower stalk grooved*, flowers 15–30mm across. Seeds in a globular cluster, hairless, finely pitted (not warted, cf Hairy Buttercup p. 24), with a short, hooked beak.

GROWTH Perennial.
HEIGHT 15–40cm.
FLOWERS Late Mar–early Jun.
STATUS Native.
ALTITUDE To 580m.

GROWTH Annual.
HEIGHT To 40cm.
FLOWERS May–Oct.
STATUS Native.
ALTITUDE Lowland.

Hairy Buttercup
Ranunculus sardous

Fairly common. Damp grassland with a maritime influence, especially coastal grazing marshes, favouring tracks, gateways and other disturbed ground. ❁ Picked out by its rather 'washed out' foliage and pale yellow flowers. As in Bulbous Buttercup, *sepals turned downwards*. DETAIL Hairy. Base of stem not swollen. Leaves 3-lobed, with the middle lobe long-stalked, densely hairy to hairless. Flowers 12–25mm across. Seeds in a globular cluster, hairless, with a few *slightly raised 'warts' in a ring towards the rim* (10× lens, but often hard to see), and a very short beak.

seed head

warts

individual seed

Small-flowered Buttercup
Ranunculus parviflorus

GROWTH Annual.
HEIGHT Stems to 20cm, usually more or less sprawling.
FLOWERS May–Jul (Dec).
STATUS Native.
ALTITUDE Lowland.

Uncommon. Dry, sparsely-vegetated habitats kept open by disturbance or drought, especially near the sea: banks, tracks, rabbit scrapes and gardens; seldom now arable fields. ❁ Small, easily overlooked, and not obviously a buttercup, with rather pale, yellowish-green foliage and small pale yellow to whitish flowers. DETAIL Softly hairy. Lower leaves 3–5 lobed, cut half way to base. Flowers 3–6 mm across, *sepals turned downwards*. Seeds in a small cluster, hairless, with *tiny hooked spines on the sides* and a short hooked beak. SIMILAR SPECIES The now rare arable weed **Corn Buttercup** *R. arvensis* has larger flowers, *spreading sepals* and seeds with a *broad rim* and *long spines*, especially on the edges.

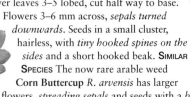

Lesser Spearwort *Ranunculus flammula*

Common in a wide variety of damp or wet places on neutral to acid soils. Will grow in shallow water and can form large mats. ✿ Low-growing, it favours sites with low levels of nutrients or shallow soils, or where other vegetation is controlled by grazing. Separated from most other buttercups by its *undivided leaves*: basal leaves pointed-oval with long stalks, *stem leaves narrow and strap-shaped*. **DETAIL** Hairless. Stems hollow, sometimes rooting at the lower nodes. *Flowers 7–20mm across, flower stalks grooved.* Seeds 1–2mm long plus a short beak, minutely pitted, not winged.

GROWTH Perennial.
HEIGHT Stems erect or spreading, to 50cm.
FLOWERS May–Oct.
STATUS Native.
ALTITUDE 0–945m.

Greater Spearwort *Ranunculus lingua*

Rather local. Native plants grow in shallow water over peat or deep mud, either still or slow-flowing, in fens, ditches and dykes, preferring alkaline conditions. Also widely introduced away from the native range into lakes, flooded gravel pits and ponds. ✿ The largest-flowered buttercup, with *long, narrow leaves.* **DETAIL** Hairless. Stems grow from long creeping runners. Stem-leaves up to 25cm long. *Flowers 20–50mm across, flower stalks not grooved.* Seeds *c.* 2.5mm long plus a short curved beak, minutely pitted and very narrowly winged.

GROWTH Perennial.
HEIGHT 50–120cm.
FLOWERS May–Sep.
STATUS Native, but so widely planted as an ornamental that its range and abundance as a wild plant are hard to work out.
ALTITUDE Lowland.

Winter Aconite *Eranthis hyemalis*

GROWTH Perennial.
HEIGHT To 10cm.
FLOWERS Jan–Mar.
STATUS Introduced to gardens from S Europe by 1596 and first recorded in the wild in 1838.
ALTITUDE Lowland.

Commonly naturalised on verges and in churchyards, parks and open woodland, usually near houses.
❋ Blooming early, the flowers are surrounded by a ruff formed by 3 deeply-cut, leaf-like bracts; they open fully only in sunshine.
DETAIL Hairless. Growing from small round tubers, the leaves appear with or after the flowers and die back by June. Flowers solitary, 20–30mm across, formed by 6 sepals (the petals are reduced to horn-shaped nectaries).

Lesser Celandine *Ficaria verna*

GROWTH Perennial, growing from small bulbs.
HEIGHT Stems to 25cm, often rooting near base.
FLOWERS Feb–May.
STATUS Native.
ALTITUDE 0–750m.

Abundant in woods, meadows and churchyards, and on verges and hedgebanks, usually on slightly damper soils. One of the earliest spring flowers, often found in large numbers. ❋ The rosettes of long-stalked, glossy, heart-shaped leaves are distinctive; they often have diffuse paler (or darker) concentric inner markings. The flowers often fade to whitish as they go over. DETAIL Flowers solitary, on long stalks, 10–30mm across, usually with 3 oval sepals and 7–12 narrowly oval petals. Seeds smooth, finely hairy.

There are three subspecies:
  F. v. fertilis Flowers relatively large, with the petals 10–20mm long. Fertile, reproducing via seed. Found throughout the range.
  F. v. verna Flowers small, petals 6–11mm long. Largely sterile, reproducing via bulbils produced in the leaf axils. Found throughout the range.
  F. v. ficariiformis Flowers very large, petals 17–26mm long. Reproduces via both ripe seed and bulbils. A garden escape sometimes naturalised in the SW. Other large-flowered forms may also escape.

Celery-leaved Buttercup
Ranunculus sceleratus

Locally common in shallow water and on bare mud around ditches, ponds and lakes, especially the 'draw-down' zone where water levels fall in summer; frequent in coastal grazing marshes. ❀ The small flowers are washed-out yellow and the whole plant is more or less hairless with a *pearly gloss to the leaves* (which only vaguely recall celery). DETAIL Very variable, from tiny, to tall, robust and often well-branched plants. Stem hollow, often rather stout. Lower leaves variably 3–5 lobed, stem leaves often divided into 3 leaflets. Flowers 5–10mm across, sepals strongly turned downwards. The rather small seeds are gathered into *elongated heads*.

GROWTH Annual.
HEIGHT To 60cm.
FLOWERS May–Sep.
STATUS Native.
ALTITUDE Lowland.

Mousetail *Myosurus minimus*

Scarce, declining, and listed as Vulnerable. Found on nutrient-rich soils subject to winter flooding, usually on bare ground around tracks and gateways on pastures, more rarely on arable. ❀ Often tiny, with *narrow, rather grass-like leaves*, all in a basal rosette. The inconspicuous flowers are grouped around a *column-like receptacle that extends in fruit to 2–7cm long.* DETAIL Hairless. Leaves slightly fleshy, 1–10cm long. Flower stalks to 10cm long. Flowers solitary, with 5 yellowish-green sepals, *c.* 4mm long, each with a short downward-pointing spur; the 5 petals are reduced to narrow nectaries, stamens 5–10. SEE ALSO Greater Plantain (p. 222), to which the flowers bear a very superficial resemblance.

GROWTH Annual.
HEIGHT 3–15cm.
FLOWERS Apr–Jul.
STATUS Native or alien.
ALTITUDE Lowland.

Stinking Hellebore
Helleborus foetidus

Nationally Scarce as a native, but also widely naturalised. Prefers open woods and scrub, hedgebanks and rocky slopes on shallow, lime-rich soils. ❀ An early-flowering evergreen, usually with many *cup-shaped yellow-green flowers,*

GROWTH Short-lived perennial.
HEIGHT To 80cm.
FLOWERS Jan–Apr.
STATUS Considered native in a band from Kent to N Wales, but commonly grown in gardens and widely naturalised.
ALTITUDE Lowland.

each edged wine-red, on robust pale stems that stand above the rest of the plant. DETAIL Leaves dark green, long-stalked, divided into 5–9 narrow, strap-shaped leaflets that arise from 2 *semi-circular splays* either side of the tip of the stalk. Flowers 10–30mm across with 5 sepals; the petals are reduced to small nectaries. Foliage said to be foul-smelling.

Green Hellebore *Helleborus viridis*

Uncommon as a native in woods and other shady places on lime-rich soils, but also found as a garden escape. ❀ *Distinctive green, saucer-shaped flowers, often held nodding downwards.* DETAIL The leaves and flowering shoots grow in clusters from a rhizome, dying down in late summer. Leaves divided into 7–11 toothed, strap-shaped leaflets that *all grow from the same central point.* Flowers 30–50mm across, up to 4 together on leafy stems, with 5 sepals

GROWTH Perennial.
HEIGHT To 40cm.
FLOWERS Mar–Apr.
STATUS Native and introduced. Recorded in the wild by 1562 but also cultivated since medieval times and sometimes naturalised.
ALTITUDE Lowland.

and many stamens; the petals are reduced to small nectaries. The sepals persist as the capsules develop, thus 'in flower' for a prolonged period.

Marsh Marigold *Caltha palustris*

Fairly common. Marshes, wet grassland, wet woodland and the margins of lakes, streams and ponds. ❋ Often somewhat stout, the whole plant is hairless. The flowers lack petals – it is the 5–8 sepals that are *butter yellow*. **DETAIL** Leaves dark,

glossy green with a network of fine paler veins, 3–30cm across, with finely toothed margins and a heart-shaped base; the basal leaves are long-stalked. Flowers 20–30mm across, occasionally up to 50mm; large- or pale-flowered forms are sometimes cultivated and may 'escape', but wild populations also throw up large-flowered plants; conversely, var. *radicans*, found in upland areas, has prostrate stems that root at the nodes and small, usually solitary, flowers.

GROWTH Perennial.
HEIGHT To 60cm.
FLOWERS Mar–Jun.
STATUS Native.
ALTITUDE 0–1100m.

Globeflower *Trollius europaeus*

GROWTH Perennial.
HEIGHT To 70cm.
FLOWERS May–Aug.
STATUS Native.
ALTITUDE 0–1090m.

Local and declining, although still common in a few regions, with a preference for cool, damp, often lightly shaded places: unimproved wet meadows, open woodland, the margins of lakes, rivers and streams and rocky ledges, mostly in the uplands and out of the reach of sheep. ❋ The beautiful globular flowers are unique. **DETAIL** Leaves cut into 3–5 deeply-lobed segments (recalling Meadow Buttercup or Meadow Cranesbill, although not downy, also Monkshood); the basal leaves are long-stalked. Flowers 25–50mm across, with 5–15 sepals curving up and inwards to conceal the petals, which are small and strap-shaped.

Wood Anemone *Anemone nemorosa*

Common. Deciduous woodland and nearby hedgebanks, as well as a variety of open habitats in the wetter N and W: scree, moorland, limestone pavement and under Bracken. ❁ The solitary, star-like flowers carpet many woodlands in the early spring; they are often flushed mauve-pink, especially on the outside. DETAIL Grows from a creeping rhizome, with long-stalked basal leaves. Flowering stems with a whorl of 3 palmately-lobed leaves. Flowers with 6–7 whitish sepals, but no petals.

GROWTH Perennial.
HEIGHT To 30cm.
FLOWERS Mar–May.
STATUS Native.
ALTITUDE 0–1190m.

GROWTH Perennial.
HEIGHT To 30cm.
FLOWERS Apr–May.
STATUS Native.
ALTITUDE Lowland.

Pasqueflower *Pulsatilla vulgaris*

Extremely local, with a few long-established colonies on old, species-rich grassland over chalk and limestone, sometimes on ancient earthworks. DETAIL The whole plant is covered with long silky hairs. Leaves finely cut, growing from the rootstock. Flowers solitary, with 6 purple sepals, each 20–50mm long; no petals. On the stem a little below the flower there is a large, finely cut bract. In fruit the style elongates into a feathery plume up to 50mm long.

Nationally Scarce and listed as Vulnerable. Even on reserves, however, a lack of grazing and nutrient enrichment can transform its preferred short, species-rich turf into rank grassland and scrub. In Britain establishment from seed is thought to be rare and colonisation of new sites unknown.

Columbine *Aquilegia vulgaris*

Very locally common on calcium-rich soils, especially on limestone: woodland glades, scrub, fens, damp grassland and scree; garden escapes are found in a wider variety of habitats. ❁ Flowers distinctive, with 5 petals extended backwards into *long, strongly hooked spurs*, and 5 petal-like sepals. Flowers usually a deep, intense blue in native populations, but can be white, purple, pink or even chocolate brown. DETAIL Stem and leaf-stalks softly hairy. Leaves divided into 3 leaflets, which in turn are divided into 3; leaflets softly hairy below. SIMILAR SPECIES Garden varieties tend to have straighter petal spurs and a hairless base to the leaf stalk, but are difficult to distinguish from native forms.

GROWTH Perennial.
HEIGHT To 100cm.
FLOWERS May–Jun.
STATUS Native, but also a frequent garden escape, thus the native range is uncertain.
ALTITUDE 0–470m.

Monkshood *Aconitum napellus*

GROWTH Perennial.
HEIGHT To 150cm.
FLOWERS May–Jun.
STATUS Recent introduction? In cultivation by 1596, it was not recorded in the wild until 1819, unlikely for such a conspicuous plant. Certainly, it is occasionally naturalised.
ALTITUDE Lowland, but naturalised up to 460m.

Nationally Scarce as a 'native', on wooded stream banks and meadows in SW England and S Wales. Widely but uncommonly naturalised in similar places; more obviously a garden throw-out on verges and waste ground. ❁ Tall, with spikes of distinctive blue to violet flowers. DETAIL Leaves hairless, finely cut (cf Globeflower). *Flower stalk hairy*, flowers 3–4cm long, with 5 sepals: the *upper sepal forms a 'hood'* that is as tall as wide and gradually tapers to a point; petals reduced to nectaries. SIMILAR SPECIES Hybrid Monkshood *A.* × *stoerkianum* is more commonly grown in gardens. Flowers, white, blue or bicoloured, *flower stalk hairless* (or almost hairless), *upper sepal distinctly taller than wide, narrowing abruptly to a point*.

Considered the most toxic wild plant in Britain, but there are very few reports of either animal or human poisoning. Unlike most poisonous plants, however, the toxins can be absorbed through the skin and it is best not to handle Monkshood.

Baneberry
Actaea spicata

GROWTH Perennial.
HEIGHT 30–60cm.
FLOWERS May–Jun.
STATUS Native.
ALTITUDE 0–450m.

Nationally Scarce. Confined to the grikes of limestone pavements, rock ledges and open woodland on limestone in N England. ✸ Spikes of wispy, creamy flowers are followed by black berries. Strong-smelling, the whole plant is poisonous. **DETAIL** Leaves glossy, hairless above, long-stalked, divided into 3–5 broad, coarsely-toothed leaflets. Flowers with 3–6 sepals, 4–6 petals and up to 15 stamens. Fruits 10–13mm across.

Alpine Meadow-rue
Thalictrum alpinum

Very locally common in mountain regions: beside streams, in flushes and on damp rock ledges, usually where the water is at least slightly alkaline. ✸ Small and very inconspicuous, with slender, upright stems and tiny flowers on well-spaced, *drooping stalks*. **DETAIL** Stems unbranched, growing from a creeping rhizome. Leaves with thread-like stalks, divided into tiny leaflets *c.* 6mm across. Flowers with 4 *purplish sepals*, no petals and dangling stamens with long purple filaments and pale brown anthers.

GROWTH Perennial.
HEIGHT 8–15cm.
FLOWERS Jun–Jul.
STATUS Native.
ALTITUDE 0–1210m (mostly above 300m).

Common Meadow-rue

Thalictrum flavum

Fairly
common
but local.
Fens, wet
meadows,
alongside ditches
and streams,
always on
alkaline soils.
✺ Usually robust,
with tiny flowers massed
together to produce *frothy, white
flower-heads*. DETAIL Near-hairless.
Stems with few or no branches,
ridged, hollow. Leaves divided into
many leaflets, each 10–20mm long,
much longer than broad, with a wedge-shaped base and 3–4-lobed tip;
upper leaves stalkless. Flowers with 4–5 small, narrow, creamy sepals and
abundant long stamens with creamy filaments and yellow anthers; petals
absent. Seeds 1.5–2.5mm long with *6 ribs*. SEE ALSO Meadowsweet (p. 71).

GROWTH Perennial.
HEIGHT 50–150cm, the
shoots coming from a network
of underground rhizomes.
FLOWERS Jul–Aug.
STATUS Native.
ALTITUDE Lowland.

Lesser Meadow-rue *Thalictrum minus*

Very locally common. Open grassy sites on dry sandy soils
or over chalk and limestone; rocky places (including
limestone pavement); stony lakeshores and
riverbanks; coastal dunes. ✺ Very variable in
size, but usually with widely-spreading branches
carrying
loose
groups
of *tiny
drooping
flowers*; stem
and flowers often tinged purplish.
DETAIL Near hairless. Stems often
zigzag, variably furrowed, hollow.
Leaves cut into many small
leaflets, 7–12mm long, *usually
broader than long* and variably
lobed or toothed at the tip (leaves
often glaucous with dense stalked
glands below in coastal dunes).
Flowers with 4–5 small greenish-
white sepals (often washed
purplish) but no petals; the long
stamens are held *more or less
drooping*, with yellowish or purple
filaments and yellow anthers.
Seeds 3–6mm long with *8–10 ribs*.

GROWTH Perennial.
HEIGHT 25–120cm.
FLOWERS Jun–Aug.
STATUS Native, but grown
in gardens and sometimes
thrown out.
ALTITUDE 0–855m.

Opposite-leaved Golden Saxifrage
Chrysosplenium oppositifolium

Locally abundant. Damp, shady places, often on acid soils, including boggy woods, streamsides, cliffs and gullies; also springs and flushes in the hills. ❀ A creeping evergreen, may form extensive patches. The short-stalked leaves are in *opposite* pairs and the umbel-like heads of tiny greenish-gold flowers are surrounded by much larger yellow-green bracts. DETAIL Stems hairy, with *leafy* non-flowering runners that root at the lower nodes and more or less erect flowering stems. Leaves sparsely hairy to hairless, 5–20mm across, more or less rounded, tapering broadly into the stalk, the edges with shallow, blunt teeth. Flowers 3–5mm across, calyx 4-lobed, no petals but a wavy-edged, nectar-secreting disk around the ovary; 8 stamens, anthers golden-yellow, ovary/stigma relatively large, 2-lobed.

GROWTH Perennial.
HEIGHT 5–15cm.
FLOWERS Feb–Jun.
STATUS Native.
ALTITUDE 0–1100m.

Alternate-leaved Golden Saxifrage
Chrysosplenium alternifolium

Scarcer and more local than Opposite-leaved Golden Saxifrage. Favours similar habitats, but usually where flushed by alkaline water, although the 2 species may occur together. ❀ Flowers slightly larger and brighter, basal leaves *long-stalked*, placed *alternately* on the stem. DETAIL Creeping, with *leafless non-flowering runners*. Flowering shoots with 2–5 sparsely hairy, basal leaves, 15–35mm across and more or less circular with the margins cut into square, shallow lobes and the base heart-shaped; leaf stalks up to 9cm long.

GROWTH Perennial.
HEIGHT To 20cm.
FLOWERS Apr–May.
STATUS Native.
ALTITUDE 0–915m.

Meadow Saxifrage
Saxifraga granulata

Locally common in old grassland such as grassy banks, verges and churchyards, also unimproved pastures and meadows (but these are now scarce); sometimes in damp woodland and along shady riverbanks. Avoids very acid soils. ❀ The near-leafless flower stems, each with

GROWTH Perennial.
HEIGHT 10–30cm.
FLOWERS Apr–Jun.
STATUS Native.
ALTITUDE 0–580m.

a loose cluster of 3–12 flowers, grow from a compact basal rosette. DETAIL Stems with abundant sticky glandular hairs. Basal leaves long-stalked, 5–30mm across, with shallow lobes and scattered hairs; stem leaves smaller and narrower with shorter stalks. The leaves die off after flowering. Flowers with 5 petals, 9–16mm long. Bulbils are produced in the axils of the rosette leaves.
SIMILAR SPECIES **Londonpride** *S.* × *urbium* is common in gardens and one of a number of similar species that may escape and turn up in damp, shady places. It has fleshy, oblong-oval leaves, 20–30mm long, with numerous blunt teeth, and smaller flowers, the petals 4–5mm long, usually marked with red spots. SEE ALSO Greater Stitchwort (p. 162).

Rue-leaved Saxifrage *Saxifraga tridactylites*

GROWTH Perennial.
HEIGHT 10–50cm.
FLOWERS Mar–May.
STATUS Native.
ALTITUDE 0–595m.

Locally common in sparse, open vegetation on thin, dry, often compacted and usually calcium-rich soils, on rocks, heaths and dunes, also walls, pavements, roofs and other man-made structures. Germinates in the late autumn, withering by midsummer. ❀ Small, sometimes tiny, with 3–5 lobed leaves and erect flower stems. The stem, leaves and calyx are often bright red and have many red-tipped glandular hairs. DETAIL Basal leaves spoon-shaped, *c.* 10mm across, but soon withering; stem leaves lobed. Flowers with 5 petals 2–3mm long, calyx bell-shaped. SEE ALSO Common Whitlowgrass (p. 139).

Purple Saxifrage *Saxifraga oppositifolia*

GROWTH Perennial.
HEIGHT Prostrate.
FLOWERS Late Feb–May.
STATUS Native.
ALTITUDE 0–1210m
(mostly 300–1000m).

Very locally common on cliffs, rock ledges, scree and stony flushes, where the rocks are base-rich and, in its more southerly sites, north-facing. ❀ Forms dense, low-growing, straggling mats that are covered in the late winter and early spring with pinkish-purple flowers. **DETAIL** Stems creeping and rooting, producing short shoots that are densely covered with *opposite* pairs of leaves, each 4–5mm long, oval, with a whitish, lime-excreting pit on the blunt, thickened tip and bristly hairs on the margins. Flowers solitary, on very short stalks, with 5 petals each 5–10mm long and 10 stamens. **SEE ALSO** Moss Campion (p. 158).

Yellow Saxifrage *Saxifraga aizoides*

Locally common in wet, rocky and stony places alongside mountain streams and flushes and on nearby areas of bare wet ground, also cliffs, ravines and, in the far north, coastal dunes. ❀ Mat-forming, with erect flower stems carrying loose clusters of 3 or more yellow, 5-petalled flowers. **DETAIL** Sprawling, with both non-flowering stems that may root at the nodes and erect leafy flower shoots; stems sparsely hairy, often reddish. Leaves almost stalkless, fleshy, strap-shaped, 6–20mm long with a few irregularly-spaced hairs along the edges and often a whitish tip (due to a lime-excreting pore). Petals 3–6mm long, often spotted red; 10 stamens and a 2-lobed ovary. **SEE ALSO** Biting Stonecrop (p. 40).

GROWTH Perennial.
HEIGHT To 25cm.
FLOWERS Jun–Aug.
STATUS Native.
ALTITUDE 0–1175m.

Mossy Saxifrage *Saxifraga hypnoides*

GROWTH Perennial.
HEIGHT To 20cm.
FLOWERS May–Jul.
STATUS Native, but also occurs as a garden escape (e.g. S and SE England), often with red or yellow flowers.
ALTITUDE 0–1210m (mostly 200–760m).

Vulnerable. Locally common, mostly in upland areas, on damp rocks and scree, by streams and flushes and in stony turf, usually on calcium-rich soils and often in partial shade. ❁ Forms *moss-like mats* from which arise *erect, near-leafless stems with a few drooping flower buds* that are often washed pink and open into white, 5-petalled flowers. **DETAIL** Grows from a short creeping rootstock, the flowering stems have sterile side shoots; also produces trailing stolons that terminate in flowerless leafy rosettes. Leaves *c.* 10mm long with sparse long hairs at the base and usually cut into 3–5 narrow, pointed lobes. Flower stems and sepals with glandular hairs, petals 8–12mm long; stamens 10, stigmas 2.

Starry Saxifrage *Saxifraga stellaris*

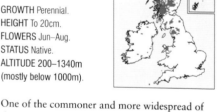

GROWTH Perennial.
HEIGHT To 20cm.
FLOWERS Jun–Aug.
STATUS Native.
ALTITUDE 200–1340m (mostly below 1000m).

One of the commoner and more widespread of the mountain specialities, growing on wet rock ledges and cliffs, alongside mountain streams and on open flushes, usually on acid soils. Plants are sometimes washed downstream to low altitudes. ❁ Compact leafy rosettes produce slender leafless reddish stems topped by loose clusters of flowers. The sepals are folded sharply downwards and each white petal has *2 yellow spots at the base*. **DETAIL** Leaves more or less stalkless, oval, toothed, 5–30mm long, often fleshy, with scattered hairs above. Stamens 10, surrounding the 2 large, pinkish carpels, each topped by a small stigma.

Orpine
Sedum telephium

Scattered on hedgebanks, verges, wood borders, limestone pavements, cliffs, rocky ravines and sometimes in ancient woodland (where it may fail to flower). ✿ *Succulent*, with upright stems, fleshy grey-green leaves and dense heads of reddish-purple flowers. **DETAIL** Hairless. *Leaves alternate*, oval, 20–100mm long, more or less flat, with toothed margins. Flowers with 5 petals, 3–5mm long, 10 stamens (about as long as the petals) and 5 erect purple ovaries / stigmas.

SIMILAR SPECIES Butterfly Stonecrop S. *spectabile* may occur as a garden throw-out. *Leaves opposite* or in whorls of 3, flowers larger (petals 5–8.5mm), with the *stamens rather longer than the petals*. **Caucasian Stonecrop** S. *spurium*, another garden throw-out, is smaller, with prostrate rooting stems, opposite leaves and larger flowers (petals 8–12mm).

GROWTH Perennial.
HEIGHT 20–60cm.
FLOWERS Jun–Sep.
STATUS Native.
ALTITUDE 0–480m.

Roseroot
Sedum rosea

Very locally common on sea cliffs and mountain crags, usually where the rocks are at least slightly basic. ✿ A fleshy rhizome produces erect shoots with many stalkless *fleshy leaves* and dense terminal clusters of yellow flowers. **DETAIL** Hairless. Leaves grey-green, 10–40mm long, more or less flat. Flowers dioecious (male and female on different plants), with 4 yellow petals 2–4mm long with an orange gland at the base; sepals and petals often washed purplish-red on outer surfaces.

GROWTH Perennial.
HEIGHT To 35cm.
FLOWERS May–Aug.
STATUS Native.
ALTITUDE 300–1165m, but to sea level in the far NW.

Reflexed Stonecrop *Sedum rupestre*

Locally common on old walls, rocky outcrops, hedgebanks and waste ground. ✿ *Succulent*, with sprawling non-flowering shoots and upright flowering stems that *droop at the tip in bud*. DETAIL Hairless. Non-flowering shoots with *evenly spaced leaves*, dead leaves not persisting. Leaves reflexed (especially on the lower part of the stem), 12–20mm long and 2–2.5mm wide, *more or less rounded in cross-section* (only slightly flattened above), pointed. Flowers with 6–7 petals, each 6–7mm long, sepals pointed.

GROWTH Perennial.
HEIGHT 10–35cm.
FLOWERS Jun–Aug.
STATUS Introduced from Europe by the 17th century (as a salad crop) and recorded in the wild by 1666.
ALTITUDE 0–365m.

Rock Stonecrop *Sedum forsterianum*

Nationally Scarce as a native, on cliffs, rocky outcrops and screes, both in dry open places and in wet woodland; also naturalised in churchyards, and on walls, hedgebanks and waste ground. ✿ Rather like Reflexed Stonecrop but less robust, stems erect in bud. DETAIL Non-flowering shoots with *tassel-like clusters of leaves* at the tip and persistent brown, shrunken, dead leaves towards the base. *Leaves held erect*, especially towards tip of stems, *c.* 1mm wide and *semi-circular in cross section (strongly flattened above)*. Sepals more or less blunt.

GROWTH Perennial.
HEIGHT To 20cm.
FLOWERS Jun–Jul.
STATUS Native to Wales and parts of the SW, but also widely introduced both there and scattered across Britain.
ALTITUDE To 680m.

Biting Stonecrop *Sedum acre*

Common. Bare places on dry, thin or virtually non-existent soils: shingle, dunes, short turf and rocks, also walls, roofs and other man-made structures. ❀ Low-growing, mat-forming evergreen *succulent* with small clusters of yellow flowers on short, erect stems. **DETAIL** Hairless. Stems green, sprawling, rooting at the nodes. Leaves alternate,

overlapping, bright green (occasionally reddish), 3–5mm long, egg-shaped, with a hot, peppery, 'biting' taste. Flowers with 5 petals, each 6–8mm long. **SEE ALSO** Yellow Saxifrage (p. 36).

GROWTH Perennial.
HEIGHT 2–10cm.
FLOWERS May–Jul.
STATUS Native.
ALTITUDE 0–550m.

Hairy Stonecrop *Sedum villosum*

GROWTH Biennial or perennial.
HEIGHT 5–10cm.
FLOWERS Jun–Aug.
STATUS Native.
ALTITUDE 0–1140m, but mostly 250–500m.

Nationally Scarce and Near Threatened. Rather local in N England and Scotland on open, stony flushes and streamsides where the water is slightly alkaline. ❀ A low-growing *succulent* with upright unbranched stems; *both stem and leaves have many glandular hairs*, flowers pink, in open clusters. **DETAIL** Leaves often reddish, *alternate, flattened on upperside*, blunt, 6–12mm long. Petals 5, more or less blunt, 3.5–5mm long; sepals separated all the way to the base. **SIMILAR SPECIES Thick-**

leaved Stonecrop *S. dasyphyllum*, a scarce garden escape on old walls in the lowlands, is also glandular-hairy. It has shorter, grey-green or pinkish *opposite leaves* that are roughly circular in cross-section, and slightly smaller flowers with 5–6 petals.

White Stonecrop *Sedum album*

Common on dry banks, rocks and shingle, but most characteristically on man-made structures such as walls, roofs, graves, paths and old concrete. ✿ Low-growing evergreen *succulent* with erect, *relatively long (7–15cm), well-branched flowering stems* carrying loose, umbel-like clusters of *at least 20 flowers*. DETAIL Hairless. Stems sprawling, rooting at the nodes, with minute, hair-like papillae. Leaves alternate, *clear green* or sometimes reddish, 6–12mm long, egg-shaped to cylindrical, blunt. Flowers with 5 more or less blunt petals, 2–4mm long, *sepals fused at the base* into a short tube; petals, ovaries / stigmas and filaments white or flushed pale pink, anthers dark red.

GROWTH Perennial.
HEIGHT 7–20cm.
FLOWERS Jun–Aug.
STATUS Ancient introduction (perhaps native in the SW).
ALTITUDE 0–570m.

English Stonecrop *Sedum anglicum*

GROWTH Perennial.
HEIGHT 2–10cm.
FLOWERS Jun–Sep.
STATUS Native.
ALTITUDE 0–1080m.

Common in W Britain but very local in the E. Acid rocks, screes, shingle, short sandy grassland, dunes, old walls and dry banks. ✿ Low-growing, mat-forming evergreen *succulent, flower spikes short (2–5cm)*, with *1–2 branches*, each with *only 3–6 flowers*. DETAIL Hairless. Stems reddish, sprawling, rooting at the nodes. Leaves pale *grey-green* to reddish, alternate, *3–6mm long*, egg-shaped. Petals 5, sharply pointed, white, tinged or streaked pink on outer side, 2.5–4.5mm long; sepals pinkish, *separate all the way to the base*. Ovaries and stigmas often heavily washed or streaked with reddish, anthers dark red.

GROWTH Perennial.
HEIGHT 3–40cm (2–50cm).
FLOWERS May–Aug.
STATUS Native.
ALTITUDE 0–550m.

Navelwort *Umbilicus rupestris*

Very common on stone walls, rocks and hedge-bottoms, often shaded, and usually on acid substrates. ❁ Spires of greenish-white flowers stand above *fleshy, circular leaves that are attached to their stalks centrally below*, and have a dimple ('navel') in the centre above. DETAIL Leaves long-stalked, hairless, with scalloped edges. Calyx with 5 narrow teeth, corolla 7–10mm long, tubular, divided at the tip into 5 teeth; stamens 10, concealed. SEE ALSO Marsh Pennywort (p. 311).

Mossy Stonecrop
Crassula tillaea

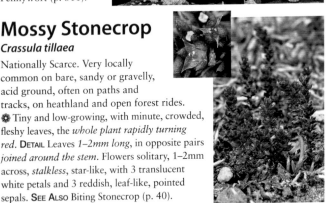

GROWTH Annual.
HEIGHT Stems to 5cm long.
FLOWERS Late Mar–Jun.
STATUS Native.
ALTITUDE Lowland.

Nationally Scarce. Very locally common on bare, sandy or gravelly, acid ground, often on paths and tracks, on heathland and open forest rides. ❁ Tiny and low-growing, with minute, crowded, fleshy leaves, the *whole plant rapidly turning red*. DETAIL Leaves *1–2mm long*, in opposite pairs *joined around the stem*. Flowers solitary, 1–2mm across, *stalkless*, star-like, with 3 translucent white petals and 3 reddish, leaf-like, pointed sepals. SEE ALSO Biting Stonecrop (p. 40).

New Zealand Pigmyweed *Crassula helmsii*

GROWTH Perennial.
HEIGHT Stems to 30cm.
FLOWERS Jun–Sep.
STATUS Introduced from Australia and New Zealand.
ALTITUDE To 345m.

An alien, sold as an oxygenator for aquaria and then dumped or planted in the wild, it has spread rapidly since 1980 and is now locally abundant. Almost impossible to eradicate once established. ❁ A small, slightly fleshy evergreen that forms *dense mats* around ditches, ponds and lakes. DETAIL Leaves 4–15mm long, fine-pointed, in *well-spaced, opposite pairs joined around the stem*, with an indistinct *fine dark ring* at the base of each junction; may root at these nodes. *Flowers stalked*, 1–2mm across, the 4 whitish petals *longer than the sepals*. SEE ALSO Water starworts (p. 348).

FAMILY FABACEAE: PEAS, VETCHES & CLOVERS

One of the larger families of British flowering plants, members of the pea family vary greatly in size, from tiny clovers, through shrubs such as broom and gorse, to *Acacia* trees. All are easily identified as members of the pea family, however, by their distinctive flower structure. The 2 lower petals are fused to form the boat-shaped *keel*, the two side petals are known as the *wings*, while the upper petal, the *standard*, is often held erect (like a flag). The 5 sepals are usually fused into a tube – the calyx – which often has 5 teeth at the mouth. The calyx teeth may be differentiated into an upper lip with 2 teeth and a lower lip with 3 teeth. In some species of pea the flowers are so small that their structure may not be apparent, while in others they are packed closely together in clusters (e.g. Red and White Clovers), disguising the structure of the individual flowers. A 10× hand lens will reveal all.

Peas produce their seeds in elongated pods. In many species the pod splits into 2 valves when ripe, releasing the seeds, but in some the pod breaks up into 1-seeded units (e.g. Horseshoe Vetch p. 45). The size and shape of the pod, and the number of seeds that it contains, may be important for identification. Some peas have modified leaves, with the midrib extended from the tip of the leaf as a tendril, which may be branched. The size and shape of the stipules, located at the junction of the leaf stalk and the main stem, are also often helpful in identification.

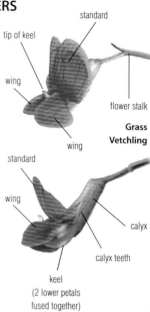

standard

tip of keel

wing

flower stalk

Grass Vetchling

wing

standard

wing

calyx

calyx teeth

keel
(2 lower petals fused together)

Purple Milk Vetch *Astragalus danicus*

Endangered. Very locally common in short, unimproved grassland on chalk and limestone, also dunes and machair in the N. ✿ A low-growing pea with *tight clusters of purple-blue flowers* and relatively small leaves that *lack tendrils*. **DETAIL** Sparsely hairy. Leaves divided into 6–13 pairs of small leaflets (usually less than 10mm long) plus a single terminal leaflet. Stipules very small, joined at the base. Flowers 15–18mm long, keel more or less blunt. Pods 7–9mm long, swollen, dark brown, with spreading white hairs. **SEE ALSO** Lucerne (p. 59).

GROWTH Perennial.
HEIGHT 5–30cm.
FLOWERS May–Jul.
STATUS Native.
ALTITUDE 0–710m.

Goat's Rue *Galega officinalis*

GROWTH Perennial.
HEIGHT 60–150cm.
FLOWERS Jun–Jul.
STATUS Introduced to cultivation from E Europe by 1568 and reported from the wild by 1640, but has only spread very recently, especially around London.
ALTITUDE Lowland.

Locally fairly common in SE England on waste ground, verges and around old gravel pits, especially on urban brownfield sites. ✹ A tall, shrub-like pea with showy, erect spikes of bluish-mauve or white flowers; *the leaves lack tendrils.* **DETAIL** Hairless to sparsely hairy. Leaves divided into 4–9 pairs of leaflets, 30–50mm long, plus a single terminal leaflet. Flowers with 5 bristle-like calyx teeth. Pods cylindrical, 20–30mm long.

Sainfoin *Onobrychis viciifolia*

GROWTH Perennial.
HEIGHT 20–60cm.
FLOWERS Jun–Aug.
STATUS Introduced from central Europe in the 17th century as a fodder crop. Little-used now, although persisting in some places, but increasingly sown in 'conservation mixes'. The prostrate form may be native.
ALTITUDE 0–375m.

Fairly common in dry grassy places on calcareous soils. ✹ The common form is an erect, showy pea, with *conical spikes of red-veined, bright pink flowers; the leaves lack tendrils.*
Smaller, prostrate forms are found on old chalk grassland; these are listed as Near Threatened. **DETAIL** Sparsely hairy. Leaves divided into 6–14 pairs of leaflets, 10–35mm long with a tiny point at the tip, plus a single terminal leaflet. Stipules papery, small and pointed. Flowers 10–14mm long, wings very short (less than half length of other petals). Pods 5–8mm across, oval, toothed, net-veined, hairy, 1-seeded.

prostrate form

Kidney Vetch *Anthyllis vulneraria*

Locally abundant. Rough grassland, usually on dry, lime-rich soils, often on sea cliffs, shingle and dunes; sometimes sown on verges.

❁ Named after its large, kidney-shaped flower heads, which may be yellow, orange or sometimes red. *The leaves lack tendrils.* DETAIL Silky-hairy. Upper leaves with 4–7 pairs of leaflets plus a single terminal leaflet. Flower heads paired,

GROWTH Perennial.
HEIGHT Stems to 60cm.
FLOWERS May–Sep.
STATUS Native.
ALTITUDE Mostly lowland, but to 945m.

2–4cm wide, with a ruff of leaf-like bracts below. Flowers 12–15mm long, each with an *inflated calyx, densely covered with woolly-white hairs.* Seed pods egg-shaped, 2–4mm long, hairless, *enclosed by the calyx.*

Horseshoe Vetch *Hippocrepis comosa*

Very locally abundant in short, unimproved, species-rich grassland on S-facing slopes on chalk and limestone.

❁ A low-growing pea with clusters of 4–8 yellow flowers *spreading star-wise* at the end of relatively long stalks. *Leaves divided into 2–8 pairs of small leaflets* (plus a single terminal leaflet); *lacks tendrils.* DETAIL More or less hairless. Leaflets 3–8mm long. Stipules 2, small and pointed. Flowers 5–10mm long. Seed pods wavy, 10–30mm long; initially several are clustered together to form a 'bird's foot', but they eventually break into *horseshoe-shaped fragments.* SEE ALSO birdsfoot trefoils (pp. 46–47) which have only 5 leaflets (2 of which are stipule-like), as well as larger flowers and different pods.

GROWTH Perennial.
HEIGHT Stems sprawling, 5–30cm.
FLOWERS May–Jul.
STATUS Native.
ALTITUDE Mostly lowland, but to 600m.

GROWTH Perennial.
HEIGHT Stems 10–50cm.
FLOWERS May–Sep.
STATUS Native.
ALTITUDE 0–915m.

Common Birdsfoot Trefoil *Lotus corniculatus*

Common in every kind of short, dry, unimproved grassland, also shingle banks and sand dunes. ✿ A sprawling, low-growing pea. Leaves divided in 5 leaflets with the basal pair, at the junction of the stem and leaf stalk, resembling stipules (the true stipules are tiny); *lacks tendrils.* Flowers yellow to orange or streaked red, often reddish in bud, in clusters at the tip of long stalks. Seed pods in groups, remarkably like a bird's foot. DETAIL Hairless to sparsely hairy. *Stems solid.* Leaflets 3–15mm × 4–9mm. Flowers 10–16mm long, in clusters of 2–8, calyx teeth *erect* in bud, the *upper 2 calyx teeth curving inwards.* Seed pods 15–30mm, multi-seeded. The introduced var. *sativus* is larger and more erect, sometimes with hollow stems; it is often sown in 'wildflower' mixes.

Greater Birdsfoot Trefoil *Lotus pedunculatus*

Common in moist grassland: marshes, wet meadows, ditches, wet woodland rides. ✿ The wetland counterpart of Common Birdsfoot Trefoil, but typically taller, more robust and often hairier, with *hollow lower stems*, broader leaves and rich yellow flowers that lack orange or red tones. DETAIL Hairless to hairy. Stems erect to sprawling. Leaflets 12–20mm long. Flowers 10–18mm long, in clusters of 5–12, the *calyx teeth curving out and down in bud*, the *upper 2 calyx teeth straight or curving outwards.*

GROWTH Perennial.
HEIGHT Shoots to 100cm.
FLOWERS Jun–Aug.
STATUS Native.
ALTITUDE Mostly lowland, but to 560m.

Narrow-leaved Birdsfoot Trefoil
Lotus tenuis

Fairly common but local. Brackish, grassy places on sea banks and coastal grazing marshes, also inland (mostly as a casual), along disused railway lines, verges and in old sand, chalk and clay pits. ❀ As Common Birdsfoot Trefoil but with *narrower and more pointed leaflets* and smaller, paler flowers. DETAIL Hairless to sparsely hairy. Stems often sprawling. Leaflets strap-shaped, 8–14mm × 1.2–2.5mm, leaflets of upper leaves usually more than 4 times as long as wide; in Common Birdsfoot Trefoil they are less then 3 times as long as wide, rarely 4 times. Flowers in clusters of 2–4 (1–6), pale yellow, 6–12mm long, with the 2 rear calyx teeth long and curving strongly inwards, like a pair of earwig's pincers.

GROWTH Perennial.
HEIGHT Stems to 90cm.
FLOWERS Jun–Aug.
STATUS Native.
ALTITUDE Lowland.

Hairy Birdsfoot Trefoil *Lotus subbiflorus*

GROWTH Annual.
HEIGHT Stems 3–30cm (80cm).
FLOWERS (May) Jun–Sep.
STATUS Native.
ALTITUDE Lowland.

Nationally Scarce. Very local in the SW in sparse, dry coastal grassland in sheltered, sunny spots amidst ranker vegetation on cliff-tops and around rocky outcrops, usually on neutral to acid, sandy soils; sometimes found inland. ❀ *Foliage greyish-green and densely hairy*, with small flowers and short pods. DETAIL Leaflets to 13mm long, narrowly oval. Flowers in clusters of 2–4 (sometimes just 1), on stalks *c.* 20mm long (longer than leaves), each flower 5–10mm long with obtuse-angled bend near base of lower edge of keel. Pods 6–15mm long, mostly less than 3 times the length of the calyx. SIMILAR SPECIES Slender Birdsfoot Trefoil *L. angustissimus* is found in similar places in the SW, although much scarcer, as well as sand and gravel workings in Hants and Kent. Not as hairy, with flowers in clusters of 1–2, on stalks shorter than leaves, with *c.* 90° bend on lower edge of keel. Pods slender, 12–30mm long, mostly more than 3 times calyx length.

GROWTH Perennial.
HEIGHT Sprawling, shoots
60–200cm.
FLOWERS Jun–Aug.
STATUS Native.
ALTITUDE Mostly lowland,
but to 560m.

Tufted Vetch
Vicia cracca

Common in undisturbed grassland on damp (but not waterlogged) soils, from road verges to river and stream banks, woodland fringes, meadows and fens. ❋ A beautiful and conspicuous pea that scrambles over other vegetation via its branched tendrils, with long-stalked, *1-sided spikes of 10–30 vivid bluish-mauve flowers* (ageing bluer). DETAIL More or less downy. Leaves divided into 5–15 pairs of narrow leaflets, 10–25mm long. Stipules *c.* 10mm long, not toothed. Flowers 8–12mm long, calyx teeth unequal. Seed pod 10–25mm long, hairless, brown, 2–6 seeded. SIMILAR SPECIES The introduced **Fodder Vetch** *V. villosa* is occasional in SE England. An annual, it has larger flowers (10–20mm), with the narrow 'stem' of the standard petal twice as long as the main part (not roughly equal) and a large bulge at the base of the calyx.

Bush Vetch *Vicia sepium*

GROWTH Perennial.
HEIGHT Shoots 20–60cm.
FLOWERS Apr–Oct.
STATUS Native.
ALTITUDE Mostly lowland,
but to 820m.

Common. Hedgebanks, shady verges, green lanes, woodland margins, and scrubby grassland. ❋ A low- to medium-growing scrambling pea with branched tendrils and clusters of 2–6 untidy-looking flowers that fade from purplish-pink to dull blue. DETAIL Stems hairless to sparsely hairy. Leaves sparsely hairy, divided into 5–9 pairs of oval-oblong leaflets 10–30mm long, more or less blunt at the tip but with a fine bristle-point (*mucro*); stipules very small, variably toothed. Flowers 12–15mm long, calyx reddish-purple, straggly-hairy, the lower teeth longer than the upper. Pods 20–35mm long, ripening to black, 3–10–seeded.

ssp. *segetalis*

Common Vetch
Vicia sativa

Very common in a wide variety of grassy places. ✻ A variable sprawling or scrambling pea with bright pinkish-purple flowers, 1–2 together on very short stalks; tendrils usually conspicuously branched. **DETAIL** Variably hairy. Leaves divided into 3–6 pairs of alternate leaflets (occasionally 2–9 pairs), each 6–15mm long, with a shallowly-notched tip containing a small point (a mucro). Stipules 4–8mm long, half arrow shaped, variably toothed. Flowers 10–30mm long, calyx teeth nearly equal. Seeds smooth. Once open, the pods coil into a spiral. There are 3 subspecies.

ssp. *nigra*

V. s. nigra 'Slender' or 'Narrow-leaved Vetch'. Slender, with the leaflets on the upper leaves (abruptly) much narrower than those on lower leaves (1–4mm wide). Flowers 14–19mm long, uniform bright pinkish-purple. Pods 23–38mm, brown to black, smooth, hairless. Native in dry, open grassland, especially on sandy soils, and commonest on the coast.

V. s. segetalis Robust, with the leaflets broader and all more or less the same width, and the flowers 9–26mm, usually 2-tone, with the standard petal much paler than the wings. Pods 28–70mm, brown to black, smooth, hairless. An ancient introduction, once cultivated for fodder, and now the commonest form over much of Britain. Rather less fussy in its choice of habitats than *nigra*.

V. s. sativa As *segetalis* but even stouter, pods 36–80mm, often hairy, light brown and 'beaded', with slight constrictions between the seeds. Formerly a fodder crop but now rare.

GROWTH Annual.
HEIGHT 15–150cm.
FLOWERS May–Sep.
STATUS Native and introduced.
ALTITUDE Mostly lowland, but to 415m.

Spring Vetch
Vicia lathyroides

Rather local. Grassy places on dry, sandy soils, often near the sea, also old walls and disturbed ground. ✻ A slender, low-growing, early-flowering pea with small flowers, very easy to overlook. Flowers almost stalkless, solitary, dull purple, ageing bluish. **DETAIL** Whole plant downy. Leaves divided into 2 *pairs of leaflets* (occasionally 3–4 pairs), usually *opposite*, 4–10mm *long*, with a shallowly-notched tip containing a minute point (mucro). *Tendrils not branched, sometimes absent*, especially on upper leaves; stipules tiny. *Flowers 6–9mm long*. Pod hairless, black, 15–30mm long; seeds closely covered with *minute pimples* (the *only certain distinction* from small Common Vetch).

GROWTH Annual.
HEIGHT 5–20cm.
FLOWERS Apr–May.
STATUS Native.
ALTITUDE Lowland.

Hairy Tare
Vicia hirsuta

Common on disturbed, well-drained soils in rough grassy places and on waste ground. ✿ An inconspicuous pea with tiny off-white flowers, trailing stems and branched tendrils. Often forms a loose tangle in and around other vegetation. **Detail** Sparsely hairy to hairless. Leaves divided into 4–10 pairs of strap-shaped leaflets, 5–13mm long. Flowers mauve-white, 3–5mm long, with the calyx teeth *equal* in length and at least as *long as* the calyx tube. Flowers in clusters of 2–7 (1–9) at the end of slender 1–3cm long stalks. Seed pod 6–11mm long, *downy, almost all 2-seeded.*

GROWTH
Annual.
HEIGHT Stems
20–80cm.
FLOWERS
May–Aug.
STATUS Native.
ALTITUDE
0–335m.

seed pods

Smooth Tare   *Vicia tetrasperma*

Fairly common. Much like Hairy Tare and found in similar habitats, but favouring heavier soils and more often in permanent grassland. ✿ *Flowers larger and more colourful, 1–2 together* (only exceptionally as many as 4), *pods hairless.* **Detail** Leaves divided into 3–6 (8) pairs of leaflets, 10–20mm long, hairless above; tendrils mostly unbranched. Flowers 4–8mm long, deep lilac, *calyx teeth unequal,* with *upper 2 shorter than calyx tube.* Pod 9–16mm long, 4-seeded (3–5).

GROWTH Annual.
HEIGHT Stems to c. 60cm.
FLOWERS May–Aug.
STATUS Native.
ALTITUDE Lowland.

Slender Tare   *Vicia parviflora*

Nationally Scarce, listed as Vulnerable. Confined to grassy places on chalky-clay soils, wet in winter but baked dry in summer; formerly also an arable weed. ✿ Rather like Smooth Tare, but with slightly larger, bluer flowers, fewer, larger leaflets, and 4–6 (8) seeds per pod. **Detail** 2–4 pairs of leaflets, 15–20mm long, hairy above; tendrils usually unbranched. Flowers 6–9mm long, pale bluish-purple, in groups of 1–4; calyx as Smooth Tare. Pods 12–17mm long. The best diagnostic feature is the *circular to oval scar* on the seed where it was attached to the placenta (seed scar 2x as long as wide in Smooth Tare; 20x lens).

GROWTH Annual.
HEIGHT Stems to
60cm.
FLOWERS
Jun–Aug.
STATUS Native.
ALTITUDE Lowland.

Wood Vetch *Vicia sylvatica*

Local and uncommon in sunny woodland, hedgerows and scrub, on scree and, in the N and W, rough cliff-top grassland and shingle. ❀ A scrambling pea with branched tendrils and loose, long-stalked, 1-sided spikes of 4–15 attractive *purple-veined white flowers*. DETAIL Near-hairless. Stems more or less square. Leaves divided into 4–12 pairs of leaflets, 5–15mm long, long-oval with a tiny point at the tip; stipules with jagged teeth. Flowers 12–20mm long. Pods 25–30mm long, 4–5-seeded, ripening to black, hairless.

GROWTH Perennial.
HEIGHT 60–200cm.
FLOWERS Jun–Aug.
STATUS Native.
ALTITUDE 0–675m.

Wood Bitter Vetch *Vicia orobus*

Nationally Scarce, listed as Near Threatened. Scarce and decidedly local on the edges of unimproved hay meadows, roads and tracks (often on banks or amongst rocks and bushes), also cliffs and crags in the north. ❀ Erect, short- to medium-sized pea, with long-stalked clusters of 6–20 stunningly attractive pink and violet flowers that fade to whitish. Not a climber, the tendril is reduced to a short point at the tip of the leaf (or occasionally a terminal leaflet). DETAIL Sparsely hairy. Stems round. Leaves divided into 6–15 pairs of leaflets, 10–15mm long with a tiny point at the tip; stipules large, toothed, half arrow-shaped, to 13mm long. Flowers 12–15mm long. Pods 20–30mm long, oblong, pointed, pale brown, hairless, with 4–5 seeds.

GROWTH Perennial.
HEIGHT 30–60cm.
FLOWERS May–Sep.
STATUS Native.
ALTITUDE 0–560m, mostly 200–300m.

GROWTH Perennial.
HEIGHT 30–60cm.
FLOWERS May–Jun.
STATUS Native.
ALTITUDE Mostly lowland,
but to 760m.

Bitter Vetch
Lathyrus linifolius

Locally common.
Rough grassland,
scrub, open woodland,
rocky places and under
Bracken, usually on
poor, acid soils. ❀ Erect
but low-growing pea
with *narrowly-winged
stems*. Not a climber,
*the tendrils are reduced
to short points at the
leaf tips*. Flowers in
loose, long-stalked
clusters of 2–6, reddish-

purple (fading to blue-green), with a *slate-blue calyx*. DETAIL Hairless.
Leaves divided into 2–4 pairs of leaflets, 10–45mm long, pointed or blunt;
occasionally has a terminal leaflet; stipules 7mm long, weakly toothed.
Flowers 10–16mm long. Seed pod 25–45mm long, pointed, hairless, brown.

GROWTH Annual.
HEIGHT 30–60cm.
FLOWERS May–Jun.
STATUS Native.
ALTITUDE Lowland.

Bithynian Vetch *Vicia bithynica*

Nationally Scarce and listed as Vulnerable.
Extremely local in grassland around under-cliffs
and hedge-bottoms, especially on clay soils near
the sea. ❀ Scrambling, with *branched tendrils*
and groups of 1–2 flowers that are *conspicuously
bicoloured*, the standard purple and the wings
and keel whitish. DETAIL More or less hairless.
Stems angled, narrowly winged. Leaves divided
into 2–3 pairs of oval leaflets, 20–55mm long.
Stipules large, 10–20mm long, with jagged teeth.
Flowers 16–20mm long. Seed pod 25–50mm,
4–8 seeded, brown to yellowish, hairy.

GROWTH Annual.
HEIGHT 15–60cm.
FLOWERS Jun–Sep.
STATUS Native.
ALTITUDE Lowland.

Yellow Vetch
Vicia lutea

Nationally Scarce, listed as Near
Threatened. Very local on coastal
grassland, especially on shingle,
and sea-cliffs; casual inland.
❀ Sprawling, with
branched tendrils. Flowers
creamy (not yellow!),
with fine darker veins,
1–2 flowers together in
the leaf axils. DETAIL
Variably hairy. Leaves
divided into 3–8

pairs of leaflets, 6–12mm long, with a fine point at the tip;
stipules tiny. Flowers 15–25mm long. Pods 20–40mm long,
4–8 seeded, black to yellow-brown, very hairy.

Meadow Vetchling *Lathyrus pratensis*

GROWTH Perennial.
HEIGHT Shoots 30–120cm.
FLOWERS May–Aug.
STATUS Native.
ALTITUDE Mostly lowland,
but to o 450m.

Very common. Rough grassy places, hedge-bottoms and scrub,
avoiding very dry soils. ❀ Scrambles via variably branched tendrils.
The leaves are cut into a *single pair of leaflets*, but the *stipules are large
and leaf-like*. Flowers in loose, long-stalked clusters of 5–12. DETAIL
Variably hairy. Stems square (but not winged). Leaflets lanceolate, sharply-
pointed, more or less grey-green, 10–35mm long; stipules arrow-shaped,
10–30mm long. Flowers 10–18mm long. Seed pods 25–35mm long,
ripening to black.

tendril   leaflets

stipules

Yellow Vetchling *Lathyrus aphaca*

Nationally Scarce and listed as
Vulnerable. Very local in dry
grassy places on chalk, chalky
clays and limestone, especially
protected road verges, sea
walls and low, earthy sea cliffs;
sometimes a casual on waste
ground etc., and formerly an
arable weed. ❀ A scrambling,
hairless pea with *waxy grey-green
foliage* and small yellow flowers
growing singly (or in pairs) at the
end of long stalks. DETAIL Hairless.
Stems slender, more or less square.
On mature plants the leaves are
reduced to *long unbranched
tendrils*, with a pair of large,
leaf-like stipules, 6–50mm long,
at their base. Flowers 10–13mm
long. Seed pods 20–30mm, more
or less curved, brown.

GROWTH Annual.
HEIGHT Stems to 40cm
(100cm).
FLOWERS May–Aug.
STATUS Native.
ALTITUDE Lowland.

Sea Pea *Lathyrus japonicus*

GROWTH Perennial.
HEIGHT Shoots sprawling, to 90cm.
FLOWERS Jun–Aug.
STATUS Native.
ALTITUDE Lowland.

Nationally Scarce. Very locally abundant on shingle beaches. ✿ Distinctive, recalling a prostrate Sweet Pea, and forming voluptuous mats of blue-green foliage. **DETAIL** Hairless. Stems 4-angled. Leaves fleshy, blue-green, divided into 2–5 pairs of oval leaflets 25–45mm long; stipules spear-shaped, 10–25mm long, tendrils sometimes branched, may be absent. Flowers 15–25mm long, in short-stalked clusters of 2–10, initially purple, fading to blue. Seed pods 30–50mm, hairless, recalling a garden pea, and reportedly eaten in Suffolk in times of famine (e.g. 1555) but toxic in large quantities.

Marsh Pea
Lathyrus palustris

Nationally Scarce and listed as Near Threatened. Very local in fens with chalky ground water, amongst tall grasses, reeds and scrub. ✿ Scrambling, using tendrils for support, with long-stemmed clusters of 2–6 pinkish-purple flowers that age to bluish. **DETAIL** Usually hairless. Stems broadly winged. Leaves divided into 2–3 pairs of narrow leaflets, each 30–70mm long; stipules 10–20mm long, half arrow-shaped. Flowers 12–20mm long. Seed pods brown, slim and flattened, 25–60mm long, opening along both sides like a garden pea.

GROWTH Perennial.
HEIGHT Stems clambering, to 120cm.

FLOWERS May–Jul.
STATUS Native.
ALTITUDE Lowland.

Wild Liquorice *Astragalus glycyphyllos*

Rather local. Woodland edges, cliffs, banks, hedge-bottoms and rough scrubby grassland, usually in a sunny, sheltered spot on calcareous soils. ❀ A stocky, *shrub-like pea*, sometimes sprawling, with subtly zigzag stems and clusters of *dull creamy-yellow flowers*. **DETAIL** More or less hairless. Leaves divided into 3–7 pairs of oval leaflets 15–20mm long, plus a terminal leaflet; stipules oval. Flowers 11–15mm long. Seed pods hairless, banana-shaped, pointed, 25–40mm long.

GROWTH Perennial.
HEIGHT 60–100cm.
FLOWERS Jun–Aug.
STATUS Native.
ALTITUDE To 365m.

The confection known as 'liquorice' is made from an extract of the roots of another member of the pea family, *Glycyrrhiza glabra*, a native of SE Europe and W Asia.

Grass Vetchling *Lathyrus nissolia*

Rather local in unimproved rough grassland, often where subject to past disturbance, including sea walls, verges, gravel pits and open woodland rides. ❀ The grass-like leaves are nearly impossible to spot when not in flower, but the small, jewel-like crimson flowers are more conspicuous.

DETAIL Hairless. Stems erect, more or less square, hollow, unwinged. Leaves 150mm long (they are technically phyllodes – flattened leaf stalks), no tendrils, stipules present, but tiny; unlike grasses the base of the leaf does not sheathe the stem and there is no ligule. Flowers 8–18mm long, 1–2 together at the end of long stalks. Pods straight and rather narrow, 30–60mm long, hairless, ripening to pale brown and opening along 2 sides.

GROWTH Annual.
HEIGHT 20–90cm.
FLOWERS Jun–early Jul.
STATUS Native.
ALTITUDE Lowland.

Narrow-leaved Everlasting Pea
Lathyrus sylvestris

Rather local. Open woodland, scrub, hedgerows, rough grassland and sea-cliffs, especially on calcareous soils. ✿ A scrambling pea, climbing via branched tendrils, with long-stalked clusters of 3–8 purplish-pink flowers (greenish-white towards base). **DETAIL** Hairless. Stems broadly winged, leaf stalks winged. Leaves with 2 opposite leaflets, 40–80mm long and narrow (more than 4 times as long as wide); stipules narrowly lanceolate, 10–30mm × 2mm, *rather less than half width of stem.* Flowers 12–20mm long, *all calyx teeth shorter than calyx tube.* Seed pods flattened, 40–70mm long.

GROWTH Perennial.
HEIGHT Stems to 200cm.
FLOWERS Jun–Aug.
STATUS Native, also introduced, especially in the north.
ALTITUDE Lowland.

Broad-leaved Everlasting Pea
Lathyrus latifolius

Occasional to locally common on verges, hedgebanks, railways and waste ground. ✿ A rampant, scrambling, large-flowered pea, climbing via branched tendrils, with long-stalked clusters of 3–12 bright pink flowers (may be white or pale pink). **DETAIL** Hairless. Stems broadly winged, leaf stalks winged. Leaves with 2 opposite, *oblong-oval leaflets,* 50–100mm long (*less than 4 times as long as wide*); stipules oval, to 70mm × 20mm, *rather more than half width of stem.* Flowers 15–30mm long, *lower calyx teeth as long as, or longer than, the calyx tube.* Seed pods brown, 50–100mm.

GROWTH Perennial.
HEIGHT Stems to 300cm.
FLOWERS Jun–Aug.
STATUS Introduced to cultivation from S Europe by the 15th century and first recorded in the wild in 1670.
ALTITUDE To 340m.

Common Restharrow *Ononis repens*

GROWTH Perennial.
HEIGHT Stems 10–60cm
long, rising to 30cm.
FLOWERS Jun–Sep.
STATUS Native.
ALTITUDE 0–365m.

Fairly common in
rough, uniproved
grassland on well-
drained, lime-rich soils,
including verges, dunes
and coastal shingle.
❀ A low-growing and
often patch-forming pea, with tough stems. Sticky-hairy all over, attracting
dust and dirt, with dusty-pink flowers. Usually lack spines, but can be
spiny (var. *horrida*). DETAIL Grows from a creeping rhizome. Stems *hairy
all round*. Leaf stalks downy, 3–5mm long, with large leaf-like stipules
clasping the stems at the base; leaves with either 1 or 3 oval leaflets,
the terminal leaflet 10–20mm long. Flowers short-stalked,
7–15mm long, the *wing equalling the keel*. Seed pods
5–8mm long, *shorter than the calyx*, with 1–2 seeds.

single leaflet

stipules

Spiny Restharrow *Ononis spinosa*

GROWTH Perennial.
HEIGHT 30–70cm.
FLOWERS Jun–Sep.
STATUS Native.
ALTITUDE Lowland.

Locally common
in rough grassland,
typically on heavy,
neutral to chalky soils,
especially boulder
clays inland and
grazing marshes and
sea banks on the coast.
❀ Similar to Common
Restharrow but lacks
rhizomes, so not
patch-forming; stems
woodier, usually more
upright, with long,
straw-coloured spines
(rarely spineless).
Best separated from
the spiny form of
Common Restharrow
by the young shoots,
which have the hairs in
1 or 2 distinct rows, and by the pods, which are generally *longer*
than the calyx. DETAIL Leaves smaller and subtly narrower
and more pointed than Common Restharrow, terminal
leaflet 4–12mm long. Flowers 10–20mm long, with the
wing shorter than the keel. Seed pods 6–10mm long
with 2 (–4) seeds.

3 leaflets

single leaflet

GROWTH Biennial.
HEIGHT 60–150cm.
FLOWERS Jun–Sep.
STATUS Introduced (via N America) *c.* 1835, probably originally from SE Europe.
ALTITUDE Lowland.

Ribbed Melilot
Melilotus officinalis

Fairly common. Verges, field margins and waste ground. ❀ A slender, erect pea with *elongated spikes of small yellow flowers.* DETAIL Hairless. Leaves divided into 3 broadly oval leaflets, 15–30mm long and up to 25mm wide. Flowers 4–7mm long, with standard and wings more or less equal in length but the *keel shorter. Seed pods 3–5mm long,* oval, *hairless, ripening to brown,* with transverse ridges, mostly with 1 seed. SIMILAR SPECIES Small Melilot M. *indicus,* a scarce casual, has smaller flowers (2–3.5mm long) with the wings and keel equal, shorter than the standard, and small, hairless, olive-green seed pods (1.5–3mm long).

GROWTH Biennial or short-lived perennial.
HEIGHT 60–150cm.
FLOWERS Jun–Aug.
STATUS Ancient introduction from Europe, first recorded in the 16th century.
ALTITUDE Lowland.

Tall Melilot
Melilotus altissimus

Fairly common in similar habitats to Ribbed Melilot (although often the commoner species in the countryside) and very similar in appearance. ❀ Best separated by the pods, which are slightly smaller, *hairy, ripen to black, and mostly contain 2 seeds.* DETAIL Leaflets long-oval, becoming strap-shaped on upper leaves, up to 8mm wide. Flowers 5–7mm long, *standard, wings and keel all more or less equal.* Seed pods 5–7mm long, with a netted or transversely-ridged surface.

GROWTH Annual or biennial.
HEIGHT 60–150cm.
FLOWERS Jun–Aug.
STATUS Introduced from Europe and first recorded in the wild in 1822.
ALTITUDE To 335m.

White Melilot
Melilotus albus

Scattered on waste ground and other disturbed places, often sporadic in its appearances. ❀ A slender, erect, pea with *elongated spikes of small white flowers.* DETAIL Leaflets 15–30mm long and up to 15mm wide. Flowers 4–5mm long, wings and keel more or less equal, standard longer. Seed pods 3–5mm long, *hairless, ripening to brown,* with a netted or transversely-ridged surface.

Lucerne *Medicago sativa* subspecies *sativa*

Scattered on verges, waste ground and arable margins, especially on light soils. ❁ An erect, bushy pea with the leaves divided into 3 leaflets, clusters of *mauve to violet flowers* and *tightly coiled seed pods*; no tendrils. DETAIL Hairy to hairless. Leaflets long-oval, each up to 30mm long and finely toothed at the tip. Stipules narrow, pointed. Flowers 8–11mm long. Seed pods smooth, hairy to hairless, spiralled in 2–4 complete tight turns, the coils meeting in the centre, with 10–20 seeds. SEE ALSO Tufted Vetch (p. 48), Purple Milk Vetch (p. 43).

GROWTH Perennial.
HEIGHT 30–90cm.
FLOWERS Jun–Oct.
STATUS Introduced in the 17th century as a fodder crop ('Alfalfa'), probably from SW Asia, and first noted in the wild in 1804.
ALTITUDE Lowland.

stipules

Sickle Medick
Medicago sativa subspecies *falcata*

Nationally Scarce as a native in Breckland (Norfolk and Suffolk), on verges and grassy heaths on dry sandy soils; scattered records elsewhere as a casual. ❁ A low-growing, somewhat bushy pea; *flowers always yellow*.

GROWTH Perennial.
HEIGHT 30–40cm.
FLOWERS Jun–Aug.
STATUS Native.
ALTITUDE Lowland.

DETAIL Leaflets smaller than Lucerne, up to 12(20)mm long. Flowers 6–9mm long. *Seed pods more or less straight to crescent-shaped*, with 2–5 seeds.

Sand Lucerne
Medicago sativa subspecies *varia*

The fertile hybrid between Lucerne and Sickle Medick, arising naturally where they meet in E Anglia (where it is often commoner than Sickle Medick), and occasionally introduced as a fodder crop elsewhere. ❁ Flower colour varies from yellow, through green, turquoise, mauve and purple to blackish, often with *mixed colours on the same plant*. Seed pods semi-circular to spiralled into 1.5 turns, the coils open in the centre. DETAIL Flowers 7–10mm long. Pods with 3–8 seeds, or seeds aborting.

Black Medick *Medicago lupulina*

Common. Short grass and disturbed places on all but the wettest or most acid soils. ❁ A low-growing clover, each leaflet with a *tiny point (mucro) in the centre of the blunt tip.* Flowers tiny, 20–40 together in compact long-stalked heads. *Seed pods tiny, black, grouped in irregular cylindrical clusters.* DETAIL Leaves variably hairy, leaflets 5–20mm long, finely toothed towards the tip; stipules pointed, finely-toothed. Flowers 2–3mm long, calyx hairy. Seed pods coiled into a disc 1.5–3mm across, net-veined, variably hairy, with 1 seed. SIMILAR SPECIES Lesser and Hop Trefoils (p. 62), but these lack a mucro at the tip of the leaflets, have a hairless calyx, and very different pods.

GROWTH Annual (dry sites), or short-lived perennial.
HEIGHT Stems ascending, 5–25cm, sometimes to 80cm.
FLOWERS May–Aug.
STATUS Native.
ALTITUDE Mostly lowland, but to 530m.

seed pods

Spotted Medick *Medicago arabica*

Locally common, especially near the sea. Short turf (including lawns) and disturbed places, often on sandy or gravelly soils. ❁ A low-growing clover, each leaflet with a characteristic *dark blotch in the centre* (spots occasionally absent). Flowers small, in short-stalked clusters of 1–5, *seed pods spiny.* DETAIL Downy when young, more or less hairless when mature. Leaflets 7–25mm long, finely toothed near tip, with a tiny point (mucro) on the blunt tip. Stipules variably but shallowly toothed. Flowers 4–6mm long. Seed pods faintly net-veined, hairless, tightly coiled into a spiral 4–7mm across with 3–5 complete turns; each turn has 2 parallel rows of slender curved spines along the outer edge, with the central raised ridge between the spines *grooved.*

GROWTH Annual.
HEIGHT Prostrate, shoots growing to 60cm.
FLOWERS May–Sep.
STATUS Native.
ALTITUDE Lowland.

Bur Medick
Medicago minima

Nationally Scarce, listed as Vulnerable, but very locally common. Dry, bare, sandy or gravelly ground, often along tracks. ❀ A small clover with clusters of 1–6 tiny flowers and spiny seed pods tightly coiled into 3–5 complete turns. DETAIL *Densely hairy.* Leaflets 3–6mm long. *Stipules more or less untoothed.* Flowers 2.5–4.5mm long. Seed pods net-veined, often sparsely hairy, 3–5mm across; each coil has 2 parallel rows of slender hooked spines along the outer edge; central raised ridge between spines *not grooved.*

GROWTH Annual.
HEIGHT Prostrate, shoots growing to 20cm.
FLOWERS May–Jul.
STATUS Native.
ALTITUDE Lowland.

Toothed Medick
Medicago polymorpha

Nationally Scarce, and very local. Short turf (including mown verges) and bare ground in dry, sandy or gravelly places near the sea; a rare casual inland. ❀ A low-growing clover with very small flowers in short-stalked clusters of 1–8 and spiny seed pods that form *open spirals* 4–6mm across. DETAIL *More or less hairless when mature.* Leaflets to 10(25)mm long. *Stipules deeply toothed.* Flowers 3–4.5mm long. Pods strongly net-veined, hairless, flattened, coiled into a spiral of 1.5–5 turns, with 2 parallel rows of slender hooked spines along the outer edge; *central raised ridge between spines not grooved.*

GROWTH Annual.
HEIGHT Prostrate, shoots growing to 60cm.
FLOWERS Jun–Sep.
STATUS Native.
ALTITUDE Lowland.

Birdsfoot
Ornithopus perpusillus

Locally common. Bare and sparsely-vegetated ground on dry, sandy or gravelly soils, usually acid, especially by tracks and paths: heathland, forestry rides and dunes. ❀ The ground-hugging leaf rosettes are often more conspicuous than the *tiny, colourful flowers.* DETAIL Finely hairy. Leaves divided into 4–13 pairs of leaflets, each 2–4mm long, plus a single terminal leaflet. Flowers short-stalked, 3–5mm long, in groups of 3–8 with leaf-like bract below; standard pale pink with red veins, wings white, keel deep yellow. Pods downy, 10–20mm long, *constricted into 4–9 sections (like a string of beads);* 2–3 pods together recall a bird's foot.

GROWTH Winter-annual.
HEIGHT Stems 2–40cm.
FLOWERS May–Aug.
STATUS Native.
ALTITUDE To 385m.

Hop Trefoil
Trifolium campestre

Fairly common. Grassy places and disturbed ground on poor, dry, neutral or alkaline soils.
❀ A low-growing clover with globular flower heads 8–15mm across, composed of *at least 20 tiny yellow flowers that turn pale brown in fruit,* resembling tiny Hop cones. DETAIL Stems densely hairy. Leaflets hairless above, sometimes hairy on midrib below, 6–10mm long, finely toothed; stalk of terminal leaflet over 1.5mm long – longer than those of lateral leaflets. Corolla 4–7mm long, *standard broad and flat,* folding forward and downward over the seed pod as the fruit matures; pod egg-shaped with a short hooked beak. SEE ALSO Black Medick (p. 60).

GROWTH Winter-annual.
HEIGHT 10–30cm.
FLOWERS May–early Oct.
STATUS Native.
ALTITUDE 0–365m.

Lesser Trefoil
Trifolium dubium

Abundant. Short turf and disturbed ground, generally on dry soils; a common lawn weed.
❀ A low-growing clover with small globular heads of yellow flowers 5–9mm across, containing 5–20 tiny flowers. DETAIL Stems sparsely hairy to hairless, leaflets may have scattered long hairs below. Leaflets 3–8mm, the terminal leaflet with a stalk longer than that of the side leaflets (0.5–1mm long). *Flower stalks c. 1mm long,* corolla 3–4mm long, turning brown in fruit, with the standard folded down on either side of its centre line over the pod like a ridged roof; pod as Hop Trefoil. SEE ALSO Black Medick (p. 60).

GROWTH Winter-annual.
HEIGHT To 25cm.
FLOWERS May–Sep.
STATUS Native.
ALTITUDE To 535m.

Slender Trefoil
Trifolium micranthum

Fairly common but inconspicuous. Sparsely vegetated places and short grassland on light, sandy or gravelly soils, neutral to moderately acid. Tolerates grazing, trampling and mowing – a lawn weed. ❀ A tiny, ground-hugging clover with clusters of 2–6 (1–10) flowers, each 1.5–3mm long, on *relatively long stalks* (to 2mm). DETAIL Leaflets and stems hairless or very sparsely hairy. Leaflets 2.5–5mm long, all very short stalked (less than 0.5mm). The standard petal is *notched at the tip and* folds down on either side of its centre line. Seed pod egg-shaped.

GROWTH Winter-annual.
HEIGHT To 15cm.
FLOWERS May–Jul.
STATUS Native.
ALTITUDE To 430m.

White Clover *Trifolium repens*

GROWTH Perennial.
HEIGHT Main stems to 50cm.
FLOWERS May–Sep.
STATUS Native.
ALTITUDE To 880m, but usually below 400m.

Abundant in most types of grassland, although scarce in tall grass and avoids very wet or acid soils; also found on waste ground and disturbed places. Horticultural varieties are frequently included in seed mixes. ❀ Main stems prostrate, *rooting at the nodes and thus patch-forming.* From these ground-hugging stems the leaves arise on long stalks, and the globular clusters of flowers are also *held upright on leafless stalks* up to 20cm long. DETAIL More or less *hairless.* Leaflets 10–30mm long, finely toothed, usually with a pale chevron near the base. Stipules translucent, oblong, *contracting abruptly to a fine point,* occasionally red-veined. Flowers 7–12mm long, usually off-white, sometimes pale pink, rarely reddish, scented when fresh. Seed pods longer than calyx.

Alsike Clover *Trifolium hybridum*

Very locally common. Often originates from commercial seed mixes and found in grassy places on farmland, verges and sometimes waste ground, but seldom persists in closed swards. In slow decline nationally. ❀ Rather similar to White Clover, with almost hairless leaves and flower heads on leafless stalks, and thus easily overlooked. DETAIL *Main stems more or less erect, not rooting* at the nodes. Stipules at the base of leaf stalks up to 25mm long, whitish at the base with green veins, *tapering gradually to a point.* Leaves usually lack whitish chevrons, with the side veins branched 1–3 times (more or less unbranched in White Clover). Flowers usually whitish with a pink base (or white, or pink; in coloration they may be closely matched by White Clover). Pods slightly longer than calyx.

GROWTH Annual or short-lived perennial.
HEIGHT Stems to 40cm (70cm).
FLOWERS Jun–Sep.
STATUS Introduced from S Europe and SW Asia and formerly commonly sown as a fodder crop. First recorded in the wild in 1762.
ALTITUDE 0–350m.

Red Clover *Trifolium pratense*

GROWTH Perennial.
HEIGHT 10–60cm.
FLOWERS May–Sep.
STATUS Native.
ALTITUDE 0–850m.

Common and familiar in all types of grassland, also waste ground. A large variety (var. *sativum*) is often included in agricultural seed mixes; conversely, the wild variety (var. *pratense*) is in decline. ❀ The tiny flowers are grouped into egg-shaped heads, 20–40mm long, at the tip of the stems; these flower heads are *more or less stalkless, with a pair of leaves immediately below.* DETAIL More or less erect. Leaves hairy below and often hairy above, divided into 3 leaflets each 15–30mm long, often with a whitish chevron. Stipules triangular, red-veined, *narrowing abruptly to a brown bristle point c. 3mm long.* Flowers pinkish-purple (rarely pale pink or white), 12–18mm long. Calyx hairy, enclosing the seed pod when mature. Var. *sativum* is larger, with hollow stems and paler flowers.

Zigzag Clover *Trifolium medium*

GROWTH Perennial.
HEIGHT 10–30cm.
FLOWERS Jun–Sep.
STATUS Native.
ALTITUDE To 610m.

Rather local. Old grassland on damp clay soils, including verges and woodland rides; in upland areas also rocky stream- and riverbanks and rocky outcrops. ❀ Very like Red Clover, but *mature flower heads obviously stalked* and *stipules taper evenly to a point* and are *entirely green or green-veined (occasionally red-veined)*; also the leaflets are narrower and the flowers a deeper, more purplish-red. On larger plants the stems are zigzag, but this is subtle and not a useful feature. DETAIL Leaves hairless on upperside, hairy below, leaflets 20–60mm long with a faint paler chevron. Flowers 12–20mm long. Calyx tube more or less hairless, but calyx teeth fringed with long hairs; calyx enclosing the seed pod when mature.

Strawberry Clover *Trifolium fragiferum*

Very locally common. Moist grassland with a brackish influence on sea banks, coastal grazing marshes and the edge of saltmarshes. Also inland on old pastures, green lanes and verges on heavy soils, especially chalky boulder clays.
❀ Patch-forming, with sprawling stems that *root at the nodes*. In the late summer the

GROWTH Perennial.
HEIGHT Sprawling, stems to 30cm.
FLOWERS Jun–Aug.
STATUS Native.
ALTITUDE Lowland.

inflated fruits, clustered into globular heads 15–20mm across, recall a raspberry as much as a strawberry. DETAIL Leaflets 8–20mm long, hairless above, sparsely hairy below, often with a paler chevron, finely-toothed, with the side veins branching several times and often curving down towards the leaf edge. Flowers 5–7mm long, pink, in long-stalked, globular heads. The calyx tube swells in fruit to form a pinkish, net-veined, inflated, downy bladder, enclosing the seed pod.

Sea Clover *Trifolium squamosum*

Nationally Scarce. Very local in grassland on salty clays: sea banks, coastal grazing marshes, the banks of tidal rivers and creeks, and the margins of saltmarshes. ❀ Like a small Red Clover, with reddish flowers in short-stalked, globular heads *c.* 10–20mm long and a pair of leaves immediately below. *Note shape of stipule and fruiting calyx.* DETAIL More or less erect. Sparsely hairy. Leaflets 10–25mm long, long-oval. Stipules green, tapering to a long point. Flowers 7–9mm long. In fruit *calyx bell-shaped*, enclosing the seed pod, with rigid teeth that spread star-wise.

GROWTH Annual.
HEIGHT 10–40cm.
FLOWERS May–mid Jul.
STATUS Native.
ALTITUDE 0–50m.

GROWTH Winter-annual.
HEIGHT 5–25cm.
FLOWERS Jun–Jul.
STATUS Native.
ALTITUDE Lowland.

Clustered Clover *Trifolium glomeratum*

Nationally Scarce. Rather local on bare, disturbed ground and short, sparse grassland, even pavement cracks, on dry, sandy or gravelly soils, mostly near the sea. ❀ A small, prostrate, *hairless* clover, with *tiny pinkish flowers*, 4–5mm long, in stalkless clusters. **DETAIL** Leaflets 4–12mm long, toothed. *Calyx teeth spreading like a star*, the calyx enclosing the seed pod.

Knotted Clover *Trifolium striatum*

GROWTH Winter annual.
HEIGHT To 30cm.
FLOWERS May–Jun.
STATUS Native.
ALTITUDE 0–320m.

Locally fairly common. Sparsely vegetated areas and open grassland on poor, dry soils, usually sandy or gravelly. ❀ Small, low-growing, *hairy* clover with *tiny pale pink flowers* 4–7mm long, in stalkless clusters. **DETAIL** Leaf stalks to 4cm long, leaflets 5–15mm long; *side veins fine, straight* or *curving slightly forward near the leaf edge. Calyx tube inflated in fruit*, enclosing the pod, with *erect teeth*.

Rough Clover *Trifolium scabrum*

GROWTH Winter annual.
HEIGHT To 20cm.
FLOWERS May–Jun.
STATUS Native.
ALTITUDE Lowland.

Locally fairly common near the sea in sparsely vegetated areas and open grassland on poor, dry soils on sand, gravel and limestone; rare inland away from Breckland. ❀ Small, low-growing, *hairy* clover with *tiny white flowers*, 4–7mm long, in stalkless clusters. **DETAIL** Leaf stalks up to 12mm long. Leaflets 4–10mm long; *side veins thickened and curving outwards and downwards towards the leaf edge. Calyx enclosing the seed pod, calyx teeth curving outwards and downwards.*

Birdsfoot Clover *Trifolium ornithopodioides*

Very locally common, mostly near the sea. Bare, disturbed ground and short turf on acid sands and gravels, usually where moist in winter and parched in summer, and often where heavily trampled on tracks and rough car parks. ✿ Easily overlooked. Prostrate, with *very small white or pale pink flowers*, 6–8mm long, in *clusters of 1–4*. **DETAIL** More or less *hairless*. Leaves with stalks up to 5cm long, leaflets 3–10mm long, toothed, sometimes with a narrow reddish-brown cross-band. Seed pods longer than calyx.

GROWTH Winter annual.
HEIGHT Shoots to 20cm.
FLOWERS May–Jul.
STATUS Native.
ALTITUDE Lowland.

Subterranean Clover *Trifolium subterraneum*

Very local on the coast in short turf and on bare ground on sandy or gravelly soils; rare inland. ✿ Similar to Birdsfoot Clover, but *flowers larger* and *leaves hairy*. The wonderful name is a reference to the globular seed pods, which are pushed into the ground as they mature. **DETAIL** Leaves with stalks 2–5cm long, leaflets 5–12mm long, often with a blackish blotch, sometimes finely toothed. Flowers 8–14mm long, in groups of 2–5; as well as the fertile flowers, there are numerous *sterile, petal-less flowers*.

GROWTH Winter annual.
HEIGHT To 20cm.
FLOWERS Apr–Jun.
STATUS Native.
ALTITUDE Lowland.

Suffocated Clover *Trifolium suffocatum*

Nationally Scarce. Very local in sparsely-vegetated places by the sea (now rare inland), on thin, dry, soils, often where trampled in car parks, picnic sites, on lawns and even pavements. ✿ Very small and inconspicuous, with the flowers in stalkless globular heads, usually densely crowded in the centre of the plant; the *whitish petals are almost hidden within the calyx*. **DETAIL** More or less hairless. Leaves on stalks 20–30mm long; leaflets 5–10mm long, with rather spine-like teeth and usually a darker chevron. Flowers 3–4mm long. Calyx teeth spreading, the calyx enclosing the seed pod.

GROWTH Winter annual.
HEIGHT To 3cm (8cm).
FLOWERS Mar–May; also Aug in wet summers.
STATUS Native.
ALTITUDE Lowland.

Sulphur Clover *Trifolium ochroleucon*

GROWTH Perennial.
HEIGHT To 50cm.
FLOWERS Jun–Jul.
STATUS Native.
ALTITUDE Lowland.

Nationally Scarce and listed as Near Threatened. Very locally common in its restricted range. Grassy places on chalky boulder clay, mostly now road verges. ✿ Short-stalked globular heads of yellowish-cream flowers distinctive. **DETAIL** Hairy. Leaves divided into 3 oblong leaflets, 15–30mm long; stipules oblong, extended into a thread-like tip, up to 12mm long. Flower clusters 2–4cm long, flowers 15–20mm long. Calyx enclosing the pod when mature.

Haresfoot Clover *Trifolium arvense*

GROWTH Annual.
HEIGHT 5–20cm.
FLOWERS Jun–Sep.
STATUS Native.
ALTITUDE Lowland.

Fairly common. Sparsely-vegetated and grassy places on dry, sandy or gravelly soils: verges, grassy heaths, dunes and fallow fields; avoids chalky soils. ✿ A distinctive clover with *softly downy, egg-shaped flower-heads* that elongate as the seeds develop into a cylinder up to 25mm long. **DETAIL** Softly hairy. Leaflets 5–15mm long, narrowly oval, often finely toothed towards tip. Stipules veined reddish, with a bristle point up to 8mm long. Flowers 3–6mm long, white to pink, the petals obscured by the long, hairy, pale brown calyx teeth. In fruit the calyx encloses the pod.

Common Milkwort *Polygala vulgaris*

GROWTH Perennial.
HEIGHT Stems to 25cm.
FLOWERS Apr–Sep.
STATUS Native.
ALTITUDE 0–730m.

Locally common. Short, unimproved grassland on neutral to basic soils, including dunes and meadows. ❀ Low growing, often threading through other vegetation, with *clusters of small flowers that may be white, pink, purple, royal blue or china blue*. Milkwort flowers are unique, with 3 tiny outer sepals, 2 large, brightly-coloured inner sepals and 3 true petals fused into a tube, with the lowest petal cut into white, finger-like lobes.
DETAIL Sparsely hairy to hairless. Stems woody at the base. *All leaves alternate, those at the base of the stem smaller than those towards the tip* (the leaves may be crowded towards the base, making it hard to see the arrangement). Usually 10–40 flowers per spike, each 4–7mm long; 2 inner sepals with well-branched veins that divide and loop around to join again, especially towards the edges.

Heath Milkwort *Polygala serpyllifolia*

GROWTH Perennial.
HEIGHT Stems to 15cm.
FLOWERS May–Sep.
STATUS Native.
ALTITUDE 0–1035m.

Locally common on heaths, moors and the drier parts of bogs, always on acid soils. ❀ Very like Common Milkwort but slightly smaller, with the *leaves on the lower part of the stem opposite* (they may have fallen, in which case examination of the leaf scars with a hand lens is necessary). DETAIL As Common Milkwort but stems only slightly woody at the base and usually 3–10 flowers per spike, each 4.5–6mm long.

Chalk Milkwort *Polygala calcarea*

Very locally fairly common to abundant. Short, species-rich turf on chalk and limestone, usually on S-facing slopes. ● Very like Common Milkwort, but *flowers bright gentian-blue* (can be white or pink) DETAIL *Leaves at base of stem larger than those towards tip and obviously crowded into a 'false rosette' above the* unbranched, *leafless stem-base.* Flowers 6–7mm long, 6–20 per spike, 2 inner sepals with little-branched veins that *do not divide, loop and re-join* (or do so only sparingly).

GROWTH Perennial.
HEIGHT Stems to 10cm.
FLOWERS Apr–Jul.
STATUS Native.
ALTITUDE Lowland.

FAMILY ROSACEAE: ROSES & ALLIES

A large and very varied family, with around 55 species native to Britain (as well as the hundreds of 'micro species' of *Alchemilla*, *Rubus* and *Sorbus*). The family contains herbaceous plants as well as numerous trees and shrubs, including plums and cherries, cotoneasters, hawthorns, apples, Rowan and whitebeams, brambles and, of course, roses (see pp. 365–376). It is not possible to produce a simple, non-technical definition of the family, but most species have alternate leaves, often with a stipule at the base of the leaf stalk, 5 petals and many conspicuous stamens.

Mountain Avens *Dryas octopetala*

GROWTH Perennial.
HEIGHT Stems to 50cm.
FLOWERS May–Jul.
STATUS Native.
ALTITUDE 0–1035m.

Nationally Scarce. Very locally common in N Scotland, in rocky places and adjacent short grassland on mountains, maritime heath and dunes on calcareous soils, also limestone pavements in W Ireland; very rare

in N England and N Wales. ● Low-growing, more or less evergreen shrub with distinctive foliage and *anemone-like flowers.* DETAIL Stems well branched, prostrate, rooting at nodes, often reddish. Leaves to 25mm long, long-oval, with 6–8 shallow lobes on each side, dark glossy green above, woolly white below. Flowers solitary, with 8 (7–10) petals, each 7–17mm long, abundant golden-yellow stamens, and glandular-hairy sepals; the many small, nut-like fruits each have a feathery plume, 20–30mm long (recalling a *Clematis*).

Meadowsweet *Filipendula ulmaria*

GROWTH Perennial.
HEIGHT To 120cm.
FLOWERS Jun–Sep.
STATUS Native.
ALTITUDE 0–880m.

Common. A wide variety of damp habitats, from wet woodland rides to roadside ditches, but avoids permanently waterlogged or very acid ground and heavy shade. May form extensive stands. ✿ The tiny flowers are gathered unto *frothy, creamy-white clusters* and are *heavily scented, recalling musk or honey*. DETAIL Stems often reddish. Leaves dark green above, pale green to silvery and variably hairy below, cut into 2–5 pairs of oval, toothed leaflets 15–80mm long that alternate with several tiny leaflets; terminal leaflet larger, often 3-lobed. Bruised stems and leaves smell of antiseptic. Flowers in clusters up to 15cm across, each 4–10mm wide, petals usually 5, many prominent stamens. Carpels 6–10, spirally-twisted, hairless.
SIMILAR SPECIES Unlike umbellifers, the stalks of the flower heads do not all come together at the tip of the stem. SEE ALSO Common Meadow-rue (p. 33).

The name refers to its use in flavouring mead and other drinks, rather than a predilection for meadows, and also used as a strewing herb, scattered on the floor to freshen up the house.

Dropwort *Filipendula vulgaris*

Rather local. Grassland on chalk and limestone, also widely planted in churchyards in W Wales, sometimes a garden escape elsewhere. ✿ The dry-ground counterpart of Meadowsweet, but usually shorter, with *much more feathery leaves*, mostly at the base of the stem. Flower heads more open, with fewer, larger, unscented flowers and conspicuous *deep pink buds*. DETAIL Leaves hairless, shiny green both sides, with 8–20 pairs of coarsely-toothed leaflets, each 5–15mm long, alternating with several tiny leaflets. Crushed stems and leaves smell of antiseptic. Flowers 10–20mm across, petals usually 6, creamy-white, 5–9mm long. Carpels 6–12, erect, hairy.

GROWTH Perennial.
HEIGHT 10–50cm (rarely to 100cm).
FLOWERS Mid May–Jul.
STATUS Native.
ALTITUDE To 365m.

Silverweed
Potentilla anserina

Very common. Bare or trampled ground and open grassy swards, on both wet and dry soils; characteristic of tracksides and the margins of drying ponds and lakes.
❂ Ground-hugging, with long, rooting runners and *distinctive silvery leaves*: uppersides range from silvery to dull green, but *undersides usually silvery-white*. DETAIL Runners often reddish. Leaves silky-hairy below, 5–25cm long, *pinnately cut* into 3–12 pairs of oval, toothed leaflets interspersed with additional tiny leaflets. Flowers with 5 petals, 7–10mm long (about twice length of sepals), growing singly on long, erect stems.

GROWTH Perennial.
HEIGHT Runners to 80cm.
FLOWERS May–Aug.
STATUS Native.
ALTITUDE Mostly lowland, but to 580m.

leaf underside

Hoary Cinquefoil
Potentilla argentea

Near Threatened.
Very locally common, especially in E Anglia, on bare ground and in open grassy swards, on dry, usually sandy soils; typical of tracksides and forest rides. ❂ Low growing. The dark green leaves contrast with whitish stems, but the leaves have to be turned over to see their white, 'hoar-frosted' undersides. DETAIL Stems woolly, arising from leaf rosette. Leaves wintergreen, up to 30mm long, *palmately-cut* into 5 leaflets, each deeply lobed with the edges narrowly turned under, undersides densely covered with felted white hairs, upperside finely hairy. Flowers in clusters, with 5 petals, 4–5mm long.

GROWTH Perennial.
HEIGHT Stems to 30cm.
FLOWERS Mid May–Sep.
STATUS Native.
ALTITUDE Lowland.

Tormentil
Potentilla erecta

Very common in rough grassy places, especially woodland rides, heaths and moors, mostly on acid soils but also on limestone and in alkaline fens. ❂ Low growing but often well branched and rather 'bushy,' most or all flowers have *4 petals and 4 sepals*. DETAIL Variably hairy. *Leaves more or less stalkless*, in a basal rosette (often withered) and on erect to sprawling, *non-rooting flowering stems*; leaves cut into 5 well-toothed leaflets (2 are actually leaflet-like stipules). Flowers 7–15mm across, in loose, stalked clusters; with 4–20 carpels.

GROWTH Perennial.
HEIGHT Stems to 45cm.
FLOWERS May–Sep.
STATUS Native.
ALTITUDE To 1110m.

Creeping Cinquefoil
Potentilla reptans

Common on short, rough grassland: verges, dry pastures, woodland rides and waste ground. ❀ Ground hugging, with *leaves palmately cut into 5 narrow leaflets*. Often flowers sparingly. **DETAIL** Sparsely hairy. A persistent basal leaf rosette gives rise to *long runners that root at the nodes*. Leaves wintergreen, on stems up to 18cm long, leaflets up to 5cm long. Flowers 15–25mm across with 5 petals, growing *singly* on long stems from the leaf axils; carpels 60–120.

GROWTH Perennial.
HEIGHT Stems to 100cm.
FLOWERS Mid May–Sep.
STATUS Native.
ALTITUDE Generally lowland.

Trailing Tormentil
Potentilla anglica

Uncommon. Well-drained acid grassland: heaths, woodland rides, sandy roadside banks and tracksides. ❀ Sprawling, *some flowers have 4 petals, others 5*. Hard to separate from Hybrid Cinquefoil: *check for developing seeds in withered flowers*. **DETAIL** A basal rosette produces fine stems that root at the nodes in late summer; *lower stem leaves with stalks 1–2cm long* and 3–5 leaflets, upper stem leaves *short-stalked* and often simpler. Flowers 12–18mm across, solitary, long-stalked; carpels 20–50. **SIMILAR SPECIES** Hybrid Cinquefoil *P.* × *mixta* (Creeping Cinquefoil × Trailing Tormentil) is rather commoner. It also has 4- and 5-petalled flowers, but *the leaf stalks are all about the same length* (not shorter towards tip of stem) and over 1cm long. Best identified by its sterility – it sets few if any seeds.

GROWTH Perennial.
HEIGHT Stems to 80cm.
FLOWERS Jun–Sep.
STATUS Native.
ALTITUDE 0–445m.

Spring Cinquefoil
Potentilla tabernaemontani

Nationally Scarce. Uncommon and very local on S-facing rocky ledges and scree on limestone, also open grassland on chalk. ❀ Mat-forming, growing from a woody rootstock and producing clusters of flowers on slender, conspicuously hairy stems. **DETAIL** *Main stems creeping and often rooting*, with rosettes of stalked leaves at the tip, each *palmately divided* into 5–7 *leaflets*, 15–35mm long, green on both sides, hairless above and sparsely hairy below. Flowering stems up to 10cm long arise from the side of these rosettes, with stalkless 3-lobed leaves and 5-petalled flowers, 10–15mm across.

GROWTH Perennial.
HEIGHT To 10cm.
FLOWERS Apr–Jun.
STATUS Native.
ALTITUDE To 355m.

Barren Strawberry
Potentilla sterilis

Common. Grassy roadside banks, scrub, open woodland and open rocky ground, usually on poor, well-drained (but not bone-dry) soils, often in partial shade in the lowlands but more frequently in

GROWTH Perennial.
HEIGHT 5–15cm.
FLOWERS Feb–May.
STATUS Native.
ALTITUDE 0–790m.

sunny places in the uplands. ❀ Fruits small, dry and inedible. Typically has *large gaps between the petals*, with the sepals clearly visible, but best separated from Wild Strawberry by the leaves. **DETAIL** Silky-hairy. Sprawling, with short runners. Leaves long-stalked, divided into 3 *dull, mid to dark green leaflets*, 5–35mm long, each with the *terminal tooth rather smaller and usually shorter than the adjacent teeth*. Flowers 8–15mm across, in clusters of 1–3 at the tip of long stems.

Wild Strawberry   **Barren Strawberry**   **Wild Strawberry**   **Barren Strawberry**

Wild Strawberry
Fragaria vesca

Common. Verges, hedgebanks, scrub and woodland on dry soils, also limestone scree and rocky outcrops; colonises old quarries and chalk pits. ❀ Fruits small but very edible. **DETAIL** Often produces long rooting

GROWTH Perennial.
HEIGHT 5–30cm.
FLOWERS Apr–Jun.
STATUS Native.
ALTITUDE 0–640m.

runners. Leaves hairy, divided into 3 *glossy, bright yellowish-green leaflets* 10–60mm long with 8–13 teeth per side and the *terminal tooth as long as, or longer than, the adjacent teeth and usually not much smaller*. Flowers 10–20mm across, in long-stemmed clusters, with the *5 petals more or less concealing the sepals*. Fruits *c.* 10mm across, the seeds standing proud of the surface. **SIMILAR SPECIES** Garden Strawberry *F. ananassa* is an occasional garden throw-out. It has more or less *hairless leaflets* and is *larger overall*: flowers 20–35mm across, fruits over 15mm wide, with the *seeds sunk into the surface*.

Water Avens
Geum rivale

Common in the N and W, local or absent in SE England and Midlands. Damp or wet places: old, wet woodland, unimproved meadows and pastures and, in the uplands, damp verges, ditches and sheltered rock ledges. ❀ Flowers *cup-shaped, drooping* at the tip of long, upright, purple stems. DETAIL Downy. Basal leaves cut into 3–6 pairs of finely toothed, unequal leaflets, terminal leaflet rather larger, typically *square-cut to rounded at base*, about as wide as long. *Petals creamy-pink with reddish-purple veins*, 8–15mm long, slightly notched at the tip and narrowing abruptly at the base; *sepals contrasting dark purple*, held upright, partly hiding petals. Stamens numerous. Seeds in a tight cluster on a short stalk in the centre of the flower, hairy, extending into a long slender style that is prominently hairy at the base and, when fresh, has a tiny plume attached to the hooked tip. SIMILAR SPECIES **Hybrid Avens** G. × *intermedium* (Water Avens × Wood Avens) is common where the parents meet, especially in disturbed habitats. Showing a mixture of characters (e.g. drooping yellow flowers), it is fertile and back-crosses with either parent to form a wide variety of intermediates.

GROWTH Perennial.
HEIGHT To 50cm.
FLOWERS Apr–Sep.
STATUS Native.
ALTITUDE 0–1050m.

Wood Avens
Geum urbanum

GROWTH Perennial.
HEIGHT To 70cm.
FLOWERS May–Nov.
STATUS Native.
ALTITUDE To 570m.

Locally common. Light shade in deciduous woodland, scrub, hedgebanks and waste ground; a persistent garden weed. ❀ Erect, slender, hairy, green stems carrying small yellow flowers with *5 well-separated petals,* the sepals clearly visible between them. DETAIL Downy. Basal leaves cut into 2–4 pairs of unequal, finely but bluntly-tooted leaflets, *terminal leaflet rather larger,* often 3-lobed, *typically tapered into the stalk* and a little longer than wide. Stem leaves 3-lobed or undivided. Petals 4–7mm long, rounded, spreading. Seeds as Water Avens but seed head stalkless, styles not as hairy and, when fresh, the tiny plume is replaced by a short side-spur; the seeds easily catch on fur or clothing.

Marsh Cinquefoil
Comarum palustre

Fairly common in the N and W, very local in the SE. Shallow water and other very wet places in fens, bogs and marshes, usually where flushed with mildly alkaline, nutrient-poor water, also wet meadows.

GROWTH Perennial.
HEIGHT To 50cm.
FLOWERS May–Jul.
STATUS Native.
ALTITUDE 0–800m.

❂ Striking star-shaped, *wine-red flowers* distinctive. **Detail** Sparsely hairy to hairless. Stems arising from a woody creeping rhizome. Leaves blue-green, cut into 3–7 coarsely toothed leaflets; upper leaves may be merely 3-lobed. Flowers 20–30mm across, in loose clusters, with 5 small, narrow petals, 5 large sepals and 20–25 stamens; the wine-red sepals persist long after the petals have fallen and, together with the dry, purple fruits, remain conspicuous.

Agrimony   *Agrimonia eupatoria*

Common. Verges, rough grassland, scrub and woodland rides; does best on chalky soils. ❂ Distinctive tall, slender, leafless flower spikes appear in midsummer. **Detail** Stems hairy, often very glandular-hairy. Leaves cut into 3–6 pairs of bluntly-toothed leaflets (alternating with 2–3 pairs of tiny leaflets), with long hairs above and often densely hairy below, but with few or no oil glands on underside; when crushed, *foliage mildly fragrant.* Flowers 5–8mm across, with 5 petals and 5 sepals. Fruit cone-shaped, *deeply grooved to the tip, with the outermost bristles spreading at 45–90°.* **See Also** Weld (p. 117).

GROWTH Perennial.
HEIGHT 30–60cm.
FLOWERS Jun–Sep.
STATUS Native.
ALTITUDE 0–365m.

fruit

Fragrant Agrimony   *Agrimonia procera*

Found in similar habitats to Agrimony, but usually much scarcer and more local; avoids chalky soils. ❂ The best distinction is the fruit, which is bell-shaped, with *grooves short or absent*, and *outermost bristles angled downwards*. When crushed, *foliage very fragrant.* **Detail** As Agrimony but stem rather shaggy-hairy (but only tiny glandular hairs), leaflets more sharply toothed, with fewer, shorter hairs on upperside but abundant stalkless shiny oil glands, especially below. Petals usually notched.

GROWTH Perennial.
HEIGHT 50–100cm.
FLOWERS Jun–Aug.
STATUS Native.
ALTITUDE 0–335m.

fruit

Salad Burnet *Poterium sanguisorba*

Locally common. Poor, dry, unimproved grassland over chalk and limestone: downs, verges, rocky slopes. ❀ The tiny greenish flowers are clustered into dense, globular heads 6–15mm across, with the uppermost flowers female, with 2 tassel-tipped reddish-purple stigmas, the central flowers bisexual, and the lower flowers male, with numerous long stamens. The leaves smell of cucumber when crushed and can be used in salads. DETAIL Sparsely hairy. Basal leaves cut into 3–12 pairs of short-stalked, toothed, rounded leaflets. Petals absent, calyx green, often tinged purple, cut into 4 lobes; in male and bisexual flowers stamens *c.* 6mm long and yellow anthers. SIMILAR SPECIES **Fodder Burnet** (subspecies *balearicum*) is larger and leafier. It was formerly grown for fodder and is still occasionally found in grassy places.

female flowers open

male flowers open

GROWTH Perennial.
HEIGHT 15–40cm.
FLOWERS May–Aug.
STATUS Native.
ALTITUDE 0–500m.

Great Burnet *Sanguisorba officinalis*

GROWTH Perennial.
HEIGHT To 120cm.
FLOWERS Jun–Sep.
STATUS Native.
ALTITUDE 0–460m.

Locally common in unimproved grassland on neutral to alkaline, often damp, soils: verges, old meadows, riverbanks; absent from most of E Anglia and S England. ❀ Relatively tall but slender.

The dark crimson flowers are crowded into dense, 1–3cm long spikes at the tip of the stems, the tiny flowers opening in succession from the bottom up. DETAIL Hairless. Basal leaves cut into 3–7 pairs of stalked, toothed leaflets. Flowers bisexual, with 4 dark crimson stamens, *c.* 4mm long, and a short, brush-topped stigma. Petals absent, sepals 4, dark crimson.

Lady's Mantle *Alchemilla vulgaris* agg.

Twelve species of Lady's Mantle are native to Britain (5 of which have very restricted distributions in N England), with another 2 introductions. They are all *apomictic*, producing seed without fertilisation, and are termed microspecies (see p. 366). Most differ only subtly in leaf shape and the distribution of hairs. Of the rest, Alpine Lady's Mantle is distinctive, while *A. mollis*, Soft Lady's Mantle, is commonly grown in gardens and is the only species likely to be found in much of lowland Britain.

GROWTH Perennial.
HEIGHT To 40cm, but often much less.
FLOWERS Jun–Aug.
STATUS Native.
ALTITUDE 0–1215m.

Soft Lady's Mantle *Alchemilla mollis*

A common garden plant, increasingly found as a garden throw-out, especially near habitation, on verges, waste-ground etc. ❀ A large, robust Lady's Mantle, with clusters of tiny greenish-yellow flowers; the softly hairy leaves often 'catch' drops of rain or dew. DETAIL Whole plant densely hairy. Flowers with 4 sepals and 4 stamens but no petals. Unlike wild Lady's Mantles, the epicalyx segments are the same length as the sepals (not rather shorter).

GROWTH Perennial.
HEIGHT To 60cm.
FLOWERS May–Sep.
STATUS Introduced to gardens in 1874 and first recorded in the wild in 1948. Has spread widely (SE Europe).
ALTITUDE To 520m.

Alpine Lady's Mantle *Alchemilla alpina*

Locally common. Grassy and rocky places on both acid and alkaline soils, mostly in the uplands, but to near sea-level in NW Scotland. DETAIL *Leaves cut to the base into 5–7 narrow leaflets that are densely silky-hairy white below.* SIMILAR SPECIES Silvery Lady's Mantle *A. conjuncta* is an occasional garden escape, sometimes in remote upland sites. Leaves silky-hairy below but cut only 60–80% of the way to the base into 7–9 lobes.

GROWTH Perennial.
HEIGHT To 20cm.
FLOWERS Jun–Aug.
STATUS Native.
ALTITUDE To 1270m.

Parsley Piert *Aphanes arvensis*

Locally common. Bare or disturbed ground and short, patchy grassland on well-drained soils, including arable fields and lawns. ❂ Small and inconspicuous, with tiny green flowers in tight clusters within a leafy cup formed by the stipules. **Detail** Hairy. Leaves 2–15mm long, fan-shaped, more or less squared-off at the base, cut into 3 deeply toothed lobes. *Stipules with oval-triangular teeth about half as long as entire portion.* Flowers c.

1mm wide. A cup-

shaped calyx conceals the reproductive parts, with the 4 erect, tooth-like sepals around its upper rim *angled slightly outwards* – the *calyx is subtly 'waisted'*; no petals. Fruits 2–2.6mm long, including sepals c. 0.6–0.8mm, their tips level with the tips of the stipule teeth.

GROWTH Winter annual.
HEIGHT 2–10cm, rarely 20cm.
FLOWERS Apr–Oct.
STATUS Native.
ALTITUDE 0–610m.

Slender Parsley Piert *Aphanes australis*

Locally common in similar habitats to Parsley Piert, but usually on poorer, drier, more acid sandy or gravelly soils. ❂ Leaves less greyish, more mid-green, tapering into the short stalk, stems more slender, and flowers even smaller, but only safely distinguished by examining the fruits (10× lens). **Detail** *Stipule lobes relatively longer and more strap-shaped, about as long as entire portion.* Fruits 1.4–1.9mm long, including sepals c. 0.3–0.5mm, their tips falling well short of the tips of the stipule lobes; the *sepals converge towards their tips and the fruit lacks a 'waist'.*

GROWTH Winter annual.
HEIGHT 2–10cm.
FLOWERS Apr–Oct.
STATUS Native.
ALTITUDE Mostly lowland, but to 450m.

Pirri-pirri-bur
Acaena novae-zelandiae

Very locally common. Dry, sparsely vegetated soil in disturbed places: heathland car parks, dunes, forestry rides, old railways, verges. ❂ Mat-forming, with erect leafy stems and *dense spherical heads of tiny flowers* that develop *hooked spines* in fruit. **Detail** Stems woody at base, reddish, hairy. Leaves cut into 4–6 pairs of deeply-toothed, oblong leaflets, glossy green above, paler green below, where hairy on the veins. Flowers with 4 greenish-white sepals, 2 stamens and 1 stigma; petals absent. Fruit with up to 4 barbed, reddish spines, the longest 6–10mm.

GROWTH Perennial.
HEIGHT To 15cm.
FLOWERS Jun–Jul.
STATUS Introduced via wool from Australia and New Zealand, first recorded 1901.
ALTITUDE Lowland.

Common Nettle
Urtica dioica

Very common in a wide range of habitats, both sunny and shaded, rich in nutrients, especially phosphates, and preferably damp: gardens, waste ground and graveyards to fens and riverbanks. ❀ One of the best-known British plants. The *tiny flowers are grouped into drooping, tassel-like clusters.*

DETAIL Patch-forming, the 4-angled purplish stems grow from a creeping rootstock. Leaves and stems with abundant long stinging hairs plus shorter non-stinging hairs. Leaves in opposite pairs, coarsely-toothed, the *terminal tooth much longer than those on either side*. Usually dioecious, with *male and female flowers on separate plants*: male flowers with 4 tiny green sepals and 4 stamens (anthers greenish-white); female flowers with a squat, pale green ovary and brush-tipped white stigma.

GROWTH Perennial.
HEIGHT To 150cm or more.
FLOWERS May–Sep.
STATUS Native.
ALTITUDE 0–850m.

Fen Nettle subspecies *galeopsifolia*

Sometimes treated as a distinct species. Frequent in fens and wet woodland in the Norfolk Broads and scattered elsewhere in S Britain. DETAIL Leaves obviously longer, narrower and more pointed, with abundant short, soft hairs, not normally stinging. Stem densely white-hairy, with the flower clusters not growing as far down on the stem. Flowers Jul–Aug.

> The tips of the stinging hairs are easily broken off, leaving a sharp point. The hair then acts as a hypodermic needle and injects a cocktail of irritants into the skin, causing both initial pain and more prolonged inflammation and tingling.

Small Nettle
Urtica urens

Locally common. Arable fields, allotments, gardens and waste ground, on fertile, well-drained, often sandy soils. ❀ As Stinging Nettle but shorter, with *longer-stalked and more coarsely and deeply-toothed leaves*, and shorter tassels of flowers. DETAIL Leaves and stems have abundant stinging hairs, and sting viciously. *Tooth at tip of leaf about as long as the teeth on either side.* Monoecious, with both male and female flowers on the same plant: the flower tassels have many female flowers but only a few male.

GROWTH Annual.
HEIGHT 10–60cm.
FLOWERS Apr–Oct.
STATUS Ancient introduction from Europe.
ALTITUDE Mostly lowland, but to §.

Pellitory of the Wall *Parietaria judaica*

Common on old walls, scarce on piles of rubble and rock and on cliffs, occasional on steep hedgebanks. Prefers sheltered, sunny spots. ❀ *Stems conspicuously reddish-purple, with very small reddish-purple flowers.* DETAIL Softly hairy. Stems well branched. Leaves alternate, short-stalked, long-oval, untoothed. Flower clusters with greenish bracts below, most flowers either male (at sides of clusters)

GROWTH Perennial.
HEIGHT 30–50cm.
FLOWERS Mar–Nov.
STATUS Native.
ALTITUDE Lowland.

or female (at centre), but some bisexual. Male and bisexual flowers with 4 short, spreading, reddish-purple calyx teeth and 4 stamens with yellowish filaments and whitish anthers; female flowers with a narrow, tubular calyx and a single brush-tipped stigma (either red or white); seed capsules reddish-purple, *c.* 2–3mm long.

Mind-your-own-business *Soleirolia soleirolii*

Fairly common garden escape (or throw-out) on bare ground in damp, shady, sheltered places, especially in frost-free areas: paths, banks, walls, churchyards. ❀ Evergreen, forming dense mats of *tiny circular leaves*, with thread-like stems and *minute flowers*. DETAIL Stems prostrate, *rooting at the nodes*, pale reddish. *Leaves alternate*, 2–7mm across, stalkless to short-stalked, sparsely hairy to hairless, with indistinct veins. Flower solitary, in leaf axils, male flowers with 4 reddish calyx teeth and 4 stamens, female flowers with a brush-tipped stigma; male and female flowers grow together on the same plant. SEE ALSO Bog Pimpernel (p. 185), Cornish Moneywort (p. 219).

GROWTH Perennial.
HEIGHT 1cm; stems to 20cm long.
FLOWERS May–Oct.
STATUS Introduced in cultivation in 1905 and recorded from the wild by 1917 (W Mediterranean).
ALTITUDE Lowland.

Grass of Parnassus
Parnassia palustris

GROWTH
Perennial.
HEIGHT
10–30cm.
FLOWERS
Jul–Oct.
STATUS
Native.
ALTITUDE 0–1005m.

Very locally common in N Britain (but extinct in much of the Midlands and S). Fens, unimproved wet meadows, dune-slacks and machair, where the ground is flushed by nutrient-poor alkaline ground water and the vegetation is short and open. ❀ The long-stalked, heart-shaped basal leaves are easily overlooked, but the exquisite flowers are unique. DETAIL Hairless. Basal leaves 1–5cm long. Flowers 15–30mm across, solitary on slender, upright stems, with 5 sepals, 5 petals, 4 stigmas and 5 stamens that alternate with 5 sterile fan-shaped staminodes, each greenish and cut into narrow lobes with a glistening golden gland at the tip. Fruit a capsule.

FAMILY OXALIDACEAE: WOOD SORRELS

Wood Sorrel
Oxalis acetosella

GROWTH Perennial.
HEIGHT 5–15cm.
FLOWERS Apr–May.
STATUS Native.
ALTITUDE 0–1160m.

Locally common. Woodland, including conifer plantations, hedgerows and other shady places on cool, moist but well-drained, humus-rich soils. In the uplands also rocky places, rough grassland and under Bracken. ❀ Patch-forming, with soft, drooping, clover-like leaves and delicate nodding flowers growing singly on leafless stems. DETAIL Thin creeping rhizomes produce both leaves and flowers on slender vertical stems up to 15cm long. Leaves sparsely hairy, divided into 3 leaflets, often purplish on underside. Flowers white with mauve veins, rarely pink or mauve, 10–25mm across, with 5 petals, 5 sepals, 10 stamens and 5 styles. Fruit a hairless, 5-angled capsule.

Procumbent Yellow Sorrel *Oxalis corniculata*

Common weed of gardens, walls, waste ground and cultivation. Spreads via seed and stem fragments and thus very hard to eradicate. ❀ Creeping, with *rooting stems*, clover-like leaves that are often dark purplish, and *long-stalked clusters of 2–8 flowers.*

GROWTH Annual or short-lived perennial.
HEIGHT Stems to 50cm.
FLOWERS May–Sep.
STATUS Introduced to cultivation by 1656 and first recorded in the wild in 1770, has spread widely in recent decades. Origin unknown.
ALTITUDE Lowland.

DETAIL Prostrate, the stems may be underground. Leaflets 6–25mm wide, sparsely hairy to hairless, leaf stalk hairy. Flowers 4–7.5mm across, *with 10 stamens, all with anthers; flower stalks spreading or bent downwards in fruit*; capsule (which may be erect) 8–20mm long, densely hairy. **SIMILAR SPECIES Least Yellow Sorrel** O. *exilis* is also a fairly common garden escape, introduced from New Zealand and Tasmania and first recorded in the wild in 1926. Overall smaller, with thread-like stems, green leaves (rarely purplish below), leaflets 2–5mm wide, hairless or sparsely hairy, and *flowers always single*, with 5 stamens with anthers and *5 stamens without anthers*. Capsule 3–4.5mm long. **Upright Yellow Sorrel** O. *stricta*, introduced by 1658 from N America and recorded from the wild by 1823, has the stems more or less erect and rarely rooting at the nodes, leaves green or purple, and flowers 5–9(15)mm across, in clusters of 2–5, the stalks erect in fruit.

Pink Sorrel *Oxalis articulata*

GROWTH Perennial.
HEIGHT To 35cm.
FLOWERS Mar–Oct.
STATUS Introduced from S America in 1870 and common in cultivation. First recorded in the wild in 1912 and probably increasing.
ALTITUDE Lowland.

Local garden escape on roadsides, waste ground and cliffs, always near houses. ❀ Pink flowers and clover-like leaves distinctive. **DETAIL** Grows from a thick, scaly-brown taproot, leaflets 25–40mm wide, densely hairy, green, with orange warts below, especially near margins; leaf stalks up to 25cm long, sparsely hairy to hairless. Flowers deep pink (rarely pale pink or white), in umbels of 3–25 on long stalks, petals 10–15mm. **SIMILAR SPECIES Large-flowered Pink Sorrel** O. *debilis* grows from a bulb, is slightly less hairy and produces bulbils at the base of the stems. **Pale Pink Sorrel** O. *incarnata* also grows from a bulb, has whorls of leaves on branched stems, more or less hairless leaflets, bulbils in the leaf axils, and solitary flowers. Both have orange warts on the leaves, and are less common garden escapes.

Family Euphorbiaceae: Spurges

In Britain, this family contains 2 species of mercury, genus *Mercurialis,* and *c.* 13 spurges, genus *Euphorbia.* The spurges are easily recognisable as such; they have simple, usually hairless, leaves that *bleed an acrid white sap that can irritate the skin* – handle with care. Spurges have very distinctive, complex 'flowers'. At the centre are several tiny male flowers, each merely a stamen tipped by 2 anthers, and a single rather larger female flower with a stalked ovary topped by 3 stigmas; this ovary develops into the seed capsule. Neither male nor female flowers have petals or sepals. The cluster of male and female flowers emerges from a cup-like *cyathium,* and attached to the rim of this cup are 2–5 rounded or crescent-shaped glands. The whole structure is usually enclosed by 2 large leaf-like bracts, often brighter and yellower than the leaves. In turn, several *cyathia* are grouped into open, umbel-like heads.

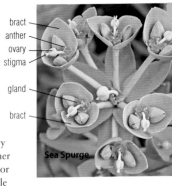

bract
anther
ovary
stigma
gland
bract

Sea Spurge

bract
ovary
stigma
gland
gland

Wood Spurge

Sun Spurge
Euphorbia helioscopia

A common weed of arable land, gardens, waste ground and other disturbed sites, especially on dry soils. ❀ The large, oval leaves are *finely toothed towards the tip,* while the greenish-yellow flowers are in large, rather *flat-topped* umbels. Detail Stems usually single, often sparsely hairy. Leaves stalkless, hairless. Flowers with *4 oval, yellowish glands. Seed capsule smooth,* hairless, *seeds with fine, netted ridges.* Similar Species Broad-leaved Spurge *E. platyphyllos* is a much scarcer arable weed, to 70cm tall, with *finely warted seed capsules* and smooth seeds.

GROWTH Annual.
HEIGHT 10–40cm.
FLOWERS Apr–Nov.
STATUS Ancient introduction.
ALTITUDE To 560m.

Cypress Spurge
Euphorbia cyparissias

A rather local garden escape or throw-out. Road- and tracksides, arable margins, grassland and scrub, especially on light and chalky soils. ❀ Patch-forming (grows from a creeping rhizome), with crowded, narrow leaves and contrasting yellow flowers. Detail Hairless. *Leaves parallel-sided, 1–2mm wide;* may turn reddish with age. Flowers with crescent-shaped glands. Seed capsule smooth with vertical bands of warts, seeds smooth.

GROWTH Perennial.
HEIGHT 10–30cm.
FLOWERS May–Jul.
STATUS Introduced to cultivation by 1640 and first recorded in the wild in 1799. May be native in SE England.
ALTITUDE Lowland.

Petty Spurge
Euphorbia peplus

An abundant weed of gardens, arable fields, pavement cracks – anywhere with disturbed soil. ✺ Usually rather small, with irregular heads of greenish flowers. DETAIL Hairless. Stems often branched. *Leaves short-stalked,* oval, blunt, *untoothed.* Flowers with 4 *crescent-shaped glands. Seed capsules smooth, ridged,* hairless; seeds pitted.

GROWTH Annual.
HEIGHT 10–30cm.
FLOWERS All year.
STATUS Ancient introduction.
ALTITUDE 0–410m.

Dwarf Spurge *Euphorbia exigua*

GROWTH Annual.
HEIGHT 5–20cm.
FLOWERS Jun–Oct.
STATUS Ancient introduction.
ALTITUDE Lowland.

Listed as Near Threatened. Rather local and usually scarce weed of arable fields on light, often base-rich soils. ✺ Short and slender but often well-branched and 'bushy', with grey-green foliage. DETAIL Hairless. *Leaves unstalked, narrow and strap-shaped, usually pointed and untoothed.* Flowers with 4 *crescent-shaped glands. Seed capsules smooth, ridged,* hairless; seeds rough.

GROWTH Perennial.
HEIGHT 20–40cm.
FLOWERS Jun–Sep.
STATUS Native.
ALTITUDE Lowland.

Sea Spurge *Euphorbia paralias*

Locally common on dunes and the strand-line of sandy beaches, less often on shingle. ❀ A robust spurge, often multi-stemmed, with *fleshy, grey-green leaves*. **DETAIL** Hairless. Stems erect, reddish. Leaves close-set, stalkless, dull grey-green above, usually shiny green below, oval to oblong, more or less pointed, *midrib obscure on underside. Seeds smooth* (but seed capsule wrinkled).

Portland Spurge
Euphorbia portlandica

GROWTH Biennial or short-lived perennial.
HEIGHT 5–30cm.
FLOWERS Apr–Sep.
STATUS Native.
ALTITUDE Lowland.

Rather local on dunes, shingle, cliffs and rocky or grassy slopes by the sea. ❀ Much like Sea Spurge but more slender and often more reddish. **DETAIL** Stems minutely hairy above, leaves only slightly succulent, grey-green both above and below, *broadest towards tip* (obovate), tapered to the base and with a *small point (mucro)* on the otherwise blunt tip; *translucent raised midrib prominent below,* glands crescent-shaped, *seeds pitted, seed capsule smooth with vertical bands of warts.*

Portland Spurge		Sea Spurge	
above	below	above	below

Wood Spurge *Euphorbia amygdaloides*

Locally common. Rides and clearings in old deciduous woodland and shaded hedgerows, mostly on neutral to basic soils, also open chalk grassland and cliffs (including sea-cliffs). ❀ A tall, conspicuous spurge with open sprays of yellowish flowers. DETAIL Stems tufted, over-wintering and flowering in their second year, *hairy*, often reddish. Leaves clustered towards tip of stem, pointed, tapering towards the base, untoothed, dull mid green or purplish, *densely hairy below. Bracts fused in pairs around the stem.* Flowers with *crescent-shaped glands. Seed capsule smooth or slightly rough*, hairless; seeds smooth. SIMILAR SPECIES The cultivated subspecies *robbiae* is a fairly frequent garden escape. Its leaves are darker, thicker, shinier and *more or less hairless.*

GROWTH Perennial.
HEIGHT 30–80cm.
FLOWERS Mar–Jul.
STATUS Native, also a garden escape (ssp. *robbiae*).
ALTITUDE To 460m.

Caper Spurge *Euphorbia lathyris*

Fairly common around towns and villages on waste ground, roadsides, old pits etc., also occasionally in open woodland. ❀ A statuesque spurge with dark blue-green foliage, the leaves with a prominent pale midrib and arranged in 4 neat ranks up the stem. DETAIL Hairless, all parts bleed bluish sap when broken. Stems usually single, over-wintering and flowering in their second year, reddish-purple near base. *Leaves opposite*, stalkless, strap-shaped, blunt. Flowers with large triangular bracts and crescent-shaped glands. *Seed capsule large, smooth, 3-sided*, hairless; seeds rough.

GROWTH Biennial.
HEIGHT 30–200cm.
FLOWERS Jun–Aug.
STATUS Ancient introduction from Europe, often grown in gardens.
ALTITUDE Lowland.

female flowers

GROWTH Perennial.
HEIGHT 10–50cm.
FLOWERS Feb–May.
STATUS Native.
ALTITUDE 0–1005m.

Dog's Mercury
Mercurialis perennis

Locally abundant in old deciduous woodland and on shady hedgebanks and verges, on moist but well-drained, often base-rich soils; in the uplands also in rocky places, especially where N-facing, and limestone pavements. Forms dense, uniform stands. ❁ Flowers small, green and inconspicuous. DETAIL Grows from creeping rhizomes, *stems unbranched, usually hairy*. Leaves opposite, oval, finely toothed, 3–8cm long, finely *hairy*. Dioecious, male and female flowers are on separate plants: male flowers in erect, catkin-like clusters up to 12cm long, growing from the leaf axils, with 3 tiny green sepals and 8–15 stamens (the anthers turn from yellow to indigo-blue after they have shed their pollen), no petals; female flowers in stalked groups of 1–3, with a bristly-hairy, 2-lobed ovary and 2 white stigmas; they develop into a hairy fruit 6–8mm across.

Annual Mercury *Mercurialis annua*

Common on disturbed ground in gardens, arable land, waste ground, beaches etc., especially on light soils. ❁ Erect, usually *branched*, with pale shiny green foliage and clusters of tiny green flowers; does not form dense, uniform stands. DETAIL Stems usually hairless. Leaves 1.5–5cm long, hairless, with 8–18 teeth per side (10–40 in Dog's Mercury). Flowers as Dog's Mercury but female flowers in almost stalkless groups of 1–3, ovary with conical warts, each tipped by a hair, fruits 3–4mm across.

GROWTH Annual.
HEIGHT 10–50cm.
FLOWERS All year.
STATUS Ancient introduction from Europe, present for at least a thousand years and probably still spreading.
ALTITUDE Lowland.

Sweet Violet *Viola odorata*

GROWTH Perennial.
HEIGHT 5–15cm.
FLOWERS Feb–May.
STATUS Native, but many
colonies are of garden origin.
ALTITUDE Lowland.

Fairly common on hedgebanks and in scrub and open woodland, especially on lime-rich soils; often naturalised in churchyards. ❂ The only *scented* violet, and the first to flower. White-flowered forms are often as common as, or commoner than, bluish-mauve forms. DETAIL *Creeping runners present*, so often patch-forming. Leaves and flowers all grow directly from a central point, with the *flowering stems leafless*. *Sepals rounded at the tip*. Leaves more or less circular in outline, with a heart-shaped base, either shiny and hairless or downy and soft to the touch, stalks with *very short hairs* (to 0.3mm long); the leaves become much larger in late summer. Flowers mauve-blue or white, spur dark mauve-blue; the common garden variety is distinctly purple with longer, narrower petals. ▶

Hairy Violet *Viola hirta*

Fairly common. Short, open turf on chalk grassland, verges, old pits and limestone pavement, also lightly shaded hedgebanks, open scrub and woodland edges; mostly on lime-rich soils. ❂ Rather like Sweet Violet, with leafless flowering stems (the leaves and flowers all coming directly from the rootstock) and blunt-tipped

GROWTH Perennial.
HEIGHT 5–15cm.
FLOWERS Late Mar–May.
STATUS Native.
ALTITUDE Generally lowland,
but to 610m.

sepals, but *not scented*. DETAIL *Lacks creeping runners.* Leaves *oval or long-oval* with a more or less *pointed* tip and heart-shaped base. Especially when fresh, the leaves and leaf-stalks are *very conspicuously hairy* (on the leaf-stalk the hairs are 0.3–1mm long and *erect*). Flower stalks sometimes hairy, capsule hairy. Flowers mauve-blue (as Sweet Violet), only *rarely white*.

GROWTH Perennial.
HEIGHT Stems to 20cm.
FLOWERS Mid Mar–May,
sometimes again Jul–Oct,
but later flowers usually lack
petals and fail to open, self-
pollinating in the bud.
STATUS Native.
ALTITUDE 0–1075m.

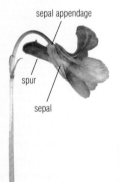

sepal appendage

spur

sepal

Common Dog Violet *Viola riviniana*

The commonest violet – very common in woodland, hedgebanks, grassland, heathland, moorland, vegetated shingle, cliffs and rock ledges, on all but very acid soils. DETAIL Leaves heart-shaped, *mostly in a central non-flowering rosette*, with the flowers carried on short *leafy stems. Leaves sparsely hairy, leaf-stalks more or less hairless.* Flowers 14–25mm across, bluish-mauve with a white throat and branched dark veins. *Spur stout* and bag-like, *obviously notched at the tip*, whitish, pale bluish-mauve or yellowish, but *always paler than the petals. Sepals pointed*, with the *sepal-appendage more than 1.5mm long*, forming a distinct short flap.

Early Dog Violet *Viola reichenbachiana*

GROWTH Perennial.
HEIGHT Flowering stems to
20cm.
FLOWERS Mar–May.
STATUS Native.
ALTITUDE Lowland.

Fairly common in woodland, hedgebanks and other shady places, on neutral to lime-rich soils. DETAIL Like Common Dog Violet but flowers smaller and more delicate, with narrower petals, and *on average paler and more washed-out*: mauve to light mauve (not so blue). Dark veins shorter, less extensive and little-branched. *Spur slimmer* (neither so obviously blunt nor so clearly notched at the tip), *as dark as, or darker than, the petals. Sepal appendage very short* (less than 1.5mm).

Heath Dog Violet *Viola canina*

GROWTH Perennial.
HEIGHT Stems to 30cm.
FLOWERS Last 10 days of
Apr–Jun.
STATUS
Native.
ALTITUDE
0–600m.

Near Threatened. Very locally common on dunes, also short turf on
heathland, the fringes of fens, rocky lake shores and stony ground by
rivers, on poor, acid soils (including clay over chalk). DETAIL Rather like
Common Dog Violet, but *flowers clear blue* (may have a hint of purple),
spur straight, blunt, creamy, pale yellow or greenish, leaves long-oval,
mostly longer than wide (1–1.5 times as long as wide), squared-off or
heart-shaped at the base. The stipules at the base of the leaf-stalk have
a few small, forward-pointing, *triangular teeth (fine and thread-like* in
Common Dog Violet), and are usually less than half the length of the leaf
stalk. Lacks the basal rosette of Common Dog Violet – all the leaves are on
the flowering shoots – but this is often hard to discern.

Pale Dog Violet *Viola lactea*

Nationally Scarce and
listed as Vulnerable.
Scarce and local on
grassy heathland, often
coastal, favouring
tracksides and similar
open ground produced
by grazing, burning or
other disturbance.
❀ As Heath Dog
Violet but *leaves
longer and narrower*
and *flowers creamy
to pale lilac* (beware
pale variants of other
violets), with narrower
petals and a short
greenish spur.
DETAIL Near hairless.
No basal rosette.
Leaves 1.4–2.5 times

GROWTH Perennial.
HEIGHT Stems to 20cm.
FLOWERS May–Jun.
STATUS Native.
ALTITUDE
Lowland.

as long as wide, more or less pointed, either rounded at the base or
tapering to the stalk; may be purplish below. Stipules relatively large and
coarsely-toothed. Hybridises with Common Dog and Heath Dog Violets.

Field Pansy *Viola arvensis*

Very common. Arable fields on light, well-drained soils, also gardens and similar places subject to soil disturbance. ❀ Flowers small but *obviously pansy-like*, almost always *creamy-white with a yellow throat*. DETAIL Stipules deeply cut, with a *large, long-oval terminal lobe, more or less lobed or toothed and resembling the leaves*. Flowers variable in size, 8–20mm top to bottom, *petals shorter than sepals*. Side petals sometimes washed or blotched violet (perhaps due to past hybridisation with Wild Pansy?). Spur 2–4mm long.

GROWTH Annual.
HEIGHT Stems 5–30cm.
FLOWERS Apr–Oct and through mild winters.
STATUS Ancient introduction.
ALTITUDE Lowland.

terminal lobe

Stipule

Wild Pansy
Viola tricolor

HEIGHT Stems to 30cm.
FLOWERS Apr–Sep.
STATUS Native.
ALTITUDE 0–460m.

Wild Pansy subspecies *tricolor*

An annual arable weed, favouring light sandy soils, but is uncommon to scarce or perhaps even rare, and listed as Near Threatened. ❀ Flowers obviously pansy-like, usually with some blue or violet, rarely all-yellow. Identification requires care, and many records probably refer to Garden Pansy, or to hybrid with Field Pansy (*V. ×contempta*). DETAIL Stipules deeply cut, *terminal lobes narrow, at most only slightly lobed*, and *not* like the leaves. *Flowers 15–25mm top to bottom* (rarely to 35mm), *petals usually longer than sepals*, not or only slightly overlapping. Spur 3–6.5mm. SIMILAR SPECIES Garden Pansy (*V. × wittrockiana*) is a fairly common garden throw-out, with flowers over 35mm top to bottom, and *strongly overlapping petals*.

Sand Pansy subspecies *curtisii*

Locally common on coastal dunes and grassland on N and W coasts, also grassy heaths in Breckland, where scarce (and similar perennial plants occur sparingly in other inland areas of dry, acid grassland). ❀ *Perennial*, with underground rhizomes. Terminal lobe of stipules rather broader than side lobes, often slightly wavy-edged. Flowers yellow or pale blue, usually less than 20mm top to bottom.

Marsh Violet *Viola palustris*

Locally fairly common. Wet woods, bogs, fens, damp heathland and moorland, especially on flushed, acid, peaty soils (often with *Sphagnum* mosses). ❀ Habitat, pale mauve flowers and rounded leaves distinctive. DETAIL Leaves and flower stalks arising a few at a time from creeping runners (1–3, sometimes 5, leaves together at each node along the runner). Leaves hairy to hairless,

GROWTH Perennial.
HEIGHT *c.* 10–15cm.
FLOWERS Apr–Jul.
STATUS Native.
ALTITUDE 0–1225m.

dull above, shiny below, *kidney-shaped, either rounded at the tip or bluntly-pointed, with a heart-shaped base,* leaf-stalks hairy to hairless. Flowers pale mauve, occasionally whitish (the dark purple veins are thus conspicuous), with a short spur and blunt petals.

Mountain Pansy *Viola lutea*

GROWTH Perennial.
HEIGHT Stems to 20cm.
FLOWERS May–Aug.
STATUS Native.
ALTITUDE To 1050m.

Locally common in unimproved, short, grazed grassland in upland areas; often common around old mine workings, but avoids more acid soils. ❀ Obviously a pansy, with yellow, blue, purple or blotched flowers. DETAIL Perennial, with slender creeping underground rhizomes and *solitary, unbranched flowering stems.* Stipules deeply cut, with the terminal lobe strap-shaped, untoothed, barely wider than side lobes. Flowers 20–35mm top to bottom (rarely only 15mm), spur 3–6mm.

Fairy Flax *Linum catharticum*

GROWTH Annual or biennial.
HEIGHT 5–25cm.
FLOWERS May–Sep.
STATUS Native.
ALTITUDE 0–840m.

Locally abundant. Short, poor, unimproved grassland, both dry turf and damp, fens and flushes, also machair and lead-mine spoil heaps. Usually found on chalk or limestone, but sometimes also mildly acidic heathland. ❀ Flowers tiny, 5-petalled, white, whole plant slender and easily overlooked, but may form mats of fine, multi-forked stems, sometimes woven through the grass. **DETAIL** Hairless. Leaves opposite, stalkless, long-oval, blunt, 5–8mm long. Sepals 5, 2–3mm long, petals 4–6mm long; stamens 5. Capsule 2–3mm long.

Allseed *Radiola linoides*

GROWTH Annual.
HEIGHT 1–8cm.
FLOWERS Jun–Aug.
STATUS Native.
ALTITUDE Lowland.

Uncommon and very local, listed as Near Threatened. Bare, acid, sand and gravel, usually where wet in winter, e.g. tracksides and the edges of ponds and ditches on heathland, forestry, grassland and dune slacks. Often found with Chaffweed (p. 185). ❀ Tiny, delicate and easily overlooked. Often well-branched, with *reddish stems contrasting with small, oval, pale green leaves*. Flowers minute. **DETAIL** Leaves opposite, to 3mm long. Sepals 4, deeply toothed at the tip, *c.* 1mm long, petals 4, about as long as sepals, stamens 4. Capsule spherical, 1mm across.

Pale Flax
Linum bienne

Rather local. Poor, dry
grassland and grassy
scrub, especially in
warm, sheltered places
near the sea: cliffs,
field margins, verges.
❋ Thin, wiry stems
support *relatively large,
satiny, 5-petalled, pale
blue flowers.*
DETAIL Hairless. *Usually
has several stems.*
Leaves alternate,
strap-shaped, pointed,
0.5–1.5(4)mm wide,
mostly 3-veined and
rather grey-green; *held
erect.* Sepals 5, pointed,
4–6mm long (as
long as ripe capsule),
petals 8–12mm long;
5 stamens, with the
stigmas more or less
as tall as the anthers.
Capsule 4–6mm long.

GROWTH Biannual or short-
lived perennial.
HEIGHT 20–30cm (60cm).
FLOWERS May–Sep.
STATUS Native.
ALTITUDE Lowland.

Flax *Linum usitatissimum*

Occasional as a relic of
cultivation or bird-seed
alien. Waste ground
and verges. ❋ *In most
places the only blue
flax, with soft-blue,
5-petalled flowers.*
DETAIL Very like Pale
Flax but larger. *Stems
usually single* (may be
branched). *Leaves 1.5–
4mm wide,* 3-veined.
Flower blue or white,
sepals 6–9mm long (as
long as ripe capsule),

GROWTH Annual.
HEIGHT To 85cm.
FLOWERS Jun–Sep.
STATUS Cultivated since at
least the Middle Ages for its
fibres, and now planted for
linseed oil. Regularly escapes.
ALTITUDE Lowland.

petals 12–20mm. Stigmas club-shaped, roughly
as tall as anthers. *Capsule 6–9mm long.* SIMILAR
SPECIES Perennial Flax *L. perenne,* rare on
chalky grassland, mostly in E England, has sky-
blue flowers. Sepals 3.5–6.5mm, *half length of
ripe capsule, inner 2 sepals rounded at tip,* and
*knob-like stigmas either shorter or taller than the
anthers.* Leaves mostly 1-veined, spreading.

Perforate St John's-wort
Hypericum perforatum

Abundant, and by far the commonest St John's-wort.
Verges, hedgebanks, meadows, scrub and open woodland.

❁ Identified by its *round stems with 2 opposite raised ridges*, hairless leaves with a few black glands on the margins below, and *abundant translucent dots in the leaves*. DETAIL Leaves 8–30mm long. Flowers 15–25mm across, petals bright yellow, at least twice as long as sepals; sepals narrow, pointed; petals and sepals with a few stalkless black glands.

GROWTH Perennial.
HEIGHT 30–80cm.
FLOWERS Jun–Sep.
STATUS Native.
ALTITUDE 0–480m.

Imperforate St John's-wort
Hypericum maculatum

Common in the N and W, scarcer elsewhere, in similar grassy places to Perforate St John's-wort, but usually damper and more shaded, including verges, rough grassland, scrub, old quarries and waste ground. ❁ Identified by its 4-lined or square stems (without wings), with few if any transparent dots in the leaves and bright yellow flowers. DETAIL Hairless. Leaves 15–40mm long, with a few black glands on margins both above and below. Flowers 15–25mm across, petals with blackish streaks and lines, sepals oval, blunt, often wavy-edged towards tip, sometimes with black glands. SIMILAR SPECIES *H. × desetangsii*, the hybrid between Perforate and Imperforate St John's-wort, is fairly common. Intermediate in the shape of the sepals and stem, and in the amount of translucent dots on the leaves, it is fertile, and can 'back-cross' with the parents to produce a range of intermediates.

GROWTH Perennial.
HEIGHT Stems to 60cm.
FLOWERS Jun–Aug.
STATUS Native.
ALTITUDE 0–510m.

Slender St John's-wort
Hypericum pulchrum

Locally common. Heathland and woodland rides and clearings on well-drained, sandy or peaty, neutral to acid soils. ❀ Dainty and attractive, especially in bud, with a slender, round, reddish stem (no ridges) and *oval leaves that are blunt at the tip and broaden to a heart-shaped base that clasps the stem.* DETAIL Hairless. Leaves to 15mm long, with translucent dots on outer part only and no black glands. Flowers 12–18mm across, petals bright yellow, tinged red below (thus bursting buds look reddish), at least twice as long as sepals, which are all equal in length; both petals and sepals have stalked black glands along the edges; anthers orange. A dwarf form with sprawling stems is found in N and W Scotland and W Ireland (var. *procumbens*), and young plants elsewhere may also be prostrate and confusable with Trailing St John's-wort.

GROWTH Perennial.
HEIGHT 20–50cm.
FLOWERS Jun–Aug.
STATUS Native.
ALTITUDE 0–820m.

Trailing St John's-wort *Hypericum humifusum*

Rather scattered and local, usually on well-drained acid soils, on heathland, moorland, woodland rides and tracksides. ❀ *More or less sprawling* with slender stems and oval-oblong leaves. DETAIL Hairless. Stems reddish, slender, round, with 2 faint raised lines. Leaves up to 10mm long, with black glands along margins on both sides; translucent dots abundant to absent. Flowers 8–12mm across, petals bright yellow, slightly longer than sepals, sepals slightly unequal in length, with 3 longer and wider than the other 2; both sepals and petals with scattered black glands along margins. SEE ALSO Slender St John's-wort – in which young plants may also be sprawling – but have more leathery leaves without black glands.

GROWTH Perennial.
HEIGHT Stems to 20cm.
FLOWERS Jun–Sep.
STATUS Native.
ALTITUDE Mostly lowland, but to 530m.

GROWTH Perennial.
HEIGHT To 40cm.
FLOWERS Jun–Sep.
STATUS Native.
ALTITUDE 0–425m.

Marsh St John's-wort *Hypericum elodes*

Locally common. Pools, flushes and the margins of streams and ponds on heathland and bogs, on poor, acid soils. Often grows in shallow water and may form a floating mat, but also terrestrial.
❀ Low-growing and patch-forming, the grey-green stems and leaves are *densely and softly hairy*, quite distinct from those of other St John's-worts, and the flowers do not open so widely.
DETAIL Stems creeping, rooting at lower nodes; flower stems erect. Leaves circular to oval, up to 30mm long, clasping the stem at the base, with many translucent dots. Flowers 12–20mm across, with reddish glandular hairs. No black glands on plant.

GROWTH Perennial.
HEIGHT 40–100cm.
FLOWERS Jun–Aug.
STATUS Native; also widely cultivated and many plants are bird-sewn from gardens.
ALTITUDE 0–630m.

Tutsan
Hypericum androsaemum

Locally frequent, especially in the S and W. Damp, shady places in woods and hedgebanks, also spread by birds to drier habitats.
❀ Shrubby, with typical St John's-wort flowers and *black berries*.
DETAIL Hairless. Erect, more or less deciduous shrub with branched, 2-ribbed, usually reddish stems. Leaves oval, stalkless, 50–120mm long. Flowers 15–25mm across, in small clusters, petals equal to or shorter than the sepals; 3 styles, stamens about as long as petals. Fruit ripens from green through red to a black berry 5–8mm across.
SIMILAR SPECIES Rose of Sharon ▶ *H. calycinum*, commonly grown in gardens, is occasionally naturalised in hedgerows, verges and railway embankments.
❀ Distinctive solitary, large flowers, 50–80mm across, have 5 styles, the stamens shorter than the petals, and the anthers reddish.

Hairy
St John's-wort
Hypericum hirsutum

Fairly common in sun or partial shade on well-drained, often lime-rich soils: woodland rides, old quarries, rough grassland, banks and verges. ❁ *Stem and leaves conspicuously hairy.* Detail Stem round. Leaves 25–50mm, *strongly veined*, with many translucent dots. Flowers 15–22mm across, petals pale yellow, with stalked black glands on petals and sepals.

GROWTH Perennial.
HEIGHT To 100cm.
FLOWERS Jul–Aug.
STATUS Native.
ALTITUDE 0–450m.

Pale
St John's-wort
Hypericum montanum

Near Threatened. Uncommon and local. Shady places on chalk or limestone: open woodland, scrub, rough grassland, rocky slopes, old quarries and limestone pavement. ❁ *Very erect*, with *exactly opposite leaves that more or less clasp the stem* and pale yellow flowers. Detail *Stems hairless, round or subtly 2-ridged.* Leaves 15–30mm long, blunt, hairless above and *finely hairy below*, with *black glands along the margins on the underside* but no translucent dots. Petals lack black glands; sepals narrow, pointed, with *stalked black glands along their margins*.

GROWTH Perennial.
HEIGHT To 100cm.
FLOWERS Jun–Aug.
STATUS Native.
ALTITUDE 0–330m.

Square-stalked
St John's-wort
Hypericum tetrapterum

Common. Marshes, wet meadows and other damp places. ❁ *Stems square*, with narrow but distinct 'wings' on the angles (0.25–0.5mm wide) and many translucent dots in the leaves. Detail Hairless. Stem reddish. Leaves to 40mm long, more or less clasping the stem, with abundant black glands along margins on both sides. Flowers 9–13mm across, petals pale yellow, as long as sepals, with few if any black glands, sepals narrow, pointed.

GROWTH Perennial.
HEIGHT To 60cm.
FLOWERS Jun–Sep.
STATUS Native.
ALTITUDE To 380m.

FAMILY GERANIACEAE: CRANESBILLS & STORKSBILLS

Cranesbills and storksbills are so-called because of the resemblance of their seed pods to cranes' and storks' bills respectively (although cranes and even more so storks have long been rare in Britain). In cranesbills (genus *Geranium*), the leaves are *palmately-lobed*, with the main veins and/or lobes all radiating from a central point at the tip of the stalk; in storksbills (genus *Erodium*), the leaves are *pinnately-lobed*, with the main veins and/or lobes radiating from the central vein.

Meadow Cranesbill | Common Storksbill

Meadow Cranesbill *Geranium pratense*

GROWTH Perennial.
HEIGHT 30–80cm.
FLOWERS Jun–Sep.
STATUS Native (but introduced in E Anglia, SW England and W Wales).
ALTITUDE To 470m.

Locally common. Rough grassland on verges, banks, stream- and riversides, mostly on damp, lime-rich soils, also unimproved meadows and pastures. ❀ Robust, with large, deeply-cut leaves and *large blue flowers*. DETAIL Clumped or almost tussock-forming, whole plant glandular-hairy. Stems often reddish. Leaves opposite, cut almost to the base into 7–9 lobes, these in turn deeply cut into long teeth. Flowers mostly paired, petals 16–24mm long, blue to violet-blue with a rounded tip. Fruit stalk angled downwards.

Dusky Cranesbill *Geranium phaeum*

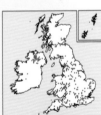

Uncommon garden escape or throw-out. Damp, shady places on verges and in woodland and churchyards, often close to houses. ❀ *Flowers dark purplish*. DETAIL Whole plant hairy, with simple hairs of various lengths and variable numbers of very short-stalked glandular hairs on leaf-stalks and on underside of leaves. Stems and leaves spotted red, the leaves cut two thirds of the way to the base into 7 lobes. Petals 8–12mm long, often with small point at tip; flowers rarely pinkish-mauve or white.

GROWTH Perennial.
HEIGHT 30–80cm.
FLOWERS Late Apr–Sep.
STATUS Introduced from C Europe and first recorded in the wild in 1724.
ALTITUDE Lowland.

Wood Cranesbill *Geranium sylvaticum*

Locally common, mostly in the uplands. Rough grassland on verges, banks, stream sides and in rocky places and open woodland; formerly common in unimproved meadows. ❁ A robust geranium with large, coarsely-toothed leaves and *purplish-pink flowers*. DETAIL Whole plant glandular-hairy. Leaves mostly alternate, deeply cut into 7 lobes, these in turn shallowly cut into teeth 1.5–2

GROWTH Perennial.
HEIGHT To 75cm.
FLOWERS Jun–Jul.
STATUS Native.
ALTITUDE To 1005m.

times longer than wide. Flowers mostly paired, petals 12–16mm long, white (and with abundant hairs) at the base and with a rounded or slightly notched tip. Fruit stalks angled upwards.

Bloody Cranesbill *Geranium sanguineum*

Locally common. Rough grassland, scrub, rocky woodland, cliffs and dunes, mostly near the sea but also inland in sheltered spots on limestone or chalk; garden throw-outs occur more widely. ❁ A mid-sized geranium with *large crimson flowers*. DETAIL Whole plant hairy, with some short glandular hairs on the

GROWTH Perennial.
HEIGHT 10–40cm.
FLOWERS May–Aug.
STATUS Native; also widely introduced in S England.
ALTITUDE 0–420m.

leaf stalks. Leaves opposite, 20–60mm across, deeply cut into 5–7 lobes. Flowers solitary, petals 14–22mm long, slightly notched at the tip.

Dovesfoot Cranesbill
Geranium molle

Very common. Dry grassland (including lawns), verges, arable fields and waste ground. ❀ A small-flowered geranium best identified by the *3 types of hair on the leaf stalks* (10× lens). **DETAIL** *Leaves alternate*, 10–40mm across, lobed up to two thirds of the way to the base. *Leaf stalk with short, erect, gland-tipped hairs (0.2–0.5mm long, unequal), abundant short glandless hairs* and *sparse long hairs* (to 1.5mm). Petals mauve-pink, 4–6mm long, narrowed at the base to a *distinct 'stem'* up to half as long as main part, tip *notched*. All stamens have anthers. Capsule hairless, usually finely ridged; seeds smooth. **SEE ALSO** Hedgerow Cranesbill (p. 104).

GROWTH Annual.
HEIGHT Stems to 40cm.
FLOWERS Apr–Sep.
STATUS Native.
ALTITUDE To 550m.

Small-flowered Cranesbill
Geranium pusillum

Fairly common. Cultivated and waste ground and open grassland, usually on sandy or gravelly soils. ❀ Much like Dove's-foot Cranesbill, but flowers smaller and often paler, *only the inner 5 stamens have anthers* (hard to see); *hairs on leaf stalks differ*. **DETAIL** *Leaves opposite*, otherwise as Dove's-foot. *Leaf-stalk with erect, very short, gland-less hairs (0.2mm, roughly equal)*, occasionally also with some stalkless glands. Petals as Dove's-foot but smaller, *2.5–4mm long*, and more shallowly notched at tip. Capsule finely downy, more or less smooth; seeds smooth.

GROWTH Annual.
HEIGHT Stems to 40cm.
FLOWERS May–Sep.
STATUS Native.
ALTITUDE Lowland.

Round-leaved Cranesbill
Geranium rotundifolium

Locally common. Dry banks, walls and waste ground, usually on sand, gravel or chalk; a garden and street weed. ❀ Small-flowered, best identified by the *red-tipped hairs on the leaf stalks* (10× lens). **DETAIL** Gland-tipped hairs abundant on most of plant. Leaves opposite, to 60mm across, lobed no more than half way to base, *often spotted red at the notches*. Leaf stalk with erect hairs (up to 0.5mm long, unequal), *many or all tipped with a red gland (like a tiny blob of sealing wax)*, as well as stalkless glands. Petals 5–7mm long, *without a narrow 'stem'*, rounded or slightly notched. Capsule hairy; seeds net-veined.

GROWTH Annual.
HEIGHT 10–40cm.
FLOWERS Apr–Aug.
STATUS Native.
ALTITUDE Lowland.

Cut-leaved Cranesbill *Geranium dissectum*

A common weed of arable land and gardens, also grassy places and waste ground. ✹ Small-flowered, with *deeply-cut leaves* (often almost to the base), but *best identified by the hairs on the leaf stalks* (10× lens). **DETAIL** Often straggling. Leaves usually opposite, at least on upper stem, 20–70mm across. *Leaf stalks densely hairy, the hairs all curved and pointing downwards* (0.3–0.8mm long, slightly

GROWTH Annual.
HEIGHT Stems to 60cm.
FLOWERS May–Aug.
STATUS Ancient introduction, present in Britain by the Roman period.
ALTITUDE To 435m.

unequal; without glands, although glandular hairs frequent on other parts of plant). *Flower stalks 25mm or less*. Petals 4.5–6mm long, *without a distinct narrow 'stem' but with a deep but blunt notch at the tip*. Sepals tipped by a fine point. Capsule long-hairy, ridged; seeds net-veined.

Long-stalked Cranesbill *Geranium columbinum*

Rather local. Dry grassy places, mostly on thin, calcareous soils, often where disturbed and/or S-facing: verges, banks, dunes, and chalk and limestone grassland. ✹ A small-flowered geranium with *deeply cut leaves* (often almost to the base) and *long-stalked, bell-shaped flowers*. **DETAIL** Erect to sprawling, often scrambling through surrounding vegetation. Stems, leaves and leaf stalks

GROWTH Annual.
HEIGHT Stems to 60cm.
FLOWERS May–Aug.
STATUS Native.
ALTITUDE Mostly lowland, but to 380m.

with long hairs, *all pressed flat*; no glandular hairs. Leaves opposite, at least on upper stem, 25–50mm across. *Most flower stalks over 25mm long*. Petals 7–10mm, reddish-pink, without a distinct narrow 'stem', rounded to shallowly notched at tip. *Sepals tipped by a long, fine point*. Capsule hairless to very sparsely hairy, seeds finely net-veined.

Shining Cranesbill *Geranium lucidum*

GROWTH Annual.
HEIGHT 10–40cm.
FLOWERS May–Sep.
STATUS Native.
ALTITUDE To 610m.

Common. Rocky places on limestone, also walls, churchyards and, especially on sandy or chalky soils, verges, banks, waste ground and gardens. ❀ Small, deep pink flowers and *obviously glossy foliage* distinctive. Whole plant often turns red. **DETAIL** Stems brittle, fleshy, hairless. Leaves cut about two thirds of the way to the base; may have some hairs on upperside. Petals 8–10mm long, with a rounded tip and long 'stem' (longer than main part). Capsules ridged, minutely hairy, at least on edges; seeds smooth.

Hedgerow Cranesbill *Geranium pyrenaicum*

Locally common, especially in SE, on verges, banks, field margins and waste ground, often near houses. ❀ A straggling geranium with *medium-sized flowers*. **DETAIL** Stems and leaves hairy. *Leaves opposite*, 35–90mm across, lobed up to half way to the base (maximum two thirds). Basal leaves long-stalked, stalks with erect hairs (a neat mixture of short gland-tipped and plain hairs as well as

GROWTH Perennial.
HEIGHT 25–60cm.
FLOWERS May–Jul.
STATUS Introduced from S Europe, first recorded in 1762 and still spreading.
ALTITUDE Lowland.

sparse rather long hairs). *Petals 7–10mm long*, notched at tip, narrowing at base to a short 'stem' (less than half length of main part). Capsules finely downy, not ridged; seeds smooth. **SEE ALSO** Dove's-foot Cranesbill (p. 102).

Herb Robert *Geranium robertianum*

GROWTH Annual or biennial.
HEIGHT 10–40cm.
FLOWERS Apr–Oct.
STATUS Native.
ALTITUDE 0–700m.

Abundant in light shade in hedgebanks, woods and gardens, and on walls, limestone pavement, scree and shingle. ✿ Strong-smelling, the whole plant may turn a beautiful red. DETAIL Whole plant hairy. Leaves deeply cut into 3–5 leaflets, these in turn deeply cut. Leaf stalk with red-tipped glandular hairs. Petals 8–14mm long, rounded at the tip and with a narrow 'stem' as long as main portion, anthers orange or purple. Capsules hairless to sparsely hairy, well ridged, seeds smooth. SIMILAR SPECIES **Little Robin** *G. purpureum*, uncommon near the sea in S and SW England, has smaller flowers (petals 5–9mm) and yellow anthers.

Common Storksbill *Erodium cicutarium*

GROWTH Annual.
HEIGHT Stems may straggle to 60cm.
FLOWERS Apr–Sep.
STATUS Native.
ALTITUDE 0–420m.

Common. Short turf and bare ground, both near the sea and inland, on light, well-drained soils on sands, gravels and leached chalk. ✿ Prostrate, with *finely cut leaves in a basal rosette*, small pink flowers and fruits with a long beak. DETAIL Whole plant rather hairy, but glandular hairs relatively sparse or absent. Leaves 4–18cm long, cut into 4–7 pairs of leaflets, these in turn deeply cut (usually to near midrib). Flowers in groups of 3–7, pinkish-purple to white, often with a black spot at the base of the 2 upper petals; 10–18mm across. Fruits 5–6.5mm long, the *pit at the tip demarcated by a sharp ridge and groove and not over-reached by hairs* (cf Sticky Storksbill, 20× lens); beak 15–40mm long.

Sticky Storksbill
Erodium lebelii

Nationally Scarce. Very local on bare ground on dunes and other sandy places. ❀ Very similar to Common Storksbill and easily overlooked, but leaves more greyish-green and overall more densely glandular-hairy (often picking up a lot of sand). Identification must be confirmed, however, by examination of the fruits. DETAIL Flowers mostly less than 10mm across, pale pink to white, without dark spots at the base of the petals. Fruits less than 5mm long with a *tiny rimless pit at the tip, over-reached by hairs* from the body of the fruit (20× lens; examine ripe dry fruits that have burst from the seed head); beak to 22mm.

GROWTH Annual.
HEIGHT Stems to 25cm.
FLOWERS May–Jul.
STATUS Native.
ALTITUDE Lowland.

Musk Storksbill
Erodium moschatum

Fairly common, and rapidly expanding its range. Open grassland, lawns, verges, waste ground and field edges, usually on light, sandy soils. ❀ *Larger and coarser* than Common Storksbill, with *less finely cut leaves*. DETAIL Whole plant glandular-hairy (the leaves minutely so), weakly musk-scented when bruised. Leaves in a basal rosette, 10–45cm long, with 4–7 pairs of leaflets, these in turn *shallowly-lobed* (usually *cut less than half way to midrib*). Flowers pinkish-purple, beak 20–45mm long.

GROWTH Annual.
HEIGHT Stems to 60cm (but often absent).
FLOWERS Mar–Jul.
STATUS Ancient introduction.
ALTITUDE Lowland.

Sea Storksbill *Erodium maritimum*

Rather local on cliffs and dunes by the sea, on short turf with bare, compacted ground created by hard grazing or trampling, also rough car parks and pavement cracks; very rare inland. ❀ Inconspicuous, with *small, petal-less flowers*, but may form patches or drape over walls and banks. DETAIL Hairy. Leaves 5–20mm long, *lobed* about half way to base, *petals usually absent* (if present, short and pink, but dropping rapidly); *beak only 8–10mm long*.

GROWTH Annual.
HEIGHT To 10cm.
FLOWERS May–Sep.
STATUS Native.
ALTITUDE Lowland.

Purple Loosestrife *Lythrum salicaria*

Locally common. Wet places on the margins of ditches, slow-flowing rivers, canals, ponds and lakes, reed swamps, fens, and light shade in willow carr; avoids acid soils. ☀ Tall, with spires of purple flowers. **DETAIL** Whole plant hairy. Stems square. Leaves stalkless, broadly spear-shaped, in opposite pairs or groups of 3 on the lower part of the stem, alternate towards the tip. Flowers in whorls around the stem, intermixed with leaf-like bracts. Calyx lobes sharply toothed at tip, reddish on inside; petals 6, strap-shaped, 8–10mm long, reddish-purple with crimson veins. Style 1 (long, medium or short), stamens 12 (6 at each of the 2 levels – long, medium or short – not occupied by the stigma), filament crimson, anther dark purple.

GROWTH Perennial.
HEIGHT 60–150cm.
FLOWERS Jun–Aug.
STATUS Native.
ALTITUDE 0–440m.

Water Purslane *Lythrum portula*

GROWTH Annual or short-lived perennial.
HEIGHT Stems 5–25cm.
FLOWERS Jun–Oct.
STATUS Native.
ALTITUDE 0–530m.

Very local. The muddy fringes of ponds, lakes and reservoirs, rutted tracks on heaths and dunes, and woodland rides that are wet in winter; avoids chalky and peaty soils. ☀ Inconspicuous, forming sprawling mats of reddish stalks with contrasting green fleshy leaves, or growing upright in shallow standing water. Tiny flowers are produced singly in the leaf axils. **DETAIL** Hairless. Stems 4-angled, rooting at nodes. Leaves opposite, 10–20mm long, spoon-shaped, with reddish rim and midrib; the whole leaf may age to red. Flowers have 6 pointed calyx teeth up to 2mm long, 6 minute purplish petals (may be absent), and 6 stamens; fruit a pea-like capsule 2mm across.

EPILOBIUM & *CHAMERION:* WILLOWHERBS

The 12 British willowherbs in the genus *Epilobium* are all perennials, with undivided leaves and flowers that have 4 notched petals in various shades of pink. They vary in height from a few cm to 180cm. Seed is produced in long, slender capsules that split open longitudinally, and the seeds have a hairy plume and are spread by the wind: several species are very successful 'weeds', able to quickly colonise disturbed ground. Willowherbs are easily recognisable as a group, but identification of the individual species requires careful examination of the shape of the stigma (either club- or cross-shaped), the presence and type of hairs on the stem, leaves and flowers (10× lens essential), and the shape of the leaves and stems. To complicate matters, they hybridise frequently, producing a range of intermediates: only plants which show *all the characters* of a species can be identified with confidence. Rosebay Willowherb (genus *Chamerion*) differs in its unequal petals and alternate (rather than opposite) leaves and is distinctive.

GROWTH Perennial.
HEIGHT 80–180cm.
FLOWERS Jul–Sep.
STATUS Native.
ALTITUDE 0–415m.

Great Willowherb
Epilobium hirsutum

Abundant in marshes and a wide variety of damp places near water, generally on fertile, non-acid soils, also drier sites such as verges and waste ground. ✿ Tall. Easily identified by the combination of a *cross-shaped stigma*, soft hairs on the stem and leaves, *large purplish-pink flowers* and clasping leaf-bases. DETAIL Stem and leaves *densely covered with soft erect hairs* (both with and without gland-tips). Petals 10–16mm long, stigma with 4 *spreading white lobes* that curl downwards. Leaves strap-shaped, unstalked, *slightly clasping the stem at their base.*

GROWTH Perennial.
HEIGHT 30–75cm.
FLOWERS Jul–Sep.
STATUS Native.
ALTITUDE 0–450m.

Hoary Willowherb
Epilobium parviflorum

Common. Marshes and other damp places near water, also drier sites such as quarries, walls, arable fields, gardens, verges and waste ground; may be a street weed. ✿ A smaller and slenderer version of Great Willowherb, separated by its *smaller, paler flowers* and *non-clasping leaves.* DETAIL Stem and leaves densely covered with soft hairs (mostly without glands), appearing greyish, sometimes reddish. *Stigma 4-lobed*, with the *lobes held upwards* (thus may appear club-shaped on a quick view, especially when fresh). Flowers pale pinkish-purple, petals 5–9mm long. Leaves strap-shaped, rounded at the base, stalkless (or very short staked) but *not clasping stem*; leaves often held erect.

Square-stalked Willowherb
Epilobium tetragonum

Common. Disturbed soils, both damp and dry, including arable, gardens, waste ground, woodland edges and streamsides. ❀ Note *club-shaped stigma, 4 ridges on the stem* (at least at base), narrow, stalkless, shiny green, hairless leaves, and *complete absence of erect glandular hairs*. DETAIL Stem, flowers and young capsules with more or less dense, flattened hairs. Leaves often held erect, strap-shaped, blunt, toothed, *stalkless or occasionally short stalked (especially in N)*, their bases running into the stem ridges.

GROWTH Perennial.
HEIGHT To 75cm.
FLOWERS Jun–Aug.
STATUS Native.
ALTITUDE Lowland.

Flowers small, petals 2.5–7mm, pale purplish-pink. Capsule 65–80mm (55–100mm). No runners.

Short-fruited Willowherb
Epilobium obscurum

Common in the W, scarcer to the SE. Damp places in marshes, woodland rides and along streams, also cultivated and waste ground. ❀ Note *club-shaped stigma*, dull green, hairless, more or less stalkless leaves (lowest may have a very short stalk), dense, short, flattened hairs on upper stem, flowers and young capsules, with a *few erect glandular hairs* on the calyx and sometimes also the capsule (can be hard to see); capsules relatively short. DETAIL Stem round, especially at base, with 2–4 raised ridges, at least towards tip. Leaves hairless (may have hairs on

GROWTH Perennial.
HEIGHT 20–80cm.
FLOWERS Jul–Aug.
STATUS Native.
ALTITUDE 0–785m.

veins), oval-lanceolate (rather broader than Square-stalked), with well-separated teeth, blunt at tip, rounded at base, which runs into the stem ridges; lower leaves joined around stem. Flowers small, petals 4–7mm, purplish-pink. Capsule 40–60mm long (30–65mm). Produces elongated runners in summer.

American Willowherb
Epilobium ciliatum

Often the commonest willowherb, especially in urban areas – gardens, waste ground and walls. Also marshes, woods and streamsides. ❀ Rather variable, but note *club-shaped stigma*, hairless, short-stalked leaves, and *glandular-hairy upper stem, flowers and young capsules*. DETAIL Whole plant usually tinged reddish. Stem either round or with 4 raised lines, with crisped hairs either all round or in 2 (sometimes 4) rows, as well as many erect gland-tipped hairs, especially towards tip. Leaves weakly toothed, strap-shaped, tapering towards a pointed tip and with either a rounded or tapering base, stalk 1.5–4mm. Flowers pinkish-

GROWTH Perennial.
HEIGHT 10–75cm (100cm).
FLOWERS Late May–Aug.
STATUS First recorded in 1891, it spread rapidly from c. 1930 and is now amongst the commonest aliens (N America).
ALTITUDE 0–790m.

purple (sometimes white), small, petals 3–6mm; flowers and young capsules with erect gland-tipped hairs and flattened glandless hairs.

Broad-leaved Willowherb
Epilobium montanum

Common. Woods, hedgerows, rocky places, gardens, walls and waste ground, on moist, free-draining soils, usually in the shade; may be a street weed. ❀ Note *4-lobed stigma*, sparsely hairy stem, and leaves that are *rounded at the base with a very short stalk*. DETAIL Stem often reddish, round, with short, inconspicuous hairs (both short, curled non-glandular hairs and erect glandular hairs). Leaves oval, pointed, sharply toothed, with a rounded base that contracts abruptly into a short stalk (1–6mm, most obvious on lower leaves); all leaves have tiny flattened hairs. Flowers pink, petals 8–10mm long.

GROWTH Perennial.
HEIGHT 20–75cm.
FLOWERS Late May–Aug.
STATUS Native.
ALTITUDE 0–790m.

Spear-leaved Willowherb
Epilobium lanceolatum

Locally fairly common in SW England, largely a garden weed elsewhere. Dry, sunny places: banks, walls, hedgerows, open woods, dunes, quarries and waste ground. ❀ Rather like Broad-leaved Willowherb (including *4-lobed stigma*) but *all leaves taper at the base into a 3–10mm long stalk* and the small flowers are *white* when first open. DETAIL Stem and leaves often reddish. Stem round, sparsely hairy, leaves with tiny flattened hairs (i.e. as Broad-leaved). Leaves narrowly oval, blunt at the tip, with a few blunt teeth. Petals 6–8mm long, white, ageing to dark pink; stigma 4-lobed.

GROWTH Perennial.
HEIGHT To 60cm.
FLOWERS Jul–Sep.
STATUS A recent colonist? First recorded 1843 and more or less confined to the SW, but may be slowly spreading.
ALTITUDE 0–400m.

Pale Willowherb
Epilobium roseum

Fairly common in the S, rarer to the N. Woods, the banks of streams and ditches, waste ground, hedgerows and gardens, on damp, disturbed soils, often shaded; may be a street weed. ❀ Note *club-shaped stigma, glandular-hairy upper stem, flowers and young capsules*, and *relatively long leaf stalks*. DETAIL Stem more or less square with 4 raised lines (2 obvious, 2 rather indistinct), very variably hairy, with both flattened non-glandular and erect glandular hairs. Leaves held at 90° or drooping, with short curled hairs on veins (occasionally hairless), toothed, almost egg-shaped, tapering to a pointed tip and a 3–20mm long stalk. Flowers very pale pink (whitish in bud), small, petals 4–7mm; flowers and young capsules with numerous erect gland-tipped hairs and flattened non-glandular hairs.

GROWTH Perennial.
HEIGHT 10–75cm.
FLOWERS Jul–Aug.
STATUS Native.
ALTITUDE To 560m.

Marsh Willowherb
Epilobium palustre

Fairly common to N, uncommon and declining in S. Marshes, bogs, around ditches, streams and lakes, and woodland flushes, usually on acid soils. ❀ A rather slender, small-flowered willowherb, note *club-shaped stigma*, sparse, erect, gland-tipped hairs on the upper stem, flowers and young capsules, short leaf stalks and *narrow, untoothed or indistinctly toothed leaves*. DETAIL Stem occasionally purplish, round, lacking raised ridges (occasionally 2 raised lines), with often sparse flattened non-glandular hairs and scattered erect glandular hairs towards tip. Leaves more or less hairless, dull green (occasionally purplish), strap-shaped, 4–10mm wide, blunt, stalkless or with short stalk (to 4mm on uppermost), tapering at the base and usually joining around the stem. Flowers pale pink, petals 4–7mm; flowers and young capsules sparsely hairy.

GROWTH Perennial.
HEIGHT 15–60cm.
FLOWERS Jul–Aug.
STATUS Native.
ALTITUDE 0–800m.

New Zealand Willowherb
Epilobium brunnescens

GROWTH Perennial.
HEIGHT Stems to 20cm.
FLOWERS May–Oct.
STATUS First recorded in the wild in 1904; now common in the N and W.
ALTITUDE 0–1100m.

Probably introduced as a rock garden plant, now locally abundant in the wetter parts of Britain on moist stony ground: old quarries, tracksides, sidings and damp walls, also more natural sites such as streamsides and damp rocks. ❀ A distinctive *small, prostrate, creeping* willowherb. **DETAIL** Stems very finely hairy, rooting at the nodes to form mats. Leaves more or less circular, small (3–7mm across), short-stalked, opposite, often purplish below. Flowers on erect, slender stalks growing from the leaf axils, white to pale pink, small (petals 2.5–4mm), stigma club-shaped.

Rosebay Willowherb *Chamerion angustifolium*

GROWTH Perennial.
HEIGHT To 150cm.
FLOWERS Jun–Sep.
STATUS Native.
ALTITUDE 0–975m.

Abundant. Disturbed ground on dunes, verges, railways, waste ground, heaths, and woodland rides and clear-fell, also rock ledges and scree in the uplands (its original habitat). ❀ Distinctive, with tall spires of rose-purple flowers. Often found in dense stands. **DETAIL** Leaves alternate, stalkless, strap-shaped, pointed. Flowers large (20–30mm across), with 4 petals, unequal in size, (upper 2 broader than lower) and 4 long, narrow, reddish sepals; style bent downwards, stigma 4-lobed. **SEE ALSO** Purple Loosestrife (p. 107).

In the early 19th century this was a rare plant and more or less confined to upland Britain, but it is now ubiquitous. Each plant produces up to 80,000 seeds and its spread was probably aided by the construction of railways and roads, along which the seeds could waft. Known as 'bombweed' following the Blitz of 1940, it grew in vast numbers amongst the bombed-out ruins.

Enchanter's Nightshade *Circaea lutetiana*

GROWTH Perennial.
HEIGHT 20–60cm.
FLOWERS Jun–Aug.
STATUS Native.
ALTITUDE To
465m.

Locally common. Damp, shady places in deciduous woodland and similar habitats, usually on base-rich soils; can be a persistent garden weed. ❀ *Patch-forming*, with leafless spikes of small, delicate white flowers. DETAIL Stems with a few glandular hairs, growing from a network of brittle, white, creeping rhizomes. Leaves opposite, oval, 4–10cm long, dull green above, shiny below, more or less hairless; leaf stalk hairy. Petals 2, deeply lobed, 2–4mm long, sepals 2, glandular-hairy, stamens 2; stigma deeply-lobed; fruits *c.* 3.5 × 2mm, covered with minute hooks. SIMILAR SPECIES **Alpine Enchanter's Nightshade** C. *alpina* averages shorter, with hairless flowers and flower stalks and the flowers more clustered. Scarce in the uplands of the N and W; the hybrid, **Upland Enchanter's Nightshade** C. × *intermedia*, although usually sterile, is commoner and *much more widespread.*

Large-flowered Evening-primrose

Oenothera glazioviana

Several species of evening-primrose have been introduced to Britain, but this is the commonest. Locally abundant on dunes, waste ground, verges and old quarries – anywhere with disturbed light soils. ❀ Tall, with large, fragrant, pale yellow flowers that mostly open in the evening. DETAIL Hairy. Stem red towards tip, and together with the fruits the green parts have numerous hairs with bulbous red bases. Leaves alternate, long-oval, often twisted. Petals 4, 30–50mm long (wider than long), sepals 4, red-striped, style longer than the 8 stamens, with the stigmas held above the anthers. SIMILAR SPECIES Several other forms of evening-primrose are found, all with smaller flowers, but identification is complicated by uncertainties about the species limits and by hybridisation.

GROWTH Biennial.
HEIGHT To 180cm.
FLOWERS Jun–Sep.
STATUS Evening-primroses were introduced to gardens from the 17th century onwards and recorded from the wild soon after; their complex genetics have given rise to several species since their arrival, probably including *O. glazioviana* (N and S America).
ALTITUDE Lowland.

GROWTH Perennial.
HEIGHT 45–90cm
(20–100cm).
FLOWERS Jun–Sep.
STATUS Ancient introduction.
ALTITUDE
Lowland.

Common Mallow
Malva sylvestris

Common in rough grassy places on well-drained soils, often sheltered spots rich in nutrients: verges, field margins and waste ground. ❀ Tall, with striking *purplish-pink flowers marked with darker veins*. DETAIL Hairy. Basal leaves long-stalked, rounded, very shallowly lobed, stem leaves cut into 5–7 rounded lobes. Flowers stalked, with 5 notched petals (12)20–30mm long, 5 large sepals, and immediately below an epicalyx of 3 small, narrow segments. Fruit a doughnut-shaped ring of nutlets, each *net-veined* but usually hairless.

GROWTH Annual.
HEIGHT Stems to 60cm.
FLOWERS Jun–Sep.
STATUS Introduction, present since at least Roman times.
ALTITUDE Lowland.

Dwarf Mallow
Malva neglecta

Fairly common. Track- and roadsides, rough car parks, waste ground, etc., on poor, disturbed soils, often around villages. ❀ Prostrate, with several stems and *small, pale lilac to whitish flowers with darker veins* that open fully only in sunshine. DETAIL Hairy. Leaves round, cut into 5–7 shallow, toothed lobes. Petals 9–13mm long. *Nutlets smooth*, downy.

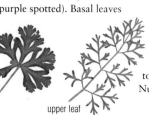

GROWTH Perennial.
HEIGHT To 80cm.
FLOWERS Late Jun–Aug.
STATUS Native.
ALTITUDE
To 305m.

Musk Mallow
Malva moschata

Locally common. Rough grassy places on well-drained soils: verges, road- and riverbanks, pastures, field margins and sea walls. ❀ Distinctive, with *rose-pink or white flowers* and *upper leaves cut almost to the base into fine lobes*. DETAIL Sparsely hairy, with swollen purple bases to the hairs on the stem (stem often purple spotted). Basal leaves

basal leaves   upper leaf

long-stalked, kidney-shaped, shallowly toothed and lobed. Petals 16mm or more. Nutlets smooth, with abundant long hairs.

Tree Mallow
Malva arborea

Local and scattered, usually within 100m of the sea, on rocks, the base of cliffs and waste ground: often around seabird roosts and dumped garden rubbish. ☼ Up to 3m tall with an unbranched stem, *woody* at the base. Flowers deep purplish-pink with dark purple veins that coalesce to form a *dark centre*. DETAIL Hairy, the younger parts with a soft, velvety texture (both simple and star-shaped hairs). Leaves long-stalked, circular, up to 25cm across, shallowly cut into 5–7 lobes. *Two or more flowers per leaf axil*. Petals 14–20mm long; epicalyx with 3 broad, oval lobes, fused at the base. Nutlets wrinkled, downy. SIMILAR SPECIES Garden Tree Mallow M. × *clementii* may escape and is more likely at inland sites. Sterile, with *1 flower per leaf axil*.

GROWTH Perennial, but monocarpic – flowering once and then dying (and killed by severe frosts).
HEIGHT To 3m.
FLOWERS Jul–Sep.
STATUS Native by the sea in the SW and W, otherwise introduced.
ALTITUDE Lowland.

Marsh Mallow
Althaea officinalis

Nationally Scarce, but locally common on the banks of brackish rivers and the dykes of coastal grazing marshes, usually out of the reach of cattle or where marshes have been converted to arable, also the upper edge of saltmarshes. ☼ *Flowers pale pastel pink*, foliage soft, velvety, and distinctively *greyish-green*.
DETAIL Whole plant hairy, with abundant star-shaped hairs. Leaves stalked, up to 11cm wide, cut into 3–5 shallow, irregularly toothed, lobes. Flowers in clusters of 1–3 in the leaf axils, petals 15–20mm long. Column (formed by fused stamens) *hairy*. *Epicalyx with 6–10 narrow segments*. SIMILAR SPECIES Hollyhock *Alcea rosea*, the familiar garden plant, has large white, yellow, red or pink flowers, with the petals 25–50mm long and the *column hairless*.

GROWTH Perennial.
HEIGHT 60–120cm.
FLOWERS Aug–Sep.
STATUS Native.
ALTITUDE Lowland.

The sweet of the same name was originally made from the roots of this plant (which contain starch, sugars, oils and plenty of gelatinous matter). Nowadays, 'marshmallows' are thoroughly artificial.

GROWTH Perennial.
HEIGHT 5–50cm.
FLOWERS May–Sep.
STATUS Native.
ALTITUDE 0–795m.

Common Rockrose
Helianthemum nummularium

Locally abundant in species-rich, short, dry grassland over chalk, limestone or basalt, also slightly acid grassland and heaths in E Scotland. ☀ More or less sprawling, with numerous wiry side branches arising from a prostrate woody stem, distinctive *drooping buds* and showy, sulphur-yellow, 5-petalled flowers. DETAIL Stems reddish when young. Leaves strap-shaped, *c.* 5mm wide, dark green and sparsely hairy above, with 1 obvious vein, white-woolly below, with *tiny strap-shaped stipules at the base.* Flowers *c.* 20–25mm across, in groups of 1–12, *style slightly kinked, longer than stamens.* SIMILAR SPECIES White Rockrose *H. apenninum* is confined to Brean Down in Somerset and Berry Head in Devon. It has white flowers.

Hoary Rockrose
Helianthemum oelandicum

Nationally Scarce. Very locally abundant in short turf and rocky places on limestone. ☀ As Common Rockrose but flowers rather smaller. DETAIL Mat-forming, stems often woody at base. Leaves oval with down-rolled margins, variably greyish-hairy above, white-woolly below; *no stipules.* Flowers to 15mm across, *style strongly S-shaped, shorter than stamens.*

GROWTH Perennial.
HEIGHT Stems to 25cm.
FLOWERS May–Jul.
STATUS Native.
ALTITUDE 0–540m.

There are 3 subspecies:
H. o. incanum Coasts of N and S Wales, NW England. Flowers usually in clusters of 3–6, rather small (often less than 10mm across), leaves usually very hairy above.
H. o. piloselloides W Ireland (the Burren). As ssp. *incanum* but flowers usually more than 10mm across, leaves larger and usually only sparsely hairy above.
H. o. levigatum Cronkley Fell (Upper Teesdale). Compact, with relatively short stems. Flowers usually in clusters of 1–3 (5), leaves more or less hairless above.

Weld
Reseda luteola

Locally common to abundant. Rough grassy places, verges, field margins, waste ground and old quarries and pits, on neutral to alkaline soils, typically where recent soil disturbance has allowed seed to germinate. ❂ *Stiffly erect*, with tall stems carrying numerous *narrow, strap-shaped leaves* and small, densely-crowded, 4-petalled flowers. DETAIL Hairless. Stem ribbed. Leaves in a basal rosette and up the stem, stalkless or short-stalked, wavy-edged. Flowers 4–5mm across, with 4 sepals and 4 pale yellowish-green petals (upper petal with several lobes, 2 lateral petals smaller, 3-lobed, lower petal insignificant, thread-like). Seed capsule 3–6mm long, short-stalked, cross-ribbed, coming to 3 points and open at the top.

GROWTH Biennial.
HEIGHT 50–150cm.
FLOWERS Jun–Sep.
STATUS Ancient introduction.
ALTITUDE To 400m.

Wild Mignonette
Reseda lutea

Locally common. Rough grassy places, field margins, dunes, old quarries and waste ground, on light, well-drained, usually alkaline soils. ❂ Rather bushy, with *well-branched stems, leaves cut into 1–2 pairs of narrow side lobes*, and spires of small, 6-petalled flowers. DETAIL Hairless. Stem ribbed. Leaves in a basal rosette and up the stems, wavy edged. Flowers 5–7mm across, with 6 sepals and 6 greenish-yellow petals (upper 2 petals with 2 large lobes, 2 lateral petals with 1 large lobe, 2 lower petals cut into 3 thread-like lobes); stamens 12–20; seed capsule oblong, 7–20mm long, indistinctly cross-ribbed, coming to 3 vague points and open at the top.

GROWTH Biennial to perennial.
HEIGHT 30–75cm.
FLOWERS May–Sep.
STATUS Recorded in the wild by 1597 but perhaps only native in the E.
ALTITUDE To 440m.

Family Brassicaceae: cabbages ('crucifers')

A large and potentially confusing family. Many are important food crops (Oil-seed Rape, Cabbage, Cauliflower, Turnip, Swede, Radish) and others are popular garden plants. Some of the more commonly naturalised garden crucifers have been treated in full, but others may be found from time to time around towns and villages (e.g. Sweet Alison *Lobularia maritima* and Aubretia *Aubrieta deltoidea*), or naturalised in wilder places (e.g. Garden Arabis *Arabis caucasica* in the Derbyshire Dales).

All have simple flowers with 4 petals and 4 sepals – the cross-like arrangement of the petals gives rise to the name 'crucifer'. Most species have 6 stamens, a few 4. One of the key identification characters in many species is the seed pod (ripe, or as near to ripe as possible). Pods come in many shapes and sizes. The main portion is made up of two *valves* that are joined together to enclose the seeds; when ripe the pod usually splits open along the join between the valves to release the seeds. In some species the *style* persists as a short, slender projection at the tip of the pod, while in others the pod has a distinctive, usually seedless, upper segment, the *beak* (tipped by a stalkless stigma). The distinction between a style and a beak is not always hard and fast.

beak
style
valves

Smith's Pepperwort   Rape

Garden Arabis

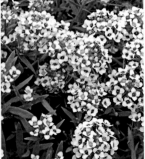

Sweet Alison

Aubretia

GROWTH Annual.
HEIGHT 7–35cm.
FLOWERS May–Aug.
STATUS Native.
ALTITUDE Lowland.

Wild Candytuft
Iberis amara

Nationally Scarce. Bare, open ground on disturbed chalky soils on S-facing slopes, including arable fields and rabbit-scrapes on grassland. ❀ Flowers white, occasionally lilac, in flat, crowded heads, the *outer flowers larger than the inner*, and the *outer pair of petals much longer than the inner*. DETAIL Stems sparsely hairy, branching. Leaves to 5cm, stalkless to broadly-stalked, shallowly-lobed, more or less hairless. Outer petals 4–8mm long, inner 2–5mm. Flower heads elongating as the fruits develop. Pods circular, flattened, 4–7mm long, narrowly winged, notched at the tip; style 0.8–2mm long. SIMILAR SPECIES Garden Candytuft *I. umbellata* occasionally escapes. The flower heads do not elongate in fruit and the pods are 7–10mm long.

Sea Rocket
Cakile maritima

Fairly common on sandy beaches, just above the line of debris left by the highest tides, less often on shingle beaches and young dunes. ● *Pale lilac flowers and shiny, fleshy leaves distinctive.* DETAIL Hairless. Stems well branched. Lower leaves stalked to stalkless, to 8cm, lobed. Petals 7.5–13mm. Pods 12–25mm, fleshy, waisted. SIMILAR SPECIES **Hoary Stock** *Matthiola incana* from Europe is very locally naturalised on cliffs and shingle on the S coast. 30–80cm tall, woody at the base, with strap-shaped, grey-green, densely hairy leaves and large, very fragrant, white, pink or purple flowers May–Aug; petals 21–31mm. Pods downy, sometimes curved. The very rare **Sea Stock** *M. sinuata*, confined to Devon and Glamorgan, has shallowly lobed leaves and conspicuous sticky oil glands on the stems.

GROWTH Annual.
HEIGHT 15–50cm, stems sprawling towards base.
FLOWERS Jun–Sep.
STATUS Native.
ALTITUDE Lowland.

Sea Kale *Crambe maritima*

GROWTH Perennial.
HEIGHT 30–50cm.
FLOWERS Late May–Jul.
STATUS Native.
ALTITUDE Lowland.

Very locally common on shingle banks and beaches. The blanched young shoots were once eaten as a delicacy. As a result, Sea Kale declined massively, and is now only slowly recovering. ● Large, well-branched clumps of thick, fleshy, cabbage-like leaves and sprays of white flowers distinctive. DETAIL Hairless. Leaves stalked, to 50cm long, oval with wavy margins, green to purple. Petals 6.5–11mm long. Pods with short, stalk-like base and pea-shaped terminal portion.

Wintercress
Barbarea vulgaris

Scattered in small numbers, mostly on damp, disturbed soils: stream- and riverbanks, ditches, waste ground, verges. ✿ *Hairless*, with *glossy dark green foliage.*
DETAIL Basal leaves to 20cm, with a large, oval terminal lobe and 1–5 pairs of narrow side lobes; uppermost stem leaves clasping the stem, *undivided* or with a *broad, large terminal lobe and 1–2 pairs of small side lobes.* Petals 4.9–7mm long. Pods held erect, slender, 15–32mm long, *style slender, 2–3.5mm long.*

GROWTH Biennial or perennial.
HEIGHT 20–90cm.
FLOWERS Apr–Jul.
STATUS Native.
ALTITUDE To 380m.

lower leaf

upper stem leaf

Medium-flowered Wintercress
Barbarea intermedia

Scattered in small numbers on disturbed ground on verges, waste ground and arable fields. ✿ Separated with care from Wintercress by more *deeply-lobed upper stem leaves* and *short, stout* style (0.6–1.7mm long) at the tip of the pod. DETAIL Hairless or with sparse hairs below. Upper stem leaves with a more or less narrow terminal lobe and 2–3 pairs of side lobes. Petals 4–6.3mm. Pods 15–35mm.

upper stem leaf

GROWTH Biennial.
HEIGHT 15–60cm.
FLOWERS Mar–Jul.
STATUS Introduced from Europe and recorded from the wild by 1849. Increasing?
ALTITUDE To 340m.

American Wintercress *Barbarea verna*

Scattered and erratic. Verges and waste ground, usually near habitation, also shingle and sandy places near the sea.
✿ Rather variable; annuals are small, perennials robust. Upper stem leaves deeply-lobed as Medium-flowered Wintercress but distinguished by *larger flowers* and *longer pods* (mostly over 40mm). DETAIL Hairless or with sparse hairs below. Basal leaves with 4–10 pairs of side lobes; upper stem leaves with 2–4 pairs of narrow side lobes. Petals 5.6–9.6mm. Pods 35–71mm, with a short, stout style.

lower leaf

upper stem leaf

GROWTH Annual or biennial.
HEIGHT 30–90cm.
FLOWERS Mar–Jul.
STATUS Originating in SW Europe and long cultivated as 'Land Cress', it was first recorded in the wild in 1803.
ALTITUDE Mostly lowland, but to 300m.

Marsh Yellow Cress *Rorippa palustris*

Fairly common. The muddy edges of rivers, lakes and ponds, and wet pastures; also a scarce weed of arable fields and waste ground. ❀ *Hairless, flowers very small,* with the *petals about as long as the sepals; pods short and squat.* DETAIL Leaves with 2–6 pairs of side lobes and a larger terminal lobe; the leaves may clasp the stem. Petals 1.4–2.8mm long. Pods held at *c.* 90° to the stem, 4–12mm (less than twice as long as stalk); style 0.2–1mm.

GROWTH Annual.
HEIGHT
5–30cm
(60cm).
FLOWERS
May–Oct.
STATUS Native.
ALTITUDE To 320m.

Creeping Yellow Cress *Rorippa sylvestris*

As Marsh Yellow Cress, and found in similar habitats, although rather commoner as an arable weed. ❀ Leaves cut into narrower lobes, flowers slightly larger and *seed pods longer* and *thinner.* DETAIL Stems erect or sometimes sprawling, rooting at the nodes and then turning up towards the tip. Petals 2.2–5.5mm long, *1.5–2 times the length of the sepals.* Pods 7–23mm (about twice as long as stalk), style 0.6–1.2mm. Self-incompatible, often setting neither fruit nor seed, and poorly developed pods make identification tricky.

GROWTH Perennial.
HEIGHT Stems 15–60cm.
FLOWERS Jun–Oct.
STATUS Native.
ALTITUDE
Lowland.

Great Yellow Cress *Rorippa amphibia*

Fairly common. Tall vegetation on the edge of slow-moving streams, rivers, ponds and ditches, usually where the water is alkaline and nutrient-rich. ❀ Robust and *hairless,* with *undivided leaves* and *short, oval seed pods.* DETAIL Stems sprawling, rooting at the nodes, then turning upwards. Lowest leaves lobed, but soon withering; stem leaves stalkless, sometimes clasping, strap-shaped with variably toothed edges. Petals 3.3–6.2mm. Pods 2.5–7.5mm (much shorter than stalk); style 0.8–1.8mm.

GROWTH Perennial.
HEIGHT 40–120cm.
FLOWERS
Jun–Sep.
STATUS Native.
ALTITUDE
Lowland.

Annual Wall Rocket *Diplotaxis muralis*

Local. Waste ground, old walls, cliffs, pavement cracks, gardens, fields and other dry, open habitats. ❁ Has a *strong, unpleasant smell* (thus 'Stinkweed'). *Flowers rich yellow,* usually not opening widely.

DETAIL *Hairless* apart from base of stem. Often branching. Rosette leaves to 10cm, long-stalked, with *broad side lobes*; stem leaves sparse: smaller, with reduced lobes and sharper teeth. *Petals 4–8mm.* Pods 15–42mm, slender, *stalk usually less than half length of pod*; style 1–3mm.

pod

no gap

GROWTH Annual or short-lived perennial.
HEIGHT 10–60cm.
FLOWERS Apr–Oct.
STATUS Introduced from S Europe in 1778 and still spreading.
ALTITUDE Lowland.

Perennial Wall Rocket
Diplotaxis tenuifolia

Local. Habitat as Annual Wall Rocket. ❁ Similarly foul-smelling, but stems *tall and leafy* and flowers paler. **DETAIL** *Hairless*. No rosette; stems often branching above. Leaves often waxy grey-green, to 12cm, long-stalked or stalkless, *unlobed or with long, narrow side-lobes*. Petals 8–15mm. Pod 14–60mm, *stalk about as long as pod, with a gap of 0.5–6.5mm between the sepal-scar and the base of the pod*; style 1.2–3mm. Largely self-incompatible, setting little seed.

pod

gap

stalk   sepal-scar

GROWTH Perennial.
HEIGHT 30–80cm.
FLOWERS May–Sep.
STATUS Ancient introduction.
ALTITUDE Lowland.

Treacle Mustard *Erysimum cheiranthoides*

Rather local, commonest in the SE. Arable fields, verges and waste ground, especially on sandy soils. ❁ *Roughly hairy* (hairs 2–3 rayed), with *long-oval leaves*, crowded flower heads and slender pods. **DETAIL** Stems usually branched, basal rosette soon dying off. Leaves short-stalked or stalkless, to 10cm, variably toothed. Petals 3–6mm. Pods 10–30mm, sparsely hairy, style 0.5–1.4mm.

GROWTH Annual.
HEIGHT 15–100cm.
FLOWERS May–Nov.
STATUS Ancient introduction.
ALTITUDE To 435m.

Wallflower *Erysimum cheiri*

Fairly common. Long naturalised in sunny spots on rocks and walls, especially on churches, castles and ruins. ❀ The combination of habitat and *large yellow, red or orange flowers* is unique. DETAIL Hairless to densely hairy. Stems tough and woody at the base. Leaves more or less stalkless, strap-shaped, pointed. Petals 16–30mm (plants with larger or even double flowers may escape from cultivation). Pods 24–70mm, style 2–3.5mm.

GROWTH Perennial.
HEIGHT 20–60cm.
FLOWERS Mar–Sep.
STATUS Cultivated since the medieval period and first recorded in the wild in 1548.
ALTITUDE Lowland.

Flixweed *Descurainia sophia*

Locally common in the E. Arable fields on light soils, occasionally also waste ground. ❀ Leaves greyish-green, *very finely divided, flowers tiny*, in a crowded head, with the *petals shorter than the sepals*. DETAIL Densely hairy, at least below, with 2–3 rayed and simple hairs. Leaves to 8cm, lobes 1–2mm wide; basal rosette soon dying. Petals 1.4–1.8mm. Pods 10–35mm, slender, slightly beaded, style 0.2–0.3mm.

GROWTH Annual.
HEIGHT 10–100cm.
FLOWERS May–Oct.
STATUS Ancient introduction.
ALTITUDE Lowland.

Bastard Cabbage *Rapistrum rugosum*

Locally common. Rough grassy places, roadsides and waste ground. ❀ Usually robust, with widely-spreading branches. *Pods, held close against the stem on short stalks, distinctive.* DETAIL Coarsely hairy on lower stem and leaf undersides. Lower leaves to 25cm, unlobed or with a large terminal lobe and several small side lobes. Upper leaves short-stalked or stalkless, strap-shaped, pointed, variably toothed. Flowers crowded, petals 5.7–11mm. Pods 3–12mm, with a cylindrical lower section and globular upper section, the latter ribbed and hairless to roughly hairy; style 1–3.5mm.

GROWTH Annual or biennial.
HEIGHT 15–100cm.
FLOWERS May–Oct.
STATUS Introduced with grain and bird seed from S Europe and SW Asia. Known in cultivation by 1739 and in the wild by 1863. Still spreading.
ALTITUDE Lowland.

Hedge Mustard
Sisymbrium officinale

One of the commonest yellow-flowered crucifers, abundant on any disturbed ground, especially on dry soils. ❀ Stems very upright, usually washed purplish, often with many branches held stiffly outward. *Flowers very small, pods small and held erect, pressed against the stem.* DETAIL Bristly-hairy to hairless. Lower leaves to 15cm, stalked, cut into more or less arrow-shaped lobes; upper leaves smaller, uppermost often unlobed. Flowers in dense clusters that elongate upwards as the lowest go to seed, petals pale yellow, 3.1–4.2mm. Pods 10–20mm, may be hairy when young, on short stalks; style 0.5–2mm.

GROWTH Annual or biennial.
HEIGHT 15–100cm.
FLOWERS Late Apr–Oct and through mild winters.
STATUS Ancient introduction.
ALTITUDE 0–570m.

Eastern Rocket
Sisymbrium orientale

Scattered. Rough ground and waste places. ❀ Flowers large, *pods long and slender, hairy when young.* DETAIL Stems hairy. Lower leaves grey-green, *softly hairy,* to 15cm, deeply cut into *1–4 pairs of triangular lobes* plus a large terminal lobe; upper leaves with a long, narrow terminal lobe and 0–2 pairs of narrow side lobes. Petals 7–10.5mm. Pods 25–120mm, held at *c.* 45°; style 0.3–3.5mm.

GROWTH Annual.
HEIGHT 10–100cm.
FLOWERS Apr–Dec.
STATUS Introduced to cultivation by 1739 and recorded in the wild by 1859 (S Europe).
ALTITUDE Lowland.

lower leaf

Tall Rocket
Sisymbrium altissimum

Scattered. Verges, tips and waste ground. ❀ Robust, with *finely cut leaves,* large, *pale yellow flowers* and *long, slender pods.* DETAIL Stems hairy towards base. Lower leaves roughly hairy, to 30cm, stalked, deeply cut with *5–10 pairs of lobes;* upper leaves stalkless or short-stalked with *2–5 pairs of very narrow lobes.* Petals 5.7–11mm. Pods 30–110mm, *hairless,* held at *c.* 45°; style 1–2.5mm. SEE ALSO Flixweed (p. 123).

GROWTH Annual.
HEIGHT 30–100cm.
FLOWERS May–Oct.
STATUS Introduced to cultivation from E Europe by 1768 and recorded in the wild from 1862.
ALTITUDE Lowland.

lower leaf

upper leaf

Hoary Mustard

Hirschfeldia incana

lower leaf

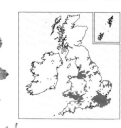

Local but increasing. Verges, waste ground, railways etc.
❀ Robust, often rather spreading, with relatively large flowers in small, pom-pom-like clusters. *Pods short, shaped like a bowling pin, held tight against the stem.* DETAIL Lower stem bristly-hairy, upper stem hairless, grey-green. Lower leaves hairy on upperside, cut into 1–5 pairs of lobes. Flowers pale yellow, *petals 5–10mm, sepals held more or less erect.* Pods sparsely hairy, 6–17mm including the beak, which is *swollen around 1–2 seeds* and tapers to a 1mm-long style. SEE ALSO Black Mustard (p. 126).

GROWTH Annual or short-lived perennial.
HEIGHT 50–130cm.
FLOWERS May–Oct.
STATUS First recorded from the wild in 1837, but has only recently increased and spread.
ALTITUDE To 360m.

Charlock

Sinapis arvensis

A common weed of roadsides (especially following soil disturbance), waste ground and arable fields. ❀ Rather variable, but *stem and leaves usually rather bristly-hairy*, dark green to purplish, with bright yellow flowers. DETAIL Stems usually well branched. Lower leaves to 20cm, stalked, *oval, with 0–2 pairs of much smaller lobes at the base*; upper leaves short-stalked or stalkless, smaller, narrower and more pointed, variably toothed. Petals 7.5–17mm, *sepals spreading at 90° or down-turned.* Pods hairy to hairless, 22–57mm, including a relatively *long (7mm plus), conical beak* with 0–1 seeds.

GROWTH Annual.
HEIGHT 20–100cm.
FLOWERS Mar–Nov.
STATUS Ancient introduction.
ALTITUDE To 450m.

White Mustard

Sinapis alba

Fairly common. Arable fields, verges, waste ground, often on lime-rich soils. Once widely grown as an ingredient for table mustard, now a casual or persistent weed, and also sown as game cover. DETAIL Rather like Charlock, but leaves more or less hairless, all leaves stalked and *more evenly and deeply lobed*, petals 7.5–14mm. Seed pods usually bristly, 20–42mm long, including the *long beak, which is curved and flattened (like a scimitar)* when mature.

GROWTH Annual.
HEIGHT 20–100cm.
FLOWERS May–Sep.
STATUS Ancient introduction.
ALTITUDE Lowland.

GROWTH Annual to biennial.
HEIGHT 30–130cm.
FLOWERS Late Mar–Jun.
STATUS Originates in
cultivation. Swede (ssp.
rapifera, with a swollen root),
is found as a rare relict of
cultivation.
ALTITUDE Lowland.

Rape *Brassica napus*

Subspecies *oleifera*, Oil-
seed Rape, is common on
roadsides, field margins and
waste ground. ❁ Tall, almost
hairless, and *waxy, grey-
green*. DETAIL Lower leaves
sometimes sparsely hairy, to
30cm, stalked, with a large
terminal lobe and several small
side lobes. Upper stem leaves
unlobed, *stalkless, clasping
the stem. Flower buds slightly
over-topping open flowers,*
forming a 'dome', petals *pale*
yellow, *mostly 13–18mm,*
sepals held more or less erect.
Pods 35–100mm long, including
a *short conical beak* with 0–1 seeds
that terminates in a stalkless style.

GROWTH Annual or biennial.
HEIGHT 30–100cm.
FLOWERS May–Jul.
STATUS Wild Turnip (ssp.
campestris) is an ancient
introduction. Turnip (ssp.
rapa) is a relict of cultivation,
and Turnip-rape (ssp. *oleifera*)
originates with bird- and
oil-seed.
ALTITUDE Lowland.

Turnip *Brassica rapa*

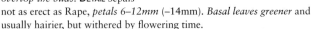

Wild Turnip is fairly common on
the banks of rivers, streams and
canals (often with Black Mustard);
Turnip and Turnip-rape are
found on verges, arable fields and
waste ground. ❁ Grey-green with
clasping stem leaves – much as
Rape – but *flowers smaller*, deeper
yellow, and held in rounded, 'pom-
pom-like' clusters; the *open flowers
overtop the buds*. DETAIL Sepals
not as erect as Rape, *petals 6–12mm* (–14mm). *Basal leaves greener* and
usually hairier, but withered by flowering time.

GROWTH Annual.
HEIGHT 30–200cm.
FLOWERS May–Sep.
STATUS Probably native, first
recorded in the wild in 1640.
ALTITUDE Lowland.

Black Mustard
Brassica nigra

Locally abundant. Riverbanks,
sea walls, cliffs and shingle,
increasingly also damp roadsides
and waste places. ❁ Tall and
often grey-green above, with
widely-spread branches. *All leaves
stalked, pods with a short, slender
beak, held pressed against the stem.* DETAIL Stem usually slightly bristly
towards base. Leaves roughly hairy to hairless, lower leaves to 20cm with
a large terminal lobe (not cut to midrib), and several side lobes; leaves
smaller and narrower higher on stem, uppermost unlobed. Flowers bright
yellow, petals 9.5–13mm, sepals held from near-erect to 90°. Pods hairless,
short-stalked, slender, subtly beaded, 8–25mm, including a *seedless beak
1.5–5mm*. SEE ALSO Hoary Mustard (p. 125).

Wild Cabbage *Brassica oleracea* var. *oleracea*

Nationally Scarce. Grows wild on sea cliffs and nearby grassland and quarries, especially on chalk and limestone. Other varieties occur inland as garden escapes. ❀ Stout, with a *woody base to the stem* marked by 'annual rings' of leaf scars. *Leaves thick and fleshy*, waxy, grey-green or purplish, flowers large, pale yellow, in a well-spaced, elongated spike. Detail Hairless. Basal leaves to 20cm, oblong, stalked, variably but often deeply-lobed, upper leaves unlobed, stalkless, clasping. Flower spike with buds greatly overtopping the open flowers. Sepals more or less erect, petals 18–30mm. Pods 35–85mm, including a short conical beak with 0–2 seeds.

GROWTH Perennial.
HEIGHT 30–60cm (200cm).
FLOWERS May–Jul.
STATUS Probably native, first recorded 1548.
ALTITUDE Lowland.

Wild Radish *Raphanus raphanistrum*

Abundant weed of arable fields and roadsides. ❀ Bristly-hairy with relatively large flowers that can be *white, mauve or yellow, often with darker lilac veins*; yellow is dominant in N and W, mauve is scarce. *Pods variably 'beaded'*. Detail Rosette leaves to 60cm, stalked, with a large oval terminal lobe and several side lobes, stem leaves progressively smaller and narrower, with fewer lobes. *Sepals erect. Petals 11–24mm*. Pods 25–90mm, variably ribbed, divided into 3–8 'beads', each longer than wide, and tapering towards a relatively long, seedless beak; the 'beads' are easily broken apart. See Also Charlock (p. 125).

GROWTH Annual.
HEIGHT 25–75cm.
FLOWERS May–Oct.
STATUS Ancient introduction.
ALTITUDE 0–540m.

Sea Radish subspecies *maritimus*

Common (although absent from most of the E and NE coasts). Coastal grassland, dunes, shingle, cliffs, muddy shores and disturbed ground by the sea. Native. ❀ As Wild Radish but biennial to perennial, the stems taller, often well branched and very leafy. Flowers yellow (rarely white), *pods more obviously 'beaded'*. Detail Petals 15–22mm long. Pods 15–55mm long, with 1–5 'beads', each as long as wide, pea-shaped, but not easily breaking up; *narrowing abruptly to a relatively short beak*.

Garlic Mustard
Alliaria petiolata

GROWTH Biennial.
HEIGHT 30–120cm.
FLOWERS Apr–Jun.
STATUS Native.
ALTITUDE 0–570m.

Common in a wide variety of habitats but does best in moist, lightly shaded sites: hedgerows, woodland margins. ❀ Tall, spring-flowering crucifer, smelling of garlic when crushed, with distinctive *kidney-shaped basal leaves*. DETAIL Stem hairy at base, otherwise plant more or less hairless. Leaves bluntly-toothed, basal leaves long-stalked, stem leaves shorter-stalked and more pointed. Petals 4.4–7.8mm. Pods 30–75mm, lightly beaded, held more or less upright.

Horseradish
Armoracia rusticana

Common. Rough, grassy places, usually near habitation: road verges, riverbanks, waste ground, railway embankments. ❀ The *tufts of large erect basal leaves* are conspicuous (but can be confused with docks), as are the *erect, often tall, flowering stems* with large open clusters of white flowers. DETAIL Hairless. Basal leaves erect, long-stalked, oval-oblong (deeply cut on young plants), toothed, to 50cm or more long, stem leaves narrower, more strap-shaped, short-stalked. Petals 5.4–7.8mm. Pods small and globular, but rarely sets seed – spreads via underground rhizomes or root fragments in dumped soil.

GROWTH Perennial.
HEIGHT 50–150cm.
FLOWERS May–Jun.
STATUS Introduced before 1500, initially for medicinal purposes but by the mid 17th century grown for hot sauces (the grated root is used), replacing Dittander. Not known in the wild, its origins are obscure.
ALTITUDE Lowland.

Dittander *Lepidium latifolium*

GROWTH Perennial.
HEIGHT 40–130cm.
FLOWERS Jun–Sep.
STATUS Native in coastal
areas. A relict of cultivation
inland (grown until the 17th
century for its fiery roots and
leaves, but replaced by Horse
Radish); also some
recent spread along
salted roads.
ALTITUDE
Lowland.

Nationally Scarce. Locally common in brackish habitats around estuaries, creeks and sea walls, increasingly also inland on waste ground and along roads, railways and canals. ✸ Often forms extensive patches. Note dock-like basal leaves and *tall, leafy flowering stems with masses of tiny flowers.* DETAIL More or less hairless. Basal leaves long-stalked, up to 30cm long, long-oval with variably toothed margins (and sometimes 2 side lobes); the leaves become smaller, narrower and shorter-stalked higher on the stem; the uppermost leaves may be stalkless. Petals 1.8–2.5mm, sometimes pinkish, sepals and ovary often purplish. Pods sometimes hairy, oval, flattened, 1.6–2.7mm, style prominent, 0.2–0.3mm.

stem leaf

Hoary Cress
Lepidium draba

GROWTH Perennial.
HEIGHT 20–60cm.
FLOWERS Apr–Jun.
STATUS Introduced
accidentally from 1802 on, it
spread rapidly (SW Asia and
S Europe).
ALTITUDE Lowland.

Locally common on rough grassy and sandy ground near the sea, also verges, waste ground, railway lines and arable fields. ✸ Patch forming, with *grey-green foliage* and *dense clusters of small white flowers.* The *upper leaves clasp the stem.* DETAIL Hairless to densely hairy. Grows from a creeping rhizome. Leaves variably toothed: lower leaves long-oval, to 15cm, blunt, tapering to the stalk; upper leaves narrower, stalkless, with pointed auricles. Petals 2.5–4.5mm. Pod usually net-veined, oval to heart-shaped, flattened, 3–4.8mm, style 0.7–1.8mm, projecting.

GROWTH Perennial.
HEIGHT 10–50cm.
FLOWERS May–Sep.
STATUS Native.
ALTITUDE To 425m.

Smith's Pepperwort *Lepidium heterophyllum*

Local. Heathy places, shingle, railway lines and arable fields on dry, acid, often gravelly soils. ❀ Branches, often towards the base, into several *crowded, parallel-sided flower spikes*, with the very small flowers on short stalks at right-angles to the stem. Foliage grey-green, the *stem leaves clasping the stem with long, pointed auricles*. DETAIL Softly hairy. Basal leaves strap-shaped, long-stalked, sometimes with 1–3 pairs of small but broad side lobes; soon withering (may regrow as the plant goes to seed). Stem leaves up to 50mm long, triangular, variably toothed, stalkless. Petals 2–3.6mm, styles 6. Pods 4.5–8.6mm, with variable numbers of tiny green blisters, *style 0.5–1.2mm, longer than the shallow notch*.

Narrow-leaved Pepperwort
Lepidium ruderale

GROWTH Annual or biennial.
HEIGHT 10–45cm.
FLOWERS Apr–Nov.
STATUS Ancient introduction.
ALTITUDE Lowland.

Local on banks and bare, waste places near the sea, also waste ground, tips and gardens, and increasingly common along salted roads. ❀ Often densely branched, forming little 'bushes'. *Foliage foul-smelling, flowers tiny, and seed pods usually shorter than their stalk*. DETAIL Stem and flower stalks often with tiny hairs, otherwise hairless. Base of stem and lower leaves often purplish. Basal leaves stalked, with 3–6 pairs of narrow, strap-shaped lobes, sometimes sparsely-toothed, usually withering early. Stem leaves progressively smaller, uppermost strap-shaped, more or less unlobed, stalkless but *not clasping*. Petals absent or *c*. 0.5mm, shorter than sepals; stamens usually 2 or 4, anthers yellow. Pods 1.5–2.7mm, oval, flattened, with a small notch at the tip; style minute.

Field Pepperwort *Lepidium campestre*

Rather local. Arable fields, open grassland, verges, walls, gardens and
waste ground, especially on sandy or gravelly soils. ❁ Very like Smith's
Pepperwort but stems more stiffly erect and usually *branching
above the middle*. Flowers slightly smaller, and seed pods with
the *style shorter to marginally longer than the shallow notch*.
DETAIL Basal leaves unlobed, soon withering. Petals 1.5–2.6mm.
Pods 4–6.8mm, oval, flattened, the centre covered with tiny
green blisters, broadly winged; style 0.1–0.7mm.

seed pod

GROWTH Annual to biennial.
HEIGHT 10–40cm.
FLOWERS Apr–Aug.
STATUS Ancient introduction.
ALTITUDE Lowland.

Swine Cress *Lepidium coronopus*

Fairly common.
Arable fields, track-
sides and waste
ground, especially
on compacted soil
(e.g. around field
entrances). ❁ Flowers
in tight clusters at
tip of main stem and
opposite the leaves,
rather small; *seed pods
irregularly wrinkled*.
DETAIL More or less
hairless. Stems sprawling. Leaves dull blue-green, slightly fleshy, to 10cm
long, basal leaves cut into 2–9 pairs of lobes (in turn irregularly-lobed)
plus a strap-shaped terminal lobe, dying off by flowering time; upper
leaves narrower, more or less unlobed. Petals 1–2mm, *longer than sepals*,
stamens 6, *all with anthers*. Pod kidney-shaped, flattened, 3–4.7mm across,
with a shortly-protruding style; stalk shorter than pod.

GROWTH Annual.
HEIGHT 5–30cm.
FLOWERS May–Oct.
STATUS Ancient introduction.
ALTITUDE Lowland.

Lesser Swine Cress *Lepidium didymum*

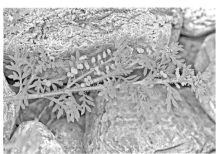

Fairly common.
Arable margins,
track- and roadsides,
waste ground and
gardens, especially in
urban areas and on
compacted soils where
water lies in winter.
❁ Smells strongly of
cress when crushed.
Flowers tiny and
inconspicuous, in short
spikes *opposite the leaves*; *seed pods shaped like a dumb-bell*.
DETAIL Stems sprawling, usually more or less hairy. Leaves hairless, to 10cm
long, cut into 2–7 pairs of lobes plus a terminal lobe, those on the basal
leaves irregularly-toothed (basal leaves dying off by flowering time). Petals
c. 0.5mm or absent, *shorter than sepals*, stamens 6, but *only 2 (sometimes
4) have anthers*. Pods notched both above and below, 2–3mm across, finely
netted when dry, style minute; stalks longer than pods.

GROWTH Annual or biennial.
HEIGHT 5–40cm.
FLOWERS Jun–Oct.
STATUS Introduced in the
early 18th century and
recorded from the wild from
1778; origin unknown.
ALTITUDE Lowland.

seed pod

Common Scurvygrass *Cochlearia officinalis*

Common. Saltmarshes and damp grassy places by the sea, also inland along salted roads, especially in SW England. ❀ Hairless, with fleshy leaves and clusters of small, whitish flowers. DETAIL Sprawling to erect. Basal leaves in a loose rosette, long-stalked, 1.5–3cm long, more or less circular or subtly angular with a *heart-shaped base*; sometimes with a few blunt teeth. Lower stem leaves short-stalked, *upper stem leaves stalkless, usually clasping stem*. Flowers 10–15mm across, petals 4–8mm. Pods spherical (or taper at the top and bottom), 3–7mm long, *not* flattened. Smaller plants, with smaller, often lilac, flowers are found on the coasts of Scotland and Ireland (subspecies *scotica*).

GROWTH Biennial to perennial.
HEIGHT 5–30cm.
FLOWERS Apr–Aug.
STATUS Native.
ALTITUDE Lowland.

seed pod in profile (left) and face-on

Pyrenean Scurvygrass *Cochlearia pyrenaica*

Locally common. Damp rocky places, flushes and old mine spoil in the uplands, usually on limestone and other base-rich rocks; absent from the coast. ❀ As Common Scurvygrass but flowers smaller and *upper stem leaves hardly clasp the stem, if at all.* DETAIL Leaves fleshy in Somerset, N Wales, Scotland and Ireland, thin and not fleshy in N England; lower leaves heart-shaped at base. Flowers 5–8mm across. Pods 3–5mm long.

basal leaf

GROWTH Biennial to perennial.
HEIGHT To 30cm.
FLOWERS Apr–Aug.
STATUS Native.
ALTITUDE To at least 960m.

English Scurvygrass *Cochlearia anglica*

Fairly common *on the coast*, on saltmarshes and the muddy margins of tidal rivers and estuaries. ❀ Note *wedge-shaped base to the basal leaves, which taper into the stalk*, and *flattened seed pods.* DETAIL Basal leaves to 3cm, stalked; stem leaves often irregularly toothed, the lower stalked, the upper often stalkless and clasping the stem. Petals 5–10mm. Pods 7–14mm long, circular to oval.

seed pod face-on (left) and in profile

GROWTH Biennial to perennial.
HEIGHT 7–40cm.
FLOWERS Apr–Jul.
STATUS Native.
ALTITUDE Lowland.

basal leaf

Danish Scurvygrass *Cochlearia danica*

Scattered on sandy and rocky shores, cliffs, walls and pavement cracks near the sea. Since the 1980s has spread along salted roads, and now *abundant on many roadsides*, forming an almost continuous fringe of low-growing lilac-white flowers on the edge of the carriageway. ❀ Slender, with *small flowers*. All but uppermost leaves *stalked*. DETAIL Stems erect, becoming sprawling in larger plants. Foliage dark green to purplish. Basal leaves long-stalked, to 1.5cm long, more or less rounded, with a heart-shaped base. *Stem leaves ivy-shaped*, 3–7 lobed, *stalked*; upper 1–2 sometimes stalkless but *not clasping* stem. Petals 2.5–4.5mm, often pale lilac. Pods 3–5mm long, egg-shaped, sometimes slightly flattened.

GROWTH Winter-annual to biennial.
HEIGHT 5–25cm.
FLOWERS Feb–Jun (Sep).
STATUS Native.
ALTITUDE 0–300m.

stem leaf

Alpine Pennycress *Noccaea caerulescens*

Nationally Scarce. Restricted to sites rich in lead or zinc, mostly old mine spoil, but also river gravels and, rarely, rocky places on limestone and other base-rich rocks. ❀ Hairless, with a persistent basal rosette, stalkless, *clasping stem leaves* and small white flowers; seed pod heart-shaped, flattened, with a *prominent style*. DETAIL Rosette leaves stalked, spoon-shaped, to 50mm, with few or no teeth. Petals 2.2–5mm; stamens 6, *anthers reddish-purple*. Pods 4.5–9.5mm, winged, the wings broadening towards the tip where the shallow notch is generally much shorter than the style.

GROWTH Perennial, sometimes biennial.
HEIGHT 5–40cm.
FLOWERS Apr–Aug.
STATUS Native.
ALTITUDE To 940m, but mostly in the uplands.

seed pod

GROWTH Biennial or perennial.
HEIGHT 5–40cm.
FLOWERS Apr–Aug.
STATUS Native.
ALTITUDE 0–1005m.

Hairy Rockcress
Arabis hirsuta

Fairly common. Rocky places, including old walls and limestone pavement, also sparse, unimproved grassland in dry, sunny places over chalk and limestone and on dunes. ✿ Distinctive slender, leafy spikes of *small white flowers* followed by very slender seed pods *held close against the stem*; as the lower pods ripen the stem elongates and is then often *pencil-thin*. DETAIL Densely hairy, with simple, forked and star-shaped hairs. Basal rosette compact, leaves long-oval, variably toothed, to 60mm; stem leaves narrower, alternate, stalkless, *clasping the stem*. Petals 4–6.2mm; stamens 6. Pods 15–50mm.

Thale Cress *Arabidopsis thaliana*

GROWTH Winter-annual.
HEIGHT 2–30cm.
FLOWERS All year, peaking Apr–Jul.
STATUS Native.
ALTITUDE 0–850m.

rosette leaf

Common on disturbed soil almost anywhere, but especially on light soils, also walls and rock ledges.
✿ Small and slender. The rosette and lower part of the stem are dark green and conspicuously hairy, but the *stem characteristically elongates* and the *upper part is hairless, often well branched, and waxy grey-green*. DETAIL Leaves variably toothed, with simple, forked and sometimes star-shaped hairs; flowering stems have a few leaves. Petals whitish, 2.5–4.5mm, sepals yellowish. Pods very slender, 6–16mm, style 0.2–0.5mm.

Field Pennycress
Thlaspi arvense

Abundant weed of arable fields and disturbed places on verges, waste ground and in gardens.
❁ Hairless, with rather waxen, grey-green foliage and small white flowers, while the *broadly-winged, penny-shaped pods* are eye-catching. DETAIL Strong-smelling when crushed. No basal rosette. Leaves broadly strap-shaped, toothed, *clasping the stem at the base* with long pointed auricles. Petals 2.7–5mm, stamens 6. Fruits 12–22mm across, with a deep cleft between the wings, *much deeper than the style.*

GROWTH Annual.
HEIGHT 10–50cm.
FLO WERS Apr–Sep.
STATUS Ancient introduction.
ALTITUDE Mostly lowland, but to 330m.

seed pod

Shepherd's Purse
Capsella bursa-pastoris

An abundant weed of gardens and arable fields, indeed anywhere where seed can germinate in loose or disturbed soil. ❁ Rather variable, but the *triangular pods (the 'purses') are always distinctive.* DETAIL Variably hairy. Leaves in a basal rosette, flowering stems branched, also leafy. *Leaves very variable, long-oval to deeply lobed*; stem leaves clasping. Petals 1.5–3.5mm long. Pods flattened, 4–10mm, style 0.2–0.8mm, located in the notch at the tip of the 'purse'.

GROWTH Annual.
HEIGHT 5–50cm.
FLOWERS All year.
STATUS Ancient introduction.
ALTITUDE 0–780m.

Shepherd's Cress
Teesdalia nudicaulis

Near Threatened. Very local in short, patchy grassland on light, sandy soils: grassy heaths, dunes, shingle, scree and spoil heaps.
❁ The tiny asymmetric flowers are less conspicuous than the *clusters of lozenge-shaped fruits.*
DETAIL Leaves long-stalked, *mostly in a basal rosette*, which may be withered by flowering time: to 5cm, slightly flshy, hairless or sparsely hairy, variably and irregularly lobed. Petals subtly unequal: outer pair 1.5–2mm, inner pair 0.8–1.5mm. Pods 3–4mm, notched at tip, flattened, style 0–0.3mm.

GROWTH Winter-annual.
HEIGHT 2–15cm.
FLOWERS Mar–Jun.
STATUS Native.
ALTITUDE To 650m.

Cuckooflower *Cardamine pratensis*

GROWTH Perennial.
HEIGHT 15–60cm.
FLOWERS Late Mar–Jun.
STATUS Native.
ALTITUDE 0–1100m.

Also known as **Lady's-smock**. Common in damp or wet grassy places: unimproved meadows, fens, woodland rides, damp verges, churchyards and sometimes lawns. ✿ Upright, the lower leaves are cut into stalked, rounded leaflets, with a larger, kidney-shaped terminal leaflet, while the upper leaves have much narrower, strap-shaped leaflets. Flowers various shades of pale pink or lilac (rarely white), often with darker veins, *anthers yellow*. **DETAIL** Usually almost hairless. A basal rosette of leaves produces an erect,

usually unbranched flower stem. Leaves divided into 2–10 pairs of leaflets. Petals 6–18mm. Pods 25–50mm, flattened, held more or less erect. Leaf fragments can root and form new plants.

Coralroot *Cardamine bulbifera*

GROWTH Perennial.
HEIGHT 25–75cm.
FLOWERS Apr–May.
STATUS Native.
ALTITUDE Lowland.

Nationally Scarce. Grows in sloping beechwoods over chalk in the Chilterns, and in wet woodland and shaded verges over wealden clay in E Sussex and adjacent Kent. Occasionally escapes from cultivation elsewhere. ✿ Vaguely similar to Cuckooflower, but the leaves and flowering stems grow *singly* from underground rhizomes, *flowers pinkish-purple*, and *purple, pea-shaped bulbils* develop in the leaf axils. **DETAIL** Hairless to sparsely hairy. Rhizome cream-coloured and coral-like (hence the name). Stems erect, unbranched, purplish at base. *No basal rosette*. Leaves cut into 1–3 pairs of bluntly-toothed leaflets plus a terminal leaflet; on flowering stems upper leaves often undivided; most plants in cultivation have broader, more coarsely-toothed leaves. Petals 10–17mm. Rarely sets seed.

Hairy Bittercress
Cardamine hirsuta

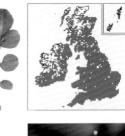

Abundant. All kinds of bare ground, including walls, rocky places, woods and stream-sides. One of the commonest garden weeds, flowering for most of the year. ✿ Short, with small white flowers that only open fully in bright weather. DETAIL *Stem usually hairless* (leaf stalks hairy). Rosette leaves with 2–6 pairs of leaflets plus a larger terminal leaflet. Stem leaves 1–4, smaller, with narrower leaflets. Flowers with 4 petals, 2.5–4mm long, and *4 stamens*. Pods 10–21mm, narrow, held upright.

GROWTH Annual.
HEIGHT 5–30cm.
FLOWERS Feb–Nov.
STATUS Native.
ALTITUDE 0–1190m.

Wavy Bittercress
Cardamine flexuosa

GROWTH Annual to short-lived perennial.	FLOWERS Mar–Sep.
	STATUS Native.
HEIGHT 7–50cm.	ALTITUDE Lowland (1190m).

Very common on bare ground in moist, shady places, but avoiding very acid soils: marshes, river- and streamsides, and gardens. ✿ Very like Hairy Bittercress, and best distinguished by having 6 *stamens* (2 are slightly shorter and can be overlooked), rather than 4. DETAIL *Stem usually hairy, with 4–10 leaves.* The flowering stems are often subtly zigzag ('wavy'), but this is not a useful feature.

Narrow-leaved Bittercress
Cardamine impatiens

Nationally Scarce, listed as Near Threatened. Rather local. Damp woodland (especially Ash), riverbanks and damp, rocky places on limestone (including scree and limestone pavement). ✿ Often tall and leafy. Leaves cut into 6–11 pairs of *conspicuously toothed leaflets*, the stem leaves with distinctive *pointed auricles at the base of the stalk, clasping the main stem*. Flowers very small, often without petals. DETAIL Hairless. Stem little- or unbranched, ridged. Basal rosette soon dying off. Petals absent or white, 1.5–3.5mm, stamens 6, anthers greenish-yellow. Pods 12–30mm, very narrow, held at 45° to the stem.

GROWTH Annual or biennial.
HEIGHT 20–80cm.
FLOWERS May–Aug.
STATUS Native.
ALTITUDE 0–610m.

Large Bittercress *Cardamine amara*

Locally common. In or around still or slow-flowing water in marshes, unimproved meadows and wet woodland, often where shaded, mostly on acid soils. ❀ Similar to Cuckooflower but *flowers usually pure white*, with *violet anthers*. **DETAIL** More or less hairless. Stems often sprawling, rooting at the nodes and turning upwards towards the tip. *No basal rosette.* Leaves wintergreen, divided into 2–5 pairs of oval leaflets and a larger terminal leaflet. *Petals 5.5–12mm*, rarely pinkish-purple. Pods 15–40mm, *flattened.*

GROWTH Perennial.
HEIGHT 15–50cm.
FLOWERS Apr–Jun.
STATUS Native.
ALTITUDE To 640m.

Watercress
Nasturtium officinale

Common in and around shallow water in streams, rivers, ditches and ponds, and in fens and wet meadows, usually on alkaline soils. ❀ *Sprawling, stems usually rooting at the nodes.* Leaves slightly glossy, dark green to purple, with 1–5 pairs of leaflets. Flowers small, with yellow anthers. **DETAIL** Hairless. No basal rosette. *Petals 3.5–6.6mm.* Pods short and squat, 11–19mm × 1.7–3.2mm, style stout, 0.5–1.7mm, stalk 7–12mm; *seeds usually in 2 parallel rows* (cut fruit lengthwise). **SIMILAR SPECIES Narrow-fruited Watercress** *N. microphyllum* is slightly scarcer in the S and commoner in the N. Found in similar places as well as more acid sites. ❀ Seeds more finely net-veined and *mostly in 1 row*, pods mostly 16–23mm × 1.3–1.8mm, with stalks 14–19mm. The hybrid between the 2 species (*R. × sterile*) is relatively common. It has longer flower spikes but shorter seed pods, often deformed, with only 0–3 well-formed seeds. **SEE ALSO** Fool's Watercress (p. 307).

GROWTH Perennial.
HEIGHT 15–60cm.
FLOWERS May–Oct.
STATUS Native. Has also been in cultivation for at least 2,000 years.
ALTITUDE Lowland.

Common Whitlowgrass
Erophila verna

GROWTH Annual.
HEIGHT 2–10cm.
FLOWERS Feb–May.
STATUS Native.
ALTITUDE 0–810m.

Locally common. Dry, bare places on sandy, gravelly and chalky soils, also walls and pavement cracks. ✿ Small, flowers early and then vanishes. *Flowering stems leafless, with small flowers that only open fully on bright days. Petals split more than halfway to base, seed pods tiny, flattened, oval.* DETAIL Variably hairy, with 1 to many stems. Rosette leaves to 20mm long, variably-toothed, tapering to a broad stalk 0.5–1 times as long as leaf blade. Petals 1.5–3.8mm; stamens 6. Pods 1.5–9mm long. SIMILAR SPECIES Glabrous Whitlowgrass *E. glabrescens* is very similar but nearly hairless, with longer leaf stalks and less deeply-cleft petals. Hairy Whitlowgrass *E. majuscula* is densely greyish-hairy, with shorter leaf stalks and less deeply cleft petals. Both are scarce but overlooked. SEE ALSO Rue-leaved Saxifrage (p. 35).

Hoary Whitlowgrass
Draba incana

Rather local in rocky places, mostly on limestone, also sparse dry turf and dunes. ✿ Usually short, with rosettes of greyish-green leaves and leafy flower stems. *The seed pods have a distinctive twist.* DETAIL Densely hairy, with some hairs star-shaped. Leaves strap-shaped (broader towards tip, which is pointed), variably toothed, up to 30mm long, more or less stalkless, those on stem sometimes slightly clasping. Petals 2.5–4.3mm, slightly notched; stamens 6. Pods 5.5–12mm, flattened. SIMILAR SPECIES Wall

GROWTH Perennial.
HEIGHT 5–40cm.
FLOWERS May–Jul.
STATUS Native.
ALTITUDE 0–1160m, but mostly in uplands.

Whitlowgrass *D. muralis* has leaves with a heart-shaped base that clasp the stem, flowers with 4 stamens, and flat pods. Nationally Scarce, and very scattered on limestone, also a garden escape.

Hutchinsia
Hornungia petraea

Nationally Scarce. Very local. Bare ground over thin soils on broken, S or SW-facing, rocky limestone slopes, also calcareous dunes. ✿ A tiny, early-flowering crucifer; in wet spring may be relatively large and well-branched. DETAIL Hairless. *Leaves both in a basal rosette and on the flowering stem.* Leaves deeply-cut into 3–10 pairs of spoon-shaped leaflets. Petals 0.5–1mm, *only slightly notched. Pods 2–3mm long, roughly oval, flattened.*

GROWTH Winter-annual.
HEIGHT 2–10cm (15cm).
FLOWERS Feb–May (Jun).
STATUS Native.
ALTITUDE Lowland (490m).

Dame's Violet *Hesperis matronalis*

GROWTH Perennial.
HEIGHT 30–100cm.
FLOWERS May–Aug.
STATUS
Introduced
to cultivation
from S Europe
by 1375 and
recorded from
the wild by 1805.
ALTITUDE
Lowland.

Commonly grown in gardens (as 'Sweet Rocket') and widely naturalised in moist, shady places, including verges and woodland borders, especially near houses. ✿ Tall and very upright, with pointed, arrow-shaped leaves *all the way up the stem* and *large, fragrant, white to deep pink or purple flowers*. DETAIL Stems branched, sparsely hairy. Leaves very hairy (slightly rough), variably toothed; lower leaves stalked, the stalks becoming shorter towards the tip of the stem, where the leaves may be stalkless. Petals 15–30mm. *Pods curving upwards*, 2–11.5cm long, *irregularly beaded*.

Honesty *Lunaria annua*

GROWTH Biennial.
HEIGHT 20–100cm.
FLOWERS Apr–Jun.
STATUS In cultivation since at least 1570, and recorded from the wild by 1597.
ALTITUDE Lowland.

Fairly common garden escape found on roadsides and waste ground, usually near houses. ✿ Large, reddish-purple flowers are followed by *flattened, coin-shaped seed pods*, whose *semi-transparent nature* is the origin of the English name. DETAIL Densely hairy. Leaves pointed-oval with a heart-shaped base, sharply-toothed, the lower stalked. Petals 15–25mm. Flowers sometimes white. Pods 25–50mm, style 4–10mm.

Bastard Toadflax
Thesium humifusum

GROWTH Perennial.
HEIGHT Stems 10–20cm.
FLOWERS Jun–Aug.
STATUS Native.
ALTITUDE Lowland.

Nationally Scarce. Rather local on short, dry, species-rich, grazed grassland over chalk or limestone, often on SW-facing slopes. ❀ Small, slender and inconspicuous, with *sprawling stems* bearing thread-like leaves and *loose spikes of tiny, white, star-like flowers*. DETAIL Hairless. Grows from a woody rootstock and partially parasitic on the roots of various other plants. Leaves alternate, stalkless, 8–25mm long × 1–2mm wide, yellowish-green with 1 vein. Flowers 2–4mm across with *3 narrow, leaf-like bracts* at base, the calyx split into 5 sepals, yellowish-green on the outside and white on the inside; no petals; 5 stamens and 1 style. Ovary inferior, ribbed, elongating in fruit.

MOSCHATEL: FAMILY ADOXACEAE

Moschatel
Adoxa moschatellina

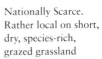

Locally common. Moist, humus-rich soils in deciduous woodland, sometimes hedgebanks and occasionally shady, rocky places in the hills. ❀ Small, delicate and low-growing, flowering in the spring and then vanishing. *Flower heads unique*: long-stalked dice-shaped clusters, 6–10mm across, with 4 greenish-yellow lateral flowers (1 to each side), and a single upwards-facing terminal flower. Aka Town-hall-clock. DETAIL Hairless. Grows from slender rhizomes. Leaves pale green, long-stalked, cut into small, 3-lobed leaflets; the flowering stems have a pair of opposite, smaller leaves. Lateral flowers with the calyx split into 3 tiny sepals, the corolla split into 5 rather larger petals, 10 stamens and usually 5 minute stigmas; terminal flower similar, but with 2 sepals, 4 petals, 8 stamens and 4 stigmas. The only British member of the family Adoxaceae.

GROWTH Perennial.
HEIGHT 5–15cm.
FLOWERS Mar–May.
STATUS Native.
ALTITUDE 0–1065m.

GROWTH Perennial.
HEIGHT To 40cm (60cm).
FLOWERS Jul–early Oct.
STATUS Native.
ALTITUDE Lowland.

Common Sea-lavender *Limonium vulgare*

Locally abundant on muddy saltmarshes (although absent from heavily grazed areas), occasionally on shingle or sea walls. ✿ Slightly fleshy, waxy, strap-shaped leaves form a *large, loose and rather upright rosette*, from which grow sprays of purplish-blue flowers. The papery, pale lilac sepals form a calyx tube which persists long after the petals have fallen. DETAIL Grows from a tough, woody rootstock, the flowering stems branching *above the middle*. Leaves long-stalked, to 20cm long, with *branched (pinnate) veins*. Flowers 8mm across, with 5 petals. Clusters of 1–5 flowers form spikelets, which in turn are grouped into *densely-packed spikes*, mostly 1–2cm long, at the end of the branches.

Lax-flowered Sea-lavender *Limonium humile*

Nationally Scarce. Locally abundant on ungrazed or lightly grazed saltmarshes, rarely on cliffs. ✿ As Common Sea-lavender but *flowering stems branch below the middle* and flower spikes longer, with the *individual spikelets well spaced*. DETAIL *Longest flower spikes 2–5cm long, with 3 or fewer spikelets per cm, the lowest 2 spikelets 4–10mm apart; anthers reddish-brown* (in Common Sea-lavender spikes mostly 1–2cm long, with more than 4 spikelets per cm, and the lowest 2 spikelets 1.5–3mm apart; anthers yellow). NB Hybrids and intermediates with Common Sea-lavender occur; in theory this species can only be identified with complete certainty by microscopic examination of the pollen.

GROWTH Perennial.
HEIGHT To 40cm.
FLOWERS Jul–Aug.
STATUS Native.
ALTITUDE Lowland.

Rock Sea-lavender *Limonium binervosum* agg.

GROWTH Perennial.
HEIGHT 8–30cm (70cm).
FLOWERS Jul–Sep.
STATUS Native.
ALTITUDE Lowland.

A complex of 9 closely related species, very similar in appearance and difficult to separate. On the S and W coasts mostly found on coastal rocks and cliffs, but on the E coast almost exclusive to a narrow zone on the drier upper saltmarsh, adjacent areas of compacted sand and gravel, and locally also shingle banks. DETAIL The flowering stems branch *below the middle* and arise from a *small, compact rosette of short leaves* (to 10cm, often less), which have short, winged stalks and *1–3 unbranched veins*.

Matted Sea-lavender *Limonium bellidifolium*

GROWTH Perennial.
HEIGHT To 30cm.
FLOWERS Late Jun–early Aug (early Sep).
STATUS Native.
ALTITUDE Lowland.

Nationally Rare. Confined to N Norfolk, where locally common on the drier, upper saltmarsh and on open areas of compacted sand and gravel where the saltmarsh meets the dunes. ❀ A delicate, low-growing, early-flowering sea-lavender, smothered in *pale lilac-pink flowers* in midsummer. DETAIL Rather *bushy*, with the stems *branching from the base*, with many *well-branched non-flowering side shoots*. Leaves oblong, 1.5–4cm long, but *withered by flowering time*. Flowers 5mm across, only on upper branches; the *white, papery bracts* remain conspicuous long after the petals have withered.

Thrift
Armeria maritima

GROWTH Perennial.
HEIGHT 5–30cm.
FLOWERS Late Apr–Jul, with smaller numbers to Oct.
STATUS Native.
ALTITUDE 0–1290m.

Common on the coast, on cliffs and clifftop grassland, well-vegetated saltmarshes and adjacent open areas of compacted sand and shingle. Very local inland: rocky places in the uplands, mountain-top moss-heath, lead mine spoil and river shingle contaminated with heavy metals. Has spread along a few salted roads. ❀ The leaves form untidy, cushion-like rosettes, with *pompom-like heads of pinkish to white flowers on unbranched, leafless stems.* **DETAIL** Grows from a tough woody rootstock. Leaves evergreen, fleshy, reddish at base, *very narrow*, to 15cm long. Flower heads 15–25mm across, with a *reddish-brown, papery collar* just below; stem downy. Flowers 8mm across, with a 5-ribbed hairy calyx, 5 petals and 5 stamens.

FAMILY FRANKENIACEAE: SEA HAETH

Sea Heath
Frankenia laevis

GROWTH Perennial.
HEIGHT Stems to 35cm.
FLOWERS Jun–Sep.
STATUS Native.
ALTITUDE 0–5m.

Nationally Scarce, and listed as Near Threatened. Very locally common on bare, open areas of compacted sand and fine gravel at the transition between saltmarsh and sand dunes, also the base of chalk cliffs in Kent. ❀ Mat-forming, with prostrate stems and tiny, strap-shaped leaves recalling a low-growing heather. The delicate, papery, *5-petalled pink flowers*, each with *6 stamens*, are distinctive. **DETAIL** Stems fine but tough and wiry. Leaves evergreen, opposite, stalkless, 2–5mm long with inrolled margins. Flowers solitary, stalkless, 5–6mm across, with 5 sepals, 5 petals and 6 stamens; stigma 3-lobed. The only British member of the family Frankeniaceae.

Common Bistort
Persicaria bistorta

Fairly common in NW England and Wales, scattered elsewhere. Damp grassy places and riverside fens, avoiding alkaline soils. ❀ *Large, cylindrical heads of pink flowers* on erect, unbranched stems, often many together in conspicuous patches. Detail Leaves oval-triangular, *square-cut to heart-shaped at the base*; lower leaves with long, winged stems. Flower spike 10–15mm across, flowers with *3 stigmas* and *stamens that project from the tepals*. Nutlet 3-angled.

GROWTH Perennial.
HEIGHT 30–80cm.
FLOWERS Late May–Aug.
STATUS Native, but often originating from garden throw-outs.
ALTITUDE To 430m.

Mountain Sorrel
Oxyria digyna

Local. Damp, ungrazed mountain ledges, the sides of streams and in wet, shady gullies. ❀ Spikes of tiny flowers grow from a basal tuft of slightly fleshy, *long-stalked, kidney-shaped leaves*. Detail Leaves to 3cm across. Flowering stems usually leafless, flowers with 4 green tepals, the inner 2 enlarging in fruit to enclose the nutlet, which is 3–4mm long, biconvex and broadly winged; stamens 6, styles 2.

GROWTH Perennial.
HEIGHT 10–30cm.
FLOWERS Late Jun–Aug.
STATUS Native.
ALTITUDE Usually 150–1240m, but sometimes along streams to near sea level.

Alpine Bistort
Persicaria vivipara

Locally common. Grassland and damp rocky places in the N Pennines and Scottish Hills, sometimes dispersing to lower levels along streams. ❀ The short, slender flower spikes, with *small whitish flowers above and tiny purple bulbils below*, are unique. Detail Flower spike 4–8mm across, flowers with *3 stigmas* and *stamens that project from the tepals*. Nutlets 3-angled. Leaves strap-shaped, the upper stalkless, clasping the stem, the lower *tapering to a stalk*.

GROWTH Perennial.
HEIGHT 6–30cm.
FLOWERS Jun–Aug.
STATUS Native.
ALTITUDE From near sea level (in the far north only) to 1210m.

GROWTH Perennial.
HEIGHT Stems to 80cm.
FLOWERS Jun–Oct.
STATUS Native.
ALTITUDE 0–560m.

Redshank *Persicaria maculosa*

A common weed of cultivation and muddy places. ✿ The tiny flowers, held in pencil-shaped clusters, are pale pink to reddish and the leaves often have a black blotch in the centre. DETAIL Stems sprawling to erect, reddish to green, swollen above leaf bases. Leaves spear-shaped, 5–12cm long, often sparsely hairy below. *Upper stem, flower stalk and tepals smooth*, with *few (if any) glands. Flowers with 2 stigmas.* Fruit a shiny, dark brown nutlet 2–3mm long, *lens-shaped or often triangular in cross-section. Ochrea* (the tubular membranous stipules at the base of the leaf stalks) *hairy*, fringed with *long, wispy, eyelash-like hairs.*

GROWTH Annual.
HEIGHT To 100cm.
FLOWERS Jun–Oct.
STATUS Native.
ALTITUDE 0–450m.

Pale Persicaria
Persicaria lapathifolia

Common on bare ground on a variety of soils, and a common weed of cultivation. ✿ Rather like Redshank and with a similar tendency to have black-blotched leaves, but often picked out by more thick-set growth, greener stems and pale flowers (pale pink to greenish). DETAIL *Upper stem and flower stalks with numerous tiny short-stalked glands* (10× lens; the tepals also have stalkless glands but these are harder to see, often also the underside of the leaves). The ochrea are *hairless* or have a *sparse fringe of short hairs. Nutlets almost all flattened* (biconcave).

Amphibious Bistort
Persicaria amphibia

Fairly common. Very variable. Aquatic forms grow in still and slow-flowing water, terrestrial forms in marshes, meadows and arable fields, especially those periodically flooded; intermediates occur, and plants can change from one to the other. ❀ The tiny, delicate, pink flowers are clustered together in a compact cylindrical head 2–4cm long.
DETAIL Aquatic plants have long-stalked, hairless, floating leaves. Terrestrial forms have smaller, slightly hairy, short-

GROWTH Perennial.
HEIGHT 30–70cm (aquatic forms may straggle to 13m).
FLOWERS Jul–Sep.
STATUS Native.
ALTITUDE 0–570m.

stalked leaves, *heart-shaped at the base*, usually with a black blotch; they flower very sparsely. Flowers 2–3mm long, with *2 stigmas*, the *stamens often projecting beyond the tepals*. The flower and flower stalk may or may not have glands. Nutlet lens-shaped, shiny brown, 2–3mm long.

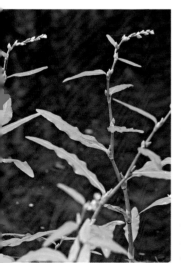

Water-pepper
Persicaria hydropiper

Fairly common. Damp places, often shady, that are wet in winter: woodland rides, pond and lake margins, and meadows. ❀ The slender spikes of tiny, well-spaced flowers often nod to one side. The English name refers to the distinctive hot, acrid taste of the leaves. **DETAIL** Leaves spear-shaped to strap-shaped, completely hairless below, tapering to a very short stalk. Flowers usually with *many stalkless yellow glands*, which may be prominent or may merely appear as small pits. Tepals greenish-white, often flushed rose-pink; *2 stigmas*. Nutlets lens-

GROWTH Annual.
HEIGHT To 75cm.
FLOWERS Jul–Oct.
STATUS Native.
ALTITUDE 0–505m.

shaped to 3-angled, 2.5–3.8mm long, *dull*, with a minutely pitted surface.
SIMILAR SPECIES Tasteless Water-pepper *P. mitis* is found in similar places but is Nationally Scarce and listed as Vulnerable. Leaves with hairs on the midrib below and not peppery, the whole plant is usually redder, with few if any glands on the flowers, and the nutlets are shiny.

Black Bindweed *Fallopia convolvulus*

GROWTH Annual.
HEIGHT Stems to 50cm (150cm).
FLOWERS Jun–Oct.
STATUS Ancient introduction, probably brought by Neolithic farmers.
ALTITUDE 0–570m.

An abundant weed of arable fields, less commonly gardens and disturbed ground on verges and waste ground. ❀ Usually prostrate, forming loose mats, but sometimes climbing. Note *heart-shaped leaves* and long-stalked clusters of *small flowers*. **DETAIL** Stems angled, mealy. Leaves 2–6cm long, mealy below, often long-stalked. Flowers short-stalked, tepals 5–6mm long, greenish, often tinged red, with whitish margins. Fruit a dull black triangular nutlet 4–5mm long, enclosed by the tepals.

Knotgrass *Polygonum aviculare* agg.

GROWTH Annual.
HEIGHT Shoots to 30cm (200cm).
FLOWERS Jun–Nov.
STATUS Native.
ALTITUDE 0–670m.

An abundant weed of disturbed soils in arable fields, also tracks, paths, pavement cracks, roadsides and beaches. ❀ Very variable, may be erect, but often prostrate and mat-forming, but always with small, simple leaves with clusters of attractive tiny flowers at their base. **DETAIL** Hairless. Leaves narrowly-oval to strap-shaped, 7–40mm long. Flowers in groups of up to 6 in leaf axils, each with 5 petals, white or pink with a green central stripe. Fruit a 3-sided pointed shiny brown nutlet 1.5–2.5mm long, enclosed by the petals. **SIMILAR SPECIES** This is a complex of four species that are so hard to separate reliably that they are often treated as mere subspecies. Two other species are scarce in coastal habitats.

Common Sorrel *Rumex acetosa*

Common. Meadows, pastures, woodland rides, verges, shingle beaches and mountain ledges, on neutral to mildly acidic soils. ❀ A slender, erect dock, the acidic leaves are popular in salads. DETAIL Stems *hollow*. Leaves 4–14cm long, long-oval with pointed, *backward-directed* basal lobes, lower leaves stalked, upper leaves *clasp* stem; chaffy brown stipules surround the stem above each leaf. Flowers in loose whorls on short side branches, reddish, tepals 2.5–4mm, circular, with tiny warts, becoming much longer than the nutlet. Flowers mostly of one sex, with male and female on separate plants (dioecious).

GROWTH Perennial.
HEIGHT 20–60cm (cultivated forms to 120cm).
FLOWERS May–Jul.
STATUS Native.
ALTITUDE 0–1210m.

Sheep's Sorrel *Rumex acetosella*

Locally abundant. Bare ground and sparse grassland on poor, acid, sandy or stony soils: heaths, dunes, shingle, rocky outcrops. ❀ A low-growing, slender dock, *conspicuous* en masse *when it carpets the ground red and gold.* DETAIL *Stems solid,*

male flowers

often rather sprawling. *Leaves stalked,* 2–4cm, with *narrow basal lobes that point forwards or sideways.* Flowers reddish, tepals without warts and not enlarging in fruit; nutlet 1–1.5mm. Dioecious (as Common Sorrel).

GROWTH Perennial.
HEIGHT 5–30cm.
FLOWERS Late Apr–Aug.
STATUS Native.
ALTITUDE 0–1050m.

Japanese Knotweed *Fallopia japonica*

A notoriously invasive weed, locally common on waste ground, verges, along railways, and on the banks of streams, rivers, canals and sea lochs; spreads via rhizomes to form dense thickets. ❀ Tall, with attractive sprays of white flowers. DETAIL Hairless. Stems stout, hollow, zigzag, often spotted reddish. Leaves oval, narrowing abruptly to a point, usually with the base at 90° to the stem, *to 12 × 10cm (21 × 14cm).* Almost all British plants are female. SIMILAR SPECIES Giant Knotweed *F. sachalinensis*, a rather scarcer garden escape, grows to 4m. Leaves to 30 × 10–20cm (40 × 28cm), heart-shaped at base and *sparsely hairy below*. *F.* × *bohemica*, the hybrid with Japanese Knotweed, also occurs.

GROWTH Perennial.
HEIGHT To 200cm (300cm).
FLOWERS Aug–Oct.
STATUS Introduced to gardens from E Asia in 1825 but not recorded from the wild until 1886.
ALTITUDE Lowland (470m).

GENUS *RUMEX*: DOCKS & SORRELS

A group of similar, rather 'weedy' species. All have small, wind-pollinated flowers, usually in well-separated whorls, with 6 sepal-like tepals, 6 stamens and 3 styles. The inner 3 tepals enlarge in fruit, and often develop swollen tubercles or 'warts'; the size and shape of the tepals *in fruit*, together with the presence or absence of warts, leaf shape and branching pattern, are key identification characters.

Broad-leaved Dock *Rumex obtusifolius*

The commonest dock, abundant on disturbed soils, sometimes in the shade. ❀ Often very robust, with large, broad leaves, distinctly *heart-shaped at the base*. Flower spike usually well branched, the branches erect. **DETAIL** Leaves stalked, oval-oblong with a blunt tip, the margins only slightly wavy; lower leaves 10–25(40)cm long × 5–15cm wide, upper leaves narrower. Flower spike leafy towards base. Tepals 3–6mm long, usually with well-developed teeth (sometimes with a few short teeth, and rarely no teeth); at least 1 tepal has a rounded wart. **SIMILAR SPECIES** Northern Dock *R. longifolius* Verges, fields and river-, stream- and lakesides, to 520m. Fairly common, and apparently spreading southwards along roads. Leaves wavy-edged, tepals *c.* 6mm × 6mm wide, without teeth or warts.

GROWTH Perennial.
HEIGHT 50–120cm.
FLOWERS May–Oct.
STATUS Native.
ALTITUDE To 700m.

tepal

wart

Curled Dock
Rumex crispus

Very common. Disturbed ground (subspecies *crispus*); shingle banks, beaches, dunes and saltmarsh fringes (subspecies *littoreus*). The second commonest dock.
❀ *Leaves narrow and strap-shaped, rounded or tapered into the stalk*, with distinctly *wavy edges*. **DETAIL** Flower spike often branched, the branches variably erect. *Tepals 3–6mm long, lacking teeth*, but with 1–3 warts: in ssp. *crispus* the warts are unequal and often only 1 develops fully; in ssp. *littoreus* the warts are slightly larger and often all 3 develop, and this subspecies also has denser sprays of fruits.

GROWTH Annual to short-lived perennial.
HEIGHT 40–100cm (120cm).
FLOWERS May–Oct.
STATUS Native.
ALTITUDE 0–845m.

fruits

littoreus

crispus

Clustered Dock
Rumex conglomeratus

Common to the S, rare in the N. Damp places besides rivers, streams, ditches and ponds, or where water stands in winter in meadows and along tracks and field margins; tolerates brackish conditions. ❀ Stems often wavy, with a very subtle zigzag, *flowering stems leafy almost to the tip. Tepals small (2–3mm long), oval, without teeth; all 3 have a large oval wart.* DETAIL Leaves more or less oblong with a rounded base. Flower spike with many long, widely-spreading branches projecting at 30–90° to the main stem.

GROWTH Biennial or short-lived perennial.
HEIGHT 30–90cm.
FLOWERS Jun–Oct.
STATUS Native.
fruit ALTITUDE 0–420m.

Wood Dock *Rumex sanguineus*

Common. Woods, hedgerows, verges, often on clay soils. ❀ Much like Clustered Dock, but prefers drier, shadier places. Flowers also in well-spaced whorls, with small, untoothed tepals, but *branches more erect* and *not leafy to the tip. Only 1 tepal has a large rounded wart, the others have either no wart or poorly-developed ones.* DETAIL Stem straight, leaves oval-oblong with a rounded base and pointed tip, the leaf stalk often reddish at the base; occasionally has reddish veins. Flower spike with many branches, arising at a *narrow angle to the stem (15–25°,* occasionally to 45°). Tepals toothless, 2–3mm long.

GROWTH Short-lived perennial.
HEIGHT 30–80cm.
FLOWERS Jun–Aug.
STATUS Native.
fruit ALTITUDE To 380m.

Fiddle Dock
Rumex pulcher

Local on lighter soils, often where disturbed by grazing or trampling: dry coastal pastures, verges, commons, village greens and churchyards. ❀ Distinctively shaped: stem rather short, zigzag, with *long branches spreading almost horizontally.* DETAIL Leaves stalked, relatively small (5–10cm × 2–4cm), usually *fiddle-shaped* – pinched in just before the heart-shaped to rounded base. Flower spike more or less leafless, with wavy branches diverging at 70°–90° to the stem. Inner tepals 4–5.5mm, usually with well-developed teeth but sometimes with a few short teeth and rarely none; all 3 usually have an *elongated rough wart.* SIMILAR SPECIES Greek Dock *R. cristatus* has a heart-shaped base to the leaves and *finely toothed, rounded tepals, 5–8mm long, with 1 large wart.* **Patience Dock** *R. patientia* also has similar tepals, but they are *untoothed, with a smaller wart,* and the lower leaves taper to the stalk. Both are introductions, mostly on waste ground around London area.

GROWTH Biennial or short-lived perennial.
HEIGHT 20–50cm.
FLOWERS Jun–Jul.
STATUS Native.
ALTITUDE Lowland.

fruit

GROWTH Perennial.
HEIGHT To 200cm.
FLOWERS Jul–Sep.
STATUS Native.
ALTITUDE Lowland.

Water Dock
Rumex hydrolapathum

Local in shallow water (to *c.* 30cm deep) on the margins of rivers, canals, ditches and lakes, including those in coastal marshes, favouring alkaline water high in nutrients, also bare, wet ground. ❁ Very large – the largest of the docks – although often half-hidden by surrounding vegetation. Foliage distinctly grey-green. **DETAIL** Leaves to 100cm long. Tepals 5–8mm long, each with a long raised wart.

fruit

GROWTH Annual to short-lived perennial.
HEIGHT 40–90cm.
FLOWERS Jun–Sep.
STATUS Native.
ALTITUDE Lowland.

Marsh Dock *Rumex palustris*

Local and uncommon. Wet, nutrient-rich mud exposed by falling water levels in summer in marshes, ponds, ditches and clay and gravel pits, also other wet disturbed ground. ❁ *Whole plant yellowish-green, when the seeds are ripe, the flower heads become pale brown or yellowish-brown.* **DETAIL** Leaves strap-shaped with wavy edges, pointed at the tip and tapering into the stalk. Flower spike diffuse, with many long, widely spreading then incurved branches. Flowers in dense, often well-separated clusters, with the flowering stems conspicuously leafy. *Flower stalks relatively thick, usually shorter than the flowers. Tepals 3–4mm long, all 3 with a large, blunt, oval wart and slender, relatively rigid teeth that are clearly shorter than the tepals (less than 2mm); anthers 0.9–1.3mm long.*

fruits

Marsh   Golden

Golden Dock
Rumex maritimus

Local and uncommon, although numbers vary greatly from year to year. Bare, nutrient-rich mud, especially on the margins of ponds, lakes, ditches and rivers as water levels fall in summer, also spoil produced by river engineering and wet hollows in marshy fields; tolerant of brackish conditions. ❁ *Whole plant initially yellowish-green, turning reddish- or yellowish-brown as the summer progresses.* Easily confused with Marsh Dock. **DETAIL** As Marsh Dock but branches shorter and straighter, individual flower stalks *very slender, longer than the flowers. Tepals slightly smaller, 2.5–3mm long, with a narrow, long-oval wart,* more or less *pointed* at the tip, and *long, flexible hair-like teeth* at least 2mm long – as long as the tepal; *anthers 0.4–0.6mm long.*

GROWTH Annual or short-lived perennial.
HEIGHT 40–90cm.
FLOWERS Jun–Sep.
STATUS Native.
ALTITUDE Lowland.

Round-leaved Sundew *Drosera rotundifolia*

Locally common. Bare peat and amongst *Sphagnum* mosses in bogs and other waterlogged ground on heaths and moors. ● The small rosettes of *long-stalked, circular leaves, covered with sticky reddish hairs* and *held pressed to the ground*, are distinctive. Flowers on long, slender stems, opening one at a time as the stem straightens, in sunny weather only and often not at all. DETAIL Leaves 7–10mm across, *narrowing abruptly into the stalk*, which is hairy and up to 40mm long. Flowers 5mm across, with 5–8 sepals, petals and stamens.

GROWTH Perennial.
HEIGHT Flower spike to 10cm (25cm).
FLOWERS Jun–Aug.
STATUS Native.
ALTITUDE 0–670m.

Sundews are carnivorous. The sticky hairs on the leaves trap small insects and then excrete enzymes to digest the victims. The resultant nutrient 'soup' is then absorbed by the leaf.

Oblong-leaved Sundew *Drosera intermedia*

Rather local. Bare, wet, acid peat on bogs and lakeshores; rarely growing amongst *Sphagnum*. ● As Round-leaved Sundew but *leaves spoon-shaped, the blade narrowing gradually into the short, more or less hairless stalk; leaves held erect.* DETAIL Leaves 4–20mm × 3–5mm, stalk 15–35mm. Flower stem arising from side of rosette, often *no longer than the leaves.*

GROWTH Perennial.
HEIGHT Flower spike to 5cm (10cm).
FLOWERS Jun–Aug.
STATUS Native.
ALTITUDE 0–335m.

Great Sundew *Drosera anglica*

Near Threatened. Very local on very wet ground in bogs, usually where not too acid; sometimes in standing water. Also stony lakeshores. ● *Leaves spatula- or strap-shaped, tapering to a hairless stalk and held upright.* DETAIL Leaves 25–40mm × 2–5mm, stalk 10–70mm long. Flower stem arising from centre of rosette, usually *much longer than the leaves,* flowers 10–13mm across.

GROWTH Perennial.
HEIGHT Flower spike to 18cm (30cm).
FLOWERS Jul–Aug.
STATUS Native.
ALTITUDE Mostly lowland, but to 915m.

Ragged Robin
Silene flos-cuculi

Common. Marshes, fens, unimproved wet grassland, wet woodland rides and ditch- and stream-banks. ❀ The large, very distinctive flowers are held in loose clusters, with the central buds the first to open. Each of the 5 petals is *deeply cut into 4 'ragged' lobes*. DETAIL Stem sparsely

GROWTH Perennial.
HEIGHT To 75cm.
FLOWERS May–Jul.
STATUS Native.
ALTITUDE 0–750m.

hairy. Leaves mostly crowded at base of stem, strap-shaped, pointed, 4–10cm long, with long hairs on the margins towards the base; opposite, their bases join around the stem. Flowers 30–40mm across, the calyx boldly ribbed and split into 5 pointed teeth at the tip. Stamens 10, styles 5, capsule with 5 teeth.

Soapwort
Saponaria officinalis

Fairly common on verges and waste ground around towns and villages, also damp woodland and along streams, especially in SW England and N Wales. ❀ Medium-tall, often clump-forming, with *dense clusters of pale pink flowers*. DETAIL *Hairless*, foliage slightly bluish-green. Leaves opposite, 5–10cm, long-oval, near-stalkless, with 3–5 bold, parallel veins. Flowers *c.* 25mm across, with the *long calyx tube* split at the tip into 5 triangular teeth, *2 styles* and 10 stamens; the double form, 'Bouncing Bett', is found quite frequently.

GROWTH Perennial.
HEIGHT 30–90cm.
FLOWERS Jul–Oct.
STATUS Ancient introduction, long grown in gardens. As the name suggests, the boiled and macerated leaves were once used to make soap.
ALTITUDE Lowland.

GROWTH Annual.
HEIGHT 30–100cm.
FLOWERS Jun–Aug.
STATUS Introduced as a contaminant of seed corn in the Iron Age, it depended on being resown each year with the crop. Once seed cleaning techniques were good enough to remove its large seeds from the corn it quickly vanished as an arable weed.
ALTITUDE Lowland.

Corncockle
Agrostemma githago

More or less extinct. Formerly as arable weed, it is now found where 'wild flower' seed mixes are used. ❀ Tall, with *solitary,*

large, reddish-purple flowers with the *sepals projecting well beyond the petals*. DETAIL Softly hairy. Leaves 5–15cm, strap-shaped, pointed, near-stalkless. Flowers on long stalks, 20–50mm across, calyx lobes longer than petals (but more or less the same length in garden varieties), and 5 styles.

Red Campion *Silene dioica*

Common in woodland clearings and rides, hedgerows and other lightly shaded places, also coastal cliffs and cliff-top grassland, shingle banks, and cliffs and scree in the uplands. ❁ One of the most familiar wild flowers, with 5-petalled red or pink flowers, each petal *deeply notched*. **Detail** Variably hairy (sometimes sticky-hairy, occasionally hairless). Leaves opposite, 4–12cm long, the lower tapered into a long, winged stalk. Flowers 18–25mm across, the calyx tube prominently ribbed, with 5 teeth at the tip. Dioecious, with male and female flowers on separate plants: male plants have 10 stamens, female plants 5 styles and a capsule with *10 strongly down-curved teeth*. Rare white-flowered forms are albinos, with no purple tints in the stem or leaves. **Similar Species** *S. × hampeana,* the hybrid with White Campion, is common. It is fertile, with pale pink flowers, but can be almost identical to pink-flowered Red Campions.

GROWTH Perennial.
HEIGHT To 100cm.
FLOWERS Apr–Oct.
STATUS Native.
ALTITUDE 0–1065m.

capsule

Bladder Campion *Silene vulgaris*

Fairly common. Rough, grassy places, including verges and woodland rides (can tolerate partial shade), also field margins, old quarries and pits, walls and waste ground, especially on chalky or sandy soils. ❁ Foliage *waxy greyish-green*, flowers with a distinctive *inflated, bladder-like calyx marked with a net-like pattern of purplish veins*. **Detail** Stems hairless or hairy, well branched. Leaves opposite, to 6cm, lanceolate, usually hairless (occasionally hairy both sides). Flowers in a loose cluster, 17–20mm across, the 5 petals deeply cut, the calyx, 10–12mm long, contracted to a narrow, toothed mouth from which the petals project. Some flowers are male, some female and some bisexual: female/bisexual flowers have 3 styles, male/bisexual 10 stamens. Capsule 6–9mm long, with 6 small, erect to spreading teeth. **Similar Species Sand Catchfly** *S. conica.* Very local, mostly Breckland and the coast of SE England. Calyx *inflated*, but green and *finely ribbed*, and flowers small (4–5mm across) and *bright pink*. **See Also** White Campion (p. 156).

GROWTH Perennial.
HEIGHT 25–80cm.
FLOWERS May–Aug.
STATUS Native.
ALTITUDE 0–360m.

GROWTH Short-lived
perennial (occasionally annual
or biennial).
HEIGHT 30–100cm.
FLOWERS May–Oct.
STATUS Introduced in the
Bronze Age.
ALTITUDE To 560m.

White Campion *Silene latifolia*

Common. Arable
fields, verges and waste
ground, generally
on light soils. ❀ A
conspicuous weed
with large, white,
5-petalled flowers.
DETAIL Softly hairy,
often with *a few* sticky
glandular hairs above.
Stem branched. Leaves
opposite, 4–12cm
long, lanceolate, the
lower stalked. Flowers
sweetly
scented in
the evening,
25–30mm
across,
petals deeply

cleft, calyx tipped by long, narrow teeth. Dioecious, with male
and female flowers on separate plants: male plants have more
flowers, each with a 10-veined calyx, 15–22mm long, and 10
stamens; females have a 20-veined calyx, 20–30mm long, and *5*
styles. Capsule with *10 erect teeth*.

Night-flowering Catchfly *Silene noctiflora*

GROWTH Spring germinating
annual.
HEIGHT 15–50cm.
FLOWERS Jul–Oct.
STATUS Ancient introduction.
ALTITUDE Lowland.

Scarce and declining
arable weed, listed as
Vulnerable, favouring
light, sandy or chalky
soils. ❀ Much like
White Campion but
obviously sticky-hairy.
Petals deeply cut,
very pale pink, lightly
washed yellowish
on the outside. The
flowers often close
during the day,
especially in bright
weather, with the
petals rolled up like a
spring. **DETAIL** May be
well branched. Leaves
5–10cm, lower oval-
lanceolate, narrowed
to a stalk-like base,
upper narrower and more pointed, clasping the
stem. Flowers *c.* 18mm across, calyx 20–30mm,
boldly veined, with long, narrow teeth; 3 styles, 10
stamens. Capsule with 6 *down-curved teeth*.

Sea Campion *Silene uniflora*

GROWTH Perennial.
HEIGHT 8–30cm.
FLOWERS Apr–Aug.
STATUS Native.
ALTITUDE 0–970m.

Common by the sea on cliffs, cliff-top grassland, dunes, shingle beaches, sea walls and waste ground. Scarce inland in rocky places in the mountains. ✿ Low-growing and mat-forming, with *fleshy, waxy, grey-green leaves* and masses of white flowers on *short, erect stalks*. DETAIL Stems woody at base, often purplish, usually hairless. Leaves 5–40mm, strap-shaped, pointed, hairless. Flowers 20–25mm across, in clusters of 1–4. Calyx inflated, cylindrical, 17–20mm long, petals deeply cut, 3 styles, 10 stamens. Capsule teeth 6, spreading or down-turned.

Nottingham Catchfly *Silene nutans*

Nationally Scarce and Near Threatened. Very local on grassy cliffs, dunes and shingle by the sea, also inland in rocky places on limestone. ✿ Has distinctive loose clusters of *drooping white flowers with the stamens usually long and prominent*. Petals deeply-cut, *rolling-up like a spring* during the day, the flowers opening at night or in dull weather when the petals *angle sharply backwards*.

GROWTH Perennial.
HEIGHT 15–115cm.
FLOWERS May–Jul.
STATUS Native.
ALTITUDE Lowland.

DETAIL Stems sticky-hairy towards tip. Leaves sparsely hairy, basal leaves spoon-shaped, 4–7cm, long-stalked, upper leaves narrower, stalkless. Flowers 17–19mm across, calyx 9–12mm, hairy, with 5 teeth, petals white, washed greenish-cream on outside, 10 stamens, 3 styles; some flowers are female, with the stamens much-reduced. Capsule with 6 spreading teeth.

Moss Campion
Silene acaulis

Rather local. Moist cliffs, ledges and other rocky places in the mountains, also dunes at sea level in the far NW. ❂ Forms *dense cushions sprinkled with solitary rose-pink flowers* (sometimes on short, erect stalks). Detail Sparsely hairy. Leaves 6–12 × 1mm, strap-shaped, pointed, with a fringe of hairs. Flowers 9–12mm across, calyx 5–9mm, hairless, with 5 teeth; styles 3, stamens 10. Capsule with 6 spreading teeth. Often dioecious, with male and female flowers on separate plants.

GROWTH Perennial.
HEIGHT To 3cm (flowering stems to 10cm).
FLOWERS Late May–Aug.
STATUS Native.
ALTITUDE 0–1305m.

Maiden Pink
Dianthus deltoides

Nationally Scarce and listed as Near Threatened. Very local in sparse grassland, in sunny, often sandy, heathy or rocky places, on dry, base-rich soils, including chalk and limestone. ❂ Flowers carnation-like (but *not* scented), often solitary, on long, leafy stalks, usually with a *dark stripe across the base of the petals* and *silvery-white spots*; they close in dull weather. Detail Stems *hairy*, especially towards base. Leaves often rather *bluish-green*, 10–25mm long, narrow, strap-shaped, joined across the stem and with a fringe of hairs at the base. Flowers 12–20mm across, in clusters of 1–3, calyx 12–18mm, slender, with 5 teeth, enclosed at the base by an *epicalyx* formed by 2–4 oval, hairy bracts; *petals irregularly toothed at tip*, styles 2, capsule with 4 teeth. Similar Species **Depford Pink** *D. armeria* is found at sacttered sites in S Britain (Nationally Scarce, Endangered, Sch 8). Annual or biennial. Leaves *green*, flowers 8–13mm across, *4 or more together*, each cluster *enclosed by leaf-like bracts*, epicalyx *almost as long as calyx*, petals often finely spotted.

GROWTH Perennial.
HEIGHT 10–20cm (45cm).
FLOWERS Jun–Aug.
STATUS Native.
ALTITUDE 0–355m.

Small-flowered Catchfly *Silene gallica*

Nationally Scarce, listed as Endangered. Very local in arable fields and sparse grassland (e.g. railway cuttings) on well-drained soils. ❂ *Sticky-hairy* and straggling, with *small* flowers (10–12mm across) – white, pale pink or *red-spotted*. Detail Leaves strap-shaped, stalkless (basal leaves die off by flowing time). Calyx 7–10mm long, with long teeth and long, erect hairs. Styles 3, capsule with 6 teeth.

GROWTH Annual.
HEIGHT 15–45cm.
FLOWERS Jun–Oct.
STATUS Ancient introduction.
ALTITUDE Lowland.

Thyme-leaved Sandwort
Arenaria serpyllifolia

Locally abundant. Dry, bare ground, especially on sandy or chalky soils: unimproved grassland, dunes, arable fields, walls, railway ballast, spoil heaps and other rocky places. ❁ Low growing but usually erect and even 'bushy', the small, white, 5-petalled flowers are inconspicuous but the *tiny pointed-oval leaves, in opposite pairs along the stem*, are distinctive. DETAIL Leaves 3–8mm long, sparsely hairy, dark grey-green. Petals undivided, shorter than sepals; styles 3, stamens 10.

GROWTH Winter-annual to biennial.
HEIGHT 2.5–20cm.
FLOWERS Apr–Sep.
STATUS Native.
ALTITUDE 0–610m.

Slender Sandwort *Arenaria leptoclados*

Very similar to Thyme-leaved Sandwort, and found in similar places, but slenderer, with the flower spike more diffuse at fruiting. Range and habit probably similar but scarcer, especially in the N.

	petals	sepals	capsule	seeds
Thyme-leaved	1.6–2.7mm	3–4 × 1.1–1.8mm	3–4 × 1.1–2.5mm flask-shaped, with a narrower neck	0.5–0.7mm
Slender	1.1–1.6mm	2.1–3.1 × 0.5–0.8mm	2.5–3 × 1.2–1.8mm conical, straight-sided	0.4–0.5mm

Spring Sandwort *Minuartia verna*

Nationally Scarce and listed as Near Threatened. Very locally abundant in short turf and rocky places on limestone, base-rich basalt and serpentine. Its occurrence on lead-mine spoil gave rise to the name 'leadwort'; also found around old zinc and copper mines. ❁ *Basal leaves very narrow, forming tufts or small cushions*, with *star-like flowers on short, upright stalks.* DETAIL Stems usually glandular-hairy, at least towards tip. Leaves *c.* 5 × 0.5mm, pointed, hairless (but glandular-hairy on the Lizard). Petals 5, *not notched*, slightly longer than sepals, which are green with a narrow white fringe and 3 veins; stamens 10, anthers dull red, styles 3. SEE ALSO Knotted and Heath Pearlworts (p. 169).

GROWTH Perennial.
HEIGHT To 15cm.
FLOWERS May–Sep.
STATUS Native.
ALTITUDE To 875m, but usually below 600m.

Three-nerved Sandwort *Moehringia trinervia*

GROWTH Annual.
HEIGHT 10–40cm.
FLOWERS Apr–Jul.
STATUS Native.
ALTITUDE 0–425m.

Locally common in relatively bare, moist places in woods and on shady hedgebanks, usually on moderately acid soils. ✿ Much like Common Chickweed, but *leaves with 3–5 prominent parallel veins* and flowers with *5 petals that are rather shorter than the sepals and not split*.
DETAIL Stems well-branched, round, densely hairy. Leaves oval, pointed, 6–25mm long, sparsely hairy, the lower stalked, the upper stalkless. Flowers with 5 sepals, 10 stamens and 3 styles.

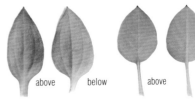

above   below   above   below

Three-nerved Sandwort   Common Chickweed

Lesser Chickweed *Stellaria pallida*

GROWTH Annual.
HEIGHT Stems to 10cm.
FLOWERS Feb–May.
STATUS Native.
ALTITUDE Lowland.

Locally common. Bare ground amongst sparse vegetation on poor, dry soils: dunes, shingle, sand pits, arable fields, forestry tracks, waste ground and sometimes lawns or walls. Flowers early then dries up and vanishes. ✿ Low-growing, often picked out by its *slightly 'sickly', pale yellow-green foliage*. Rather like Common Chickweed but leaves and flowers smaller, with *no petals* and usually *only 2 stamens* (Common Chickweed

occasionally lacks petals). DETAIL Stems sprawling, much-branched, round, with a single line of hairs. Leaves hairless, stalked, 3–7mm long, oval with a pointed tip. Flowers 3–6mm across, sepals 2–3mm long, usually conspicuously hairy (sometimes hairless), stamens 1–2, occasionally 3; anthers dull purple. Capsules on short stalks. SEE ALSO Thyme-leaved Sandwort (p. 159).

Common Chickweed *Stellaria media*

An abundant weed of disturbed, fertile soils, including arable fields, gardens, waste ground, roadsides and pavements. By far the commonest chickweed. ❀ Chickweeds have a rounded stem with a fine line of hairs along alternate sides (sometimes 2 opposite lines), more or less hairless leaves, the upper stalkless, the lower stalked, and small flowers with 5 deeply-cut petals and 3 stigmas. DETAIL Stems more or less straggling. Leaves sparsely hairy to hairless, pointed-oval, 6–25mm long, with a *single vein*. Flowers 5–9mm across, petals 1.1–3.1mm long (no longer than sepals), sepals 2.7–5.2mm long, variably hairy with a fine white margin, 3–5 stamens (but up to 8) with reddish-violet anthers, 3 styles. Seed pods on long stalks that curve downwards or are wavy.

GROWTH Annual or short-lived perennial.
HEIGHT Stems to 30cm.
FLOWERS All year.
STATUS Native.
ALTITUDE 0–950m.

Greater Chickweed *Stellaria neglecta*

Locally common. Damp, shady places along hedgerows, woodland edges and streams. ❀ Very like Common Chickweed but subtly larger (although large Common Chickweed can match it), and the flowers have *8–10 stamens*. A good spotting feature is the ripening capsules, which are on long stalks that are *angled sharply downwards* at first, eventually becoming erect. DETAIL Leaves hairless, 20–30mm long; when viewed against the light, they show a fine network of *subtly darker lines between the translucent veins*. Flowers 9–11mm across, petals 2.5–4mm, sepals 5–6.5mm (often slightly longer than petals); anthers reddish-purple, becoming blackish once the pollen is shed.

GROWTH Annual or short-lived perennial.
HEIGHT 20–100cm.
FLOWERS Apr–Jul.
STATUS Native.
ALTITUDE To 440m.

GROWTH Perennial.
HEIGHT 15–60cm.
FLOWERS Late Mar–Jun.
STATUS Native.
ALTITUDE 0–915m.

Greater Stitchwort *Stellaria holostea*

Abundant. Verges, hedgebanks and woodland, usually on moist but well-drained soils in light shade. ❀ Conspicuous in the spring and often patch-forming, growing tangled through the grass. Leaves narrow and pointed, often grey-green, stalkless. *Bracts uniformly green, and flowers relatively large (20–30mm across), the 5 petals split about halfway to the base.* DETAIL Stems growing from creeping rhizomes, 4-sided, slightly rough on the angles towards the base, and sometimes finely hairy towards the tip. Leaves in opposite pairs, 40–80mm × 5mm, hairy above or hairless, with a comb-like fringe of tiny hairs on the margins. *Petals much longer than the 5 sepals*; stamens 10, styles 3. SEE ALSO Wood Stitchwort (p. 164).

GROWTH Perennial.
HEIGHT Shoots to 40cm.
FLOWERS May–Jun.
STATUS Native.
ALTITUDE To 1005m.

Bog Stitchwort *Stellaria alsine*

Locally common in damp grassy places including river-, stream- and ditchsides, fens and wet woodland rides, often on acid soils. Low-growing and inconspicuous. Flower small, the *5 petals shorter than the 5 sepals, and split almost to the base.* DETAIL Stems growing from creeping rhizomes, straggling, mat-forming, brittle, 4-sided but smooth on the angles, hairless. Leaves 10–20mm × 5mm, pointed-oval, slightly greyish green, *hairless* apart from a fringe of hairs along the margins towards the base; stalkless on flowering shoots, but *short-stalked on non-flowering stems.* Bracts thin and papery in texture, whitish, with a narrow green midrib. Stamens 10, styles 3.

Lesser Stitchwort *Stellaria graminea*

Common in rough grassy places on neutral to acid soils. ❀ The 5-petalled flowers are very variable in size, and *not always small* (5–18mm across). *Petals split almost to the base* and about as long as sepals. Bracts *thin and papery in texture, whitish, with a green midrib.* Leaves narrow and pointed, stalkless. DETAIL Stems growing from creeping rhizomes, straggling, brittle, 4-sided, hairless. Leaves in opposite pairs, 10–40mm × 5mm,

GROWTH Perennial.
HEIGHT Stems to 80cm.
FLOWERS May–Oct.
STATUS Native.
ALTITUDE 0–740m.

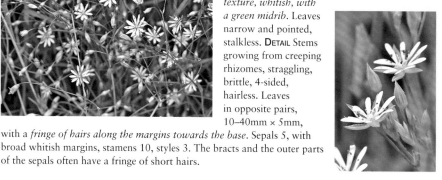

with a *fringe of hairs along the margins towards the base.* Sepals 5, with broad whitish margins, stamens 10, styles 3. The bracts and the outer parts of the sepals often have a fringe of short hairs.

Marsh Stitchwort *Stellaria palustris*

Listed as Vulnerable. Rather local in marshes, fens, dyke margins and wet pastures on alkaline soils, especially where water stands in the winter. ❀ *Leaves hairless,* even at the base, and usually (but not always) *waxy blue-green.* Flowers relatively large (12–20mm across), 5-petalled, the petals split to the base and about as long as the sepals. DETAIL Stems growing from creeping rhizomes, often sprawling then turning up near the tip, slender, brittle, 4-sided, hairless. Leaves stalkless, opposite, 15–50mm × 2–3mm, strap-shaped with a pointed tip. Bracts thin and papery, whitish, with a narrow green midrib; neither bracts nor sepals have a fringe of hairs. Stamens 10, styles 3. Beware Lesser Stitchwort growing in wet grassland, which may have hairless margins to the bracts and sepals, but not the leaves.

GROWTH Perennial.
HEIGHT Stems to 60cm.
FLOWERS May–Aug.
STATUS Native.
ALTITUDE To 360m.

Wood Stitchwort *Stellaria nemorum*

Rather local in damp, shady places on fertile soils: stream- and ditchsides, wet woods and damp hedgebanks. ❁ Has Greater Stitchwort-like flowers on a relatively tall, straggling, chickweed-like frame. Lower leaves *long-stalked.* **DETAIL** Stems

GROWTH Perennial.
HEIGHT
15–60cm.
FLOWERS
Apr–Jun.
STATUS
Native.
ALTITUDE
To 915m.

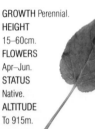

growing from creeping rhizomes, weak, swollen above nodes, *sparsely hairy all round*. Leaves 40–60mm long with a *heart-shaped base*, hairless to hairy; upper leaves clasping stem, lower leaves with 20–40mm stalk (with long wavy hairs). Flowers *c.* 13–18mm across, petals split to near base, *twice length of sepals*; sepals hairless on outer half, with narrow whitish margins; 10 stamens, 3 styles.

Water Chickweed *Myosoton aquaticum*

Local in damp and wet places, including marshes, wet woodland and stream-, river- and ditchsides. ❁ Resembles Greater or especially Wood Stitchwort, but *lower leaves short-stalked or stalkless* and *flowers with sticky-hairy stalks and sepals and 5 styles.* **DETAIL** Stems erect to straggling, branched, brittle, usually glandular-hairy towards tip. Leaves 20–50mm long, pointed-oval, heart-shaped at the base, in opposite pairs, more or less hairy, and short-

GROWTH Perennial.
HEIGHT 20–100cm.
FLOWERS Jun–Oct.
STATUS Native.
ALTITUDE Lowland.

stalked (less than 10mm) or stalkless. Flowers 12–15mm across, petals 5, split almost to base; 10 stamens, 5 styles.

Sea Sandwort *Honckenya peploides*

GROWTH Perennial.
HEIGHT Stems to 25cm.
FLOWERS May–Jul.
STATUS Native.
ALTITUDE Lowland.

Common on beaches and dunes, on bare sand and sandy shingle, just above the high-water mark. ❀ One of the

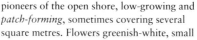

pioneers of the open shore, low-growing and *patch-forming*, sometimes covering several square metres. Flowers greenish-white, small and inconspicuous, but the *yellowish-green leaves, fleshy but stiff and arranged into pagoda-like tiers*, are distinctive, as are the conspicuous *yellowish-green, pea-like fruits*. DETAIL *Hairless.* Stems growing from creeping rhizomes, rounded. Leaves 5–18mm long, oval, pointed, stalkless. Flowers 6–10mm across, with 5 sepals and 5 rounded petals. Flowers hermaphrodite or, more often, with male and female flowers on different plants (dioecious): male flowers with 10 stamens and petals about as long as sepals, female with 3 styles and petals much shorter than sepals.

Field Mouse-ear *Cerastium arvense*

Rather local on dry, sandy soils: verges and other rough, grassy places, dunes, sand pits and arable margins. ❀ *Mat-forming*, with *both non-flowering and flowering shoots* and *softly hairy* leaves. Flowers in groups of 3–5, *relatively large* (12–20mm across),

GROWTH Perennial.
HEIGHT Shoots to 30cm.
FLOWERS Apr–Aug.
STATUS Native.
ALTITUDE 0–300m.

with the 5 petals deeply notched and twice as long as the sepals, and 5 styles. DETAIL Whole plant with short, soft hairs as well as some glandular hairs near the tip of the shoots and upper leaves. Stems growing from creeping rhizomes, more or less sprawling, with the shoots often weaving up through the grass. Leaves in opposite pairs, stalkless, joining around the stem, 10–25mm long, strap-shaped. 5 sepals, 10 stamens. SIMILAR SPECIES **Snow-in-summer** *C. tomentosum.* A garden escape or throw-out on verges, waste ground, dunes and shingle. *Commoner and rather more widespread than Field Mouse-ear.* Also mat-forming, with similar flowers and leaves, but *densely-hairy*, with short greyish-white felted hairs, the *whole plant appearing greyish or white.* Introduced from Italy and in cultivation since 1648. First recorded in the wild in 1915.

GROWTH Annual.
HEIGHT To 25cm (45cm).
FLOWERS Apr–Sep.
STATUS Native.
ALTITUDE 0–715m.

Sticky Mouse-ear
Cerastium glomeratum

Common in sunny places on disturbed soils: gardens, tracksides, waste ground etc. ❀ Short, erect and *sticky-hairy*. Flowers in *compact clusters*, with 5 *deeply-notched petals* and *uniformly green bracts*. Most flower early then die off. DETAIL Hairy, with some long hairs that *project beyond the tips of the sepals and leaves*; glandular hairs mostly on upper stem and sepals. Leaves 20–30mm, stalkless. Flowers 8–10mm across, sepals as long as petals, narrowly fringed white. Stamens 10 (5 short, 5 long), styles 5; capsules 7–10mm, slightly curved, with 10 teeth.

GROWTH Annual.
HEIGHT 1–4cm (20cm).
FLOWERS Mar–Jun.
STATUS Native.
ALTITUDE 0–485m.

Little Mouse-ear
Cerastium semidecandrum

Locally common on dry, bare ground and short turf, typically on sandy soils: dunes, pathsides, walls; also rocky places on limestone. ❀ *Early-flowering, sticky-hairy,* and usually very small. Flowers often on drooping stalks, with 5 *shallowly-notched petals. Bracts broadly fringed white,* with at least the *terminal third white*. DETAIL Abundant glandular and non-glandular hairs all over, but *no long hairs projecting past tip of sepals or leaves*. Leaves 5–18mm, stalkless, often rimmed purple. Flowers 5–7mm across, petals *c.* 0.67 times sepal length, sepals broadly rimmed white. Stamens 5, capsule 4.5–6.5mm, straight, with 10 teeth.

GROWTH Annual.
HEIGHT To 8cm (30cm).
FLOWERS Mar–Jul.
STATUS Native.
ALTITUDE 0–455m.

Sea Mouse-ear
Cerastium diffusum

Local on sand, shingle and open grassland by the sea. Formerly found inland on railway ballast, now sometimes along sandy or salted roadsides. ❀ Low-growing, *sticky-hairy,* with loose clusters of flowers, each with 4 *deeply-notched petals, 4 sepals* and 4 *stamens* (occasionally 5 of each). Bracts *uniformly green*. DETAIL Abundant glandular and some non-glandular hairs throughout (but *no long hairs projecting past tip of sepals*); rarely hairless. Leaves 5–18mm long, often rimmed purple. Flowers 3–6mm across, petals *c.* 0.75 times sepal length, sepals fringed white. Capsule 5–7.5mm, straight, with 8(10) teeth.

Common Mouse-ear *Cerastium fontanum*

Common. A variety of grassy places, also dunes, shingle, waste ground and walls. ✿ Low-growing, *perennial*, sometimes mat-forming, with straggling non-flowering stems and erect flowering shoots. Usually *softly hairy* (but *few or no glandular hairs*) with *loose clusters* of small white flowers, 5 deeply notched petals (about the same length as the sepals), and narrow white margins to the sepals and often also *the bracts*. DETAIL Stem and leaves usually hairy, but may be nearly hairless (the least hairy plants are found in wet habitats, especially in the N). Leaves 10–25mm long, stalkless. Flowers 6–10mm across, with 10 stamens (5 short, 5 long) and 5 styles; capsules 7–16mm long, slightly curved, with 10 teeth.

GROWTH Perennial.
HEIGHT To 50cm.
FLOWERS Late Mar–Sep.
STATUS Native.
ALTITUDE 0–1220m.

Upright Chickweed *Moenchia erecta*

Rather local. Dry, close-cropped turf on heaths, dunes, sandy shingle, quarries and sand pits. ✿ Very small, stiffly-erect and *waxy greyish-green*, but very inconspicuous, as the flowers *usually open only partially*, even in bright sunshine. *Petals 4, rounded, sepals 4, pointed, longer than the petals, and with neat white margins*. DETAIL *Hairless*. Stems erect to sprawling. *Leaves rigid*, strap-shaped, *joined at the base across the stem*. Flowers held erect on long stalks, a maximum of 7–9mm across, with 4 stamens and 4 styles; capsule with 8 teeth.

GROWTH Annual.
HEIGHT 2–12cm.
FLOWERS Apr–Jun.
STATUS Native.
ALTITUDE To 455m.

GROWTH Perennial.
HEIGHT Stems to 20cm.
FLOWERS Apr–Sep.
STATUS Native.
ALTITUDE 0–1150m.

Procumbent Pearlwort *Sagina procumbens*

Abundant. Bare and trampled ground on paths, walls, and rocky places, also short turf and plant pots. ❀ Forms vivid green, rather *moss-like mats of irregular crisscross stems.* **DETAIL** Usually hairless. A central *non-flowering rosette* produces trailing stems that *root at the nodes to produce more rosettes.* Leaves 5–15mm long, needle-like, with a bristle-tip 0.2–0.3mm long. Flowers very small, stigma 4-lobed, stamens 4 (rarely 5), *petals either absent* or *whitish and usually no more than half length of sepals;* the 4 (rarely 5) *blunt sepals* spread cross-wise around the capsule, which is 2–3mm tall and opens at the top to reveal minute black seeds.

GROWTH Annual.
HEIGHT 2–15cm.
FLOWERS Apr–Aug.
STATUS Native.
ALTITUDE 0–535m.

Annual Pearlwort
Sagina apetala

Common. Walls, paths, bare sand, gravel and clinker, and other open ground. ❀ Insignificant, low-growing and spindly. **DETAIL** *Basal rosette tiny, the leaves usually shorter than the stem leaves and quickly withering,* leaving *more or less erect, branching stems,* often hairy, *all with flowers.* Leaves 3–10mm long, with a bristle-tip 0.1–0.4mm and a margin of fine hairs at the base. Sepals 4, green, *pointed,* either hairless or glandular-hairy, stamens 4, *petals minute* (less than 1mm long) or *absent.* In fruit the sepals are held more or less erect around the capsule, which opens at the top to reveal tiny black seeds. **SIMILAR SPECIES Slender Pearlwort** *S. filicaulis* is probably commoner in man-made habitats. Outer sepals blunt or 'hooded' and often fringed red rather than white; in fruit they often spread crosswise.

GROWTH Annual.
HEIGHT To 15cm.
FLOWERS May–Sep.
STATUS Native.
ALTITUDE Lowland.

Sea Pearlwort
Sagina maritima

Fairly common by the sea in rocky and sandy places, dune-slacks, on shingle, walls and in pavement cracks. ❀ Like Annual Pearlwort but *leaves blunt and often fleshy.* **DETAIL** Stems erect to spreading, occasionally purplish or sparsely hairy. *Bristle-point at tip of leaves minute (<0.1mm) or absent.* Petals minute or absent, sepals 4, blunt.

Knotted Pearlwort
Sagina nodosa

Local in damp places flushed with alkaline water: mires, flushes, dune slacks, sometimes also drier calcareous grassland. ❀ Slender and low-growing, with *relatively large* 5-petalled satin-white flowers. *Leaves in distinct clusters* ('knots') along the stem. DETAIL Hairless to glandular-hairy. Stems often somewhat sprawling. Lower leaves 4–11 × 0.5–1mm, upper leaves much shorter, all with a minute bristle tip. Flowers 5–10mm across, *petals rounded, twice as long as sepals*; sepals 5, styles 5, stamens 10.

GROWTH Perennial.
HEIGHT 5–15cm.
FLOWERS Jul–Sep.
STATUS Native.
ALTITUDE To 850m.

Heath Pearlwort *Sagina subulata*

GROWTH Perennial.
HEIGHT Shoots to 10cm.
FLOWERS Jun–Aug.
STATUS Native.
ALTITUDE 0–940m.

Local in dry, sparsely-vegetated places on sandy and gravelly soils: heaths, moors and short turf, often along tracks and paths. ❀ Mat-forming, with short, non-flowering shoots as well as *erect, 2–4cm-long, glandular-hairy, leafless* flowering stems. Flowers 4–5mm across, the 5 *petals about as long as the sepals.* DETAIL Glandular-hairy. Stems sprawling, forming a rosette but shoots usually *not rooting at nodes*. Leaves 3–12 × 0.5mm, with a bristle-point 0.3–0.6mm long. Sepals 5, *glandular-hairy* (sometimes hairless in S), stamens 10, styles 5; capsule 2.5–3mm.

Annual Knawel
Scleranthus annuus

Nationally Scarce and listed as Endangered. Local and declining on disturbed sandy soils on heaths, waste places and, rarely, arable fields. ❀ Low-growing and inconspicuous, the *well-branched,* straggling stems sometimes form tight mats. Annual, *all shoots with clusters of small greenish flowers* that lack petals, but the *pointed green sepals* have a *neat white border*. DETAIL Stems variably hairy. Leaves pointed, 10–18 × 0.6mm, with a fringe of hairs near the base, in stalkless, *opposite pairs that join across the stem*. Flower clusters in leaf axils and at shoot tips, flowers 4mm across, calyx split into 5 sepals, stamens 5–10, styles 2.

GROWTH Annual.
HEIGHT Stems to 20cm.
FLOWERS Late May–Aug.
STATUS Native.
ALTITUDE To 365m.

Greater Sea Spurrey *Spergularia media*

Fairly common. Saltmarshes and other damp, more or less muddy, saline places near the sea, also locally on wet cliffs. Very rare inland. ❀ Low-growing, usually *more or less hairless*, with small, *cylindrical, fleshy leaves* and small, whitish, 5-petalled flowers, the *petals slightly longer than the sepals*. DETAIL Stems usually hairless or with glandular hairs close to the flowers, often purplish, *not woody at base*. Leaves opposite, 10–25mm, pointed, usually hairless. *Flower buds held erect*, 3 styles and 10 stamens (rarely fewer).

GROWTH Perennial.
HEIGHT Stems to 30cm.
FLOWERS May–Sep.
STATUS Native.
ALTITUDE Lowland.

Lesser Sea Spurrey *Spergularia marina*

Locally common in the drier parts of saltmarshes and other muddy, saline places near the sea. Since the late 1970s has also spread extensively along the margins of salted main roads inland. ❀ As Greater Sea Spurrey but *flowers rather smaller, usually lilac-pink*, with the *petals usually shorter than the sepals*. DETAIL Stem hairless at base, *usually glandular-hairy near flowers and often a little way below them*. Leaves hairless to glandular-hairy. Flowers with 2–7 (0–10) stamens.

GROWTH Annual, sometimes perennial.
HEIGHT Stems to 20cm.
FLOWERS May–Aug.
STATUS Native.
ALTITUDE 0–590m.

Rock Sea Spurrey *Spergularia rupicola*

GROWTH Perennial.
HEIGHT Stems 5–15cm.
FLOWERS Jun–Aug.
STATUS Native.
ALTITUDE Lowland.

Locally common on cliffs and other rocky places near the sea, including breakwaters and walls, also short cliff-top turf. ❀ As Greater Sea Spurrey but *very sticky-hairy;* flowers slightly smaller. DETAIL *Stem woody at base*, densely glandular-hairy. Leaves sparsely glandular-hairy. *Flower buds nodding, seeds not winged.*

Sand Spurrey *Spergularia rubra*

GROWTH Annual or biennial.
HEIGHT Stems 5–25cm.
FLOWERS May–Jul.
STATUS Native.
ALTITUDE 0–590m.

Local. Bare, disturbed ground on acid sands and gravel: heaths, forestry tracks, old quarries and pits, waste ground and occasionally coastal shingle and dunes. ❁ Rather like Lesser Sea Spurrey but *leaves flatter* and not as fleshy and *flowers smaller*; not in saline habitats. Flowers only opening in sunshine. DETAIL Stem hairless at base, glandular-hairy near flowers and usually a little way below them. Leaves usually glandular-hairy, at least some with a *0.2–0.4mm long bristle tip*. Flowers pale lilac. Sepals usually longer than petals. Capsule long-stalked, seeds not winged.

	Greater Sea	Lesser Sea		Rock Sea	Sand
flower size	10–12mm (7–13mm)	(4)5–8mm		8–10mm	3–5mm
sepal length	(3.5)4–6mm	2.5–4mm		4–4.5mm	3–4(5)mm
capsule length	(4)7–9mm	3–6mm		4.5–7mm	3.5–5mm
seeds	0.7–1mm wide, most or all also have a broad, flat wing	0.6–0.8mm wide; a variable proportion (sometimes none) also have a wing		not winged	<0.6mm, not winged

Corn Spurrey *Spergula arvensis*

GROWTH Annual.
HEIGHT 10–40cm.
FLOWERS Jun–Oct.
STATUS Ancient introduction.
ALTITUDE To 450m.

Due to recent declines, listed as Vulnerable, but still a locally abundant weed of arable fields and other disturbed places on light, usually sandy, acid soils, also river shingle. ❁ Slender, but much-branched and somewhat 'bushy', sticky-hairy all over with clusters of thread-like leaves and small satin-white flowers. DETAIL Stems swollen around nodes. Leaves often greyish-green, 10–30mm long. Flowers 4–7mm across, with 5 styles and 10 stamens. Seeds not winged or with a very narrow wing.

FAMILY AMARANTHACEAE: GOOSEFOOTS, ORACHES ETC.

There are around 12 species of goosefoot, genus *Chenopodium*, and 10 species of orache, genus *Atriplex*, in Britain. Some are rather hard to identify, and only the commoner species are treated here. All are drab plants that love areas rich in nutrients: inland they are arable weeds and some are frequently found around manure heaps; on the coast they favour the strand-line. Their foliage is often *mealy*, especially when young, as if excreting tiny grains of salt, due to bladder-like hairs. The tiny, wind-pollinated flowers are bisexual in goosefoots, with both stigmas and stamens, and unisexual in oraches: either male, with 5 green sepals and 5 stamens (but no petals), or female, lacking both petals and sepals but enclosed by bracteoles – more or less swollen, triangular flaps.

Good King Henry *Chenopodium bonus-henricus*

GROWTH Perennial.
HEIGHT 25–75cm.
FLOWERS May–Aug.
STATUS Introduced as a pot herb from the mountains of central Europe and present since at least the Roman period.
ALTITUDE 0–455m.

Rather local and declining, listed as Vulnerable. Disturbed, nutrient-rich soil on verges, waste ground, especially around villages, farmsteads and ruins. ❀ Leaves large, triangular, with wavy edges and sharp corners but otherwise *unlobed or only sparsely lobed. Flower spikes almost leafless.* **DETAIL** Stems variably but sparsely mealy, often reddish when old. Leaves 5–10cm long, *net-veined*, mealy when young, especially below. Flowers with 5 tepals, 5 stamens and 2–3 *relatively long, projecting stigmas (0.8–1.5mm).*

Red Goosefoot
Chenopodium rubrum

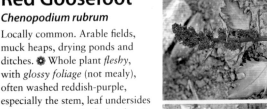

Locally common. Arable fields, muck heaps, drying ponds and ditches. ❀ Whole plant *fleshy*, with *glossy foliage* (not mealy), often washed reddish-purple, especially the stem, leaf undersides and flowers, particularly when in fruit. **DETAIL** Stems variably erect. Leaves 4–8cm long, diamond-shaped, deeply but irregularly toothed. Flower spike leafy, flowers with 2–4 tepals, fused below the middle.

GROWTH Annual.
HEIGHT 30–80cm.
FLOWERS Jul–Oct.
STATUS Native.
ALTITUDE Lowland.

Fat Hen
Chenopodium album

GROWTH Annual.
HEIGHT 15–150cm.
FLOWERS Late Jun–Oct.
STATUS Native.
ALTITUDE To 570m.

Abundant weed of arable fields, gardens and waste ground. ❀ Very variable, leaves lanceolate to oval or diamond-shaped, variably toothed or shallowly-lobed, and tapering into the stalk; flowers in *roughly globular clusters* that are gathered into dense, erect spikes, *leafless towards the tip.* DETAIL Stems often washed or striped reddish. Leaves 3–6cm long, greyish-green, variably mealy (especially when young). Flowers 2mm wide, tepals not toothed. SEE ALSO Common Orache (p. 174).

Fig-leaved Goosefoot *Chenopodium ficifolium*

Local. Arable fields, muck heaps. ❀ As Fat Hen but *lower leaves clearly 3-lobed.* DETAIL Leaves 2.5–5(8) cm long, central lobe roughly parallel-sided, 2–3 times as long as side lobes.

GROWTH Annual.
HEIGHT 20–50cm (100cm).
FLOWERS Jul–Sep.
STATUS Introduced in the Iron Age.
ALTITUDE Lowland.

Many-seeded Goosefoot
Chenopodium polyspermum

GROWTH Annual.
HEIGHT 15–50cm (100cm).
FLOWERS Jul–Oct.
STATUS Ancient introduction.
ALTITUDE Lowland.

Locally common. Arable, waste ground, drying ponds, especially on boulder clay and avoiding sandy or chalky soils. ❀ Superficially like Fat Hen but stems often spreading or prostrate (especially the long outer stems), leaves long-oval, *always more or less untoothed,* blunt or pointed, tapering into the stalk, not mealy, but *edged purplish and often purplish below.* Flowers in irregular, contorted 'strings', gathered into long, *leafy spikes,* tepals often reddish and *seeds obvious, contrastingly dark.* DETAIL More or less hairless. Stems mostly reddish, square, usually much-branched. Leaves 1.5–6cm long, occasionally sparsely mealy when young. Tepals not toothed.

Spear-leaved Orache
Atriplex prostrata

Common. The strand-line on beaches, sea walls and along the saltmarsh edge, also inland on arable fields, disturbed ground and salted roads. ❀ *Very variable*: upright and robust or small and prostrate, but *lower leaves always triangular*, the *base at 90° to the stalk*. DETAIL Leaves to 10cm, toothed, often mealy. Bracteoles 2–6mm long, triangular with the *base at c. 90° to the stalk*, variably spongy but mostly thin and leafy, variably toothed and spiked. The whole plant may turn red in autumn. SIMILAR SPECIES Babington's Orache *A. glabriuscula* is confined to beaches and waste places by the sea. Prostrate, with the bracteoles *diamond-shaped*, thick, spongy and rather spiny.

GROWTH Annual.
HEIGHT Stems
10–100cm.
FLOWERS Jul–Oct.
STATUS Native.
ALTITUDE 0–570m.

hinge

Babington's   **Spear-leaved**

Common Orache
Atriplex patula

Common. Arable fields and other disturbed ground, but largely absent from the seashore. ❀ Varies from upright and robust to slender and prostrate. Lower leaves roughly diamond-shaped, *tapering into the stalk*, sometimes with a pair of *forward-pointing lobes* at the base, sometimes irregularly toothed. DETAIL Stems well branched. Leaves to 9cm, upper leaves diamond- to strap-shaped. Bracteoles 3–7mm long, diamond-shaped, with a *thin, leafy upper section*, toothed, especially at angle, hinged at angle.

GROWTH Annual.
HEIGHT Stems 15–100cm
(150cm).
FLOWERS Jul–Oct.
STATUS
Native.
ALTITUDE
To 570m.

Grass-leaved Orache
Atriplex littoralis

Common. The strand-line on beaches, sea walls and along saltmarshes, and increasingly inland on salted roads. ❀ Always *erect*, with *narrow, strap-shaped leaves*. DETAIL Leaves 4–8cm, sometimes slightly toothed, but *never* with triangular basal lobes. Bracteoles 3–6mm long, *spongy at base, squat*, triangular to diamond-shaped, *well-toothed/spined*, hinged at angle (bracteoles are the best distinction from a narrow-leaved Common Orache).

GROWTH Annual.
HEIGHT Stems 50–100cm
(30–100cm).
FLOWERS Jul–Oct.
STATUS Native.
ALTITUDE
Lowland.

bracteole

Frosted Orache *Atriplex laciniata*

GROWTH Annual.
HEIGHT 6–30cm (50cm).
FLOWERS Jul–Oct.
STATUS Native.
ALTITUDE 0–5m.

Fairly common on the strand-line of undisturbed sandy beaches. ❀ 'Sugar-frosted' with distinctive mealy, *silvery-white foliage* and contrasting *reddish stems*. DETAIL Stems well branched, sprawling. Leaves 1.5–4cm, diamond-shaped, wavy-edged. Bracteoles 6–7mm, diamond-shaped, variably toothed.

Sea Purslane *Atriplex portulacoides*

GROWTH Perennial.
HEIGHT To 80cm (100cm).
FLOWERS Jul–Sep.
STATUS Native.
ALTITUDE 0–5m.

bracteole

Abundant on saltmarshes, often forming dense stands along the fringes of pools and creeks, also scattered on rocky coasts in W Britain. ❀ A small shrub with distinctive pale, *mealy, grey-green foliage* and *fleshy, spoon-shaped leaves*. DETAIL Stems woody, well branched. Leaves evergreen, to 5cm, lower leaves elliptical, upper narrower. Bracteoles 2.5–5mm long, 3-lobed near tip, hinged near tip.

Prickly Saltwort *Salsola kali*

Common on the strand-line of undisturbed sandy beaches. ❀ Distinctive *spiny, grey-green foliage*. DETAIL Stems branched, usually somewhat prostrate, bristly, often striped with red. Leaves succulent, strap-shaped, 1–4cm long, *tapering to a spine at the tip*. Flowers tiny, enclosed by a pair of leaf-like bracteoles, with 5 sepals and both stigmas and stamens.

GROWTH Annual.
HEIGHT 20–60cm.
FLOWERS Jul–Sep.
STATUS Native.
ALTITUDE 0–5m.

Glassworts *Salicornia* spp.

GROWTH Annual.
HEIGHT 10–45cm.
FLOWERS Aug–Sep.
STATUS Native.
ALTITUDE Lowland.

Abundant on saltmarshes, forming extensive stands on the lower shore, the short green shoots often turning red, orange or yellow in autumn, and also grows mingled with other saltmarsh vegetation. ❀ Highly specialised, glassworts appear to lack leaves and be merely jointed, succulent stems, but leaves are present in the form of pairs of triangular, scale-like lobes tightly clasping opposite sides of the stem. As the season progresses, the lower scale-leaves fall off to reveal an increasingly tough and woody stem. There are 7 species of glasswort in Britain, together with numerous hybrids, but they are very hard to distinguish without prolonged study. Glassworts are edible, and are more usually known as 'Samphire'. DETAIL *Stems erect*, constricted or even 'beaded', often branched. Flowers tiny, comprising a fleshy disc emerging from behind the 'leaves' with the tiny anther extruded from a central pore (the stigma remains hidden). Most species have flowers in groups of 3, with the central flower the largest, but One-flowered Glasswort *S. pusilla* has a single flower.

Perennial Glasswort *Sarcocornia perennis*

GROWTH Perennial.
HEIGHT Stems to 30cm.
FLOWERS Aug–Sep.
STATUS Native.
ALTITUDE Lowland.

Nationally Scarce. Uncommon and local on firm, raised areas of saltmarsh, where sand or gravel are mixed with the mud, and usually where well-vegetated saltmarsh gives way abruptly to bare ground, often near a channel. ❀ Both flowering and non-flowering shoots arise from a creeping rootstock to form *large tussocks*. DETAIL Stems woody at base, often *sprawling*, rooting at nodes. Mature plants turn yellowish to reddish. Flowers in groups of 3, about equal in size. Known as 'Sheep's Samphire', this species is not edible.

Annual Seablite *Suaeda maritima*

Abundant on saltmarshes, where often a pioneer species, growing on relatively bare mud together with a variety of glassworts. Also found on muddy shingle and other bare, wet, brackish ground. ✿ Very variable, sometimes prostrate, sometimes upright, and often superficially similar to glassworts and similarly turning red in the autumn, but unlike glassworts the leaves are clearly distinct from the wiry stems. DETAIL Stem often rather woody at base. Leaves semicircular in cross-section, 3–25mm long, pointed at tip. Flowers tiny, 1–2mm across, with 5 succulent green sepals, 5 stamens and 2 stigmas. SIMILAR SPECIES Young plants of Shrubby Seablite (p. 394) have leaves that are circular in cross-section.

GROWTH Annual.
HEIGHT Stems to 30cm.
FLOWERS Jul–Sep.
STATUS Native.
ALTITUDE Lowland.

Sea Beet *Beta vulgaris* subspecies *maritima*

Common. The margins of saltmarshes, tidal rivers and sea walls, particularly around the nutrient-rich debris marking the limit of the highest tides. ✿ The large, glossy, leathery leaves, red-striped stems and slender, leafy spikes of tiny, fleshy flowers are distinctive.

DETAIL Hairless. Stems often tinged red, usually well branched, longer stems often collapsing. Lower leaves 4–40cm long, broadly triangular, blunt, tapering into a long stalk; upper leaves becoming smaller, narrower and stalkless. Flowers bisexual, with 2 stigmas and 5 green to reddish tepals. SIMILAR SPECIES Sugar Beet subspecies *vulgaris* is widely cultivated for its large, swollen root.

GROWTH Perennial.
HEIGHT To 100cm.
FLOWERS Jun–Sep.
STATUS Native.
ALTITUDE Lowland.

Indian Balsam
Impatiens glandulifera

Locally abundant alien. The banks of streams, rivers and canals, wet woodland and marshes. ❂ Tall, with large pinkish-purple to white flowers. The seed capsules explode when ripe to shoot the seeds up to 7m, and popping the nearly ripe seed heads is a favourite amongst children. DETAIL Hairless, with rather fleshy, reddish stems, swollen at the nodes. Leaves opposite or in whorls of 3, oval, pointed, 6–15cm long, with finely-toothed edges (24–75 teeth per side), the teeth often tipped with red glands. Flowers 25–40mm long, enclosed at the base by a brightly-coloured, pouch-like lower sepal that ends in a short spur that is bent through at least 90°.

GROWTH Annual.
HEIGHT 100–200cm.
FLOWERS Jul–Oct.
STATUS Introduced to gardens from the Himalayas in 1839. First recorded in the wild in 1855. Has increased rapidly in recent decades.
ALTITUDE Lowland (320m).

Touch-me-not Balsam
Impatiens noli-tangere

Nationally Scarce. Very local in central Wales and the Lakes, introduced elsewhere. Shady streamsides and seepages in woodland. ❂ Large yellow flowers, marked with small brown spots and with a *long spur bent at c. 90°*, distinctive. DETAIL Stems yellowish-green to reddish, often swollen at nodes. Leaves alternate, 4–12cm long, with 6–11 teeth per side. Flowers mostly 20–35mm long, with the pouch-like lower sepal tapered to the 6–12mm-long spur.

GROWTH Annual.
HEIGHT 40–70cm.
FLOWERS Jul–Sep.
STATUS Native.
ALTITUDE Lowland.

Small Balsam
Impatiens parviflora

Local. Dry, shady places beside paths and tracks in woodland and along rivers. ❂ The small, pale yellow, spurred flowers are distinctive. DETAIL Stems green, swollen at nodes. Leaves alternate, 5–15cm long, finely toothed with 20–30 teeth per side. Flowers mostly 6–18mm long, with the pouch-like lower sepal gradually tapered to a straight spur.

GROWTH Annual.
HEIGHT 30–60cm (100cm).
FLOWERS Jul–Oct.
STATUS Introduced in 1823 from central Asia. First recorded in the wild in 1851 and still spreading.
ALTITUDE Lowland.

BALSAMS: FAMILY BALSAMINACEAE **179**

Orange Balsam *Impatiens capensis*

GROWTH Annual.
HEIGHT 40–70cm (150cm).
FLOWERS Jul–Sep.
STATUS Introduced from N
America. First recorded in the
wild in 1822.
ALTITUDE Lowland.

Locally common, and slowly expanding its range. Marshy ground, sometimes shaded, by streams, rivers and canals. ❁ The orange flowers, heavily blotched reddish, are unique. Detail Stems brownish, swollen at the nodes. Leaves 3–9cm long with 8–12 teeth per side. Flowers mostly 20–35mm long, the pouch-like lower sepal ending in a spur that is bent double.

JACOB'S-LADDER, PHLOX: FAMILY POLEMONIACEAE

Jacob's Ladder
Polemonium caeruleum

Nationally Rare, but very locally common on lightly shaded, N-facing, steep rocky slopes on limestone in the Peak District and Yorkshire Dales, also riverbank clays in Northumberland. Scarce but more widespread as a garden escape. ❁ Tall, with clusters of blue flowers with prominent orange stamens. Detail Stems erect, unbranched, angled, glandular-hairy towards tip. Leaves sometimes sparsely hairy, alternate, 10–40cm long, cut into 6–12 pairs of leaflets. Flowers 2–3cm across, calyx cut into 5 sepals, corolla cut into 5 petals, 5 stamens and a single style with 3 stigmas at the tip; flowers occasionally white.
Similar Species
Garden varieties have shorter, more oval leaflets, no more than 2.7 times as long as wide (at least 3 times as long in wild plants).

GROWTH Perennial.
HEIGHT 30–100cm.
FLOWERS Jun–Jul (Sep).
STATUS Native.
ALTITUDE 190–580m.

Springbeauty
Claytonia perfoliata

Locally common on disturbed ground on sandy soils, sometimes in light shade: arable fields, gardens, verges. ❀ The 2 stem leaves are fused to form a *unique, saucer-like structure below the cluster of small white flowers.* DETAIL Hairless, slightly fleshy. Basal leaves 1–3cm long, oval, long-stalked. Flowers 5–8mm across, petals 5, sometimes slightly notched at tip, sepals 2, stamens 5, fruit a 1-seeded capsule.

GROWTH Perennial.
HEIGHT 10–30cm.
FLOWERS Apr–Jul.
STATUS Introduced to gardens from N America in 1794. First recorded in the wild in 1849.
ALTITUDE Lowland.

Pink Purslane
Claytonia sibirica

Locally common. Damp, bare ground, usually in light shade. ❀ Low-growing and rather fleshy, with long-stalked, long-oval leaves, often tinted bronze, and stalked clusters of pink (occasionally white), 5-petalled flowers. DETAIL Hairless. Basal leaves long stalked, 1–5cm long with 3(5) parallel veins; stem leaves 2, opposite, stalkless. Flowers 15–20mm across. Sepals 2, petals notched at tip, stamens 5, fruit a 1-seeded capsule.

GROWTH Annual to perennial.
HEIGHT 15–40cm (60cm).
FLOWERS Apr–Jul.
STATUS Introduced to gardens by 1768. First recorded in the wild in 1838, still spreading (E Asia and N America).
ALTITUDE To 425m.

Blinks *Montia fontana*

Locally common. Bare, muddy ground on pastures and around tracks, ditches, streams, lakeshores, flushes and springs, also sandy and gravelly places that are wet in winter; avoids chalky soils. ❀ Varies from spreading mats to small, compact tufts; in water has long, floating stems. Rather *fleshy*, with pairs of small, stalked, oval leaves, often joined around the stem at their base, and clusters of 1–3 *tiny flowers, 2–3mm across*, each with *2 sepals forming a cup* and 5 white petals; *flowers seldom open widely.* DETAIL Stems often reddish. Leaves 4–15mm long. Petals less than 2mm long. The sepals persist around the ripe capsule, which has 3 blackish seeds variably covered with minute warts. SEE ALSO Water starwort (p. 348).

GROWTH Annual to perennial.
HEIGHT 1–20cm (50cm).
FLOWERS Mar–Oct.
STATUS Native.
ALTITUDE 0–995m.

Chickweed Wintergreen *Trientalis europaea*

Rather local. Damp, mossy woods of oak, birch and pine, also moorland. Seed rarely produced and seldom colonises new sites. ❀ *Short and delicate with 1–2 star-like flowers.* DETAIL Leaves 5–8, hairless, shiny, well veined, 2–8cm long, arranged in a ring at the tip of the stem (there may be 1–2 smaller leaves lower down). Flowers on long reddish stalks, 15–18mm across, corolla divided almost to the base into 7(5–9)

GROWTH Perennial.
HEIGHT To 25cm.
FLOWERS Jun–Jul.
STATUS Native.
ALTITUDE To 1100m.

petals; anthers orange-yellow. A member of the primula family and not related to other 'wintergreens'.

Cyclamen ('Sowbread') *Cyclamen hederifolium*

A popular garden plant that is an increasing garden throw-out, a relict of old gardens or deliberately planted: woods, churchyards and other dry, shady places. ❀ Flowers appear in early autumn and are held 'upside-down' on long bare stalks. DETAIL Grows from an underground corm. Leaves 4–10cm long, 5–9 sided with a heart-shaped base, a band of silvery

GROWTH Perennial.
HEIGHT To 10cm.
FLOWERS Aug–Sep, before the leaves.
STATUS Introduced from S Europe sometime before 1596 and recorded from the wild from 1597.
ALTITUDE Lowland.

marbling above and purplish below; summer dormant, leaves are only present Sep–May. Flowers nodding, divided nearly to the base into 5 pale pink, twisted petals, 3–8m long that point backwards (i.e. upwards); the petals have conspicuous bosses at the base. After flowering the flower stalk coils onto the soil's surface to release the ripe seeds, which may be carried long distances by ants. SIMILAR SPECIES Other garden cyclamens are spring flowering.

Primrose
Primula vulgaris

GROWTH Perennial.
HEIGHT To 12cm.
FLOWERS Mar–May
(Dec–Jun).
STATUS Native.
ALTITUDE 0–850m.

Common. Deciduous
woodland, hedgerows,
roadside banks and
old grassland. Especially in the E, does best in light shade on moist, heavy
soils. ✿ An iconic spring flower, slightly scented and sometimes patch-
forming. **DETAIL** Leaves hairless above except along midrib, evergreen,
strongly wrinkled, more or less unstalked and *tapering gradually to the
base*. Flowers borne *singly*, on *shaggy-hairy stems*, saucer-shaped, *c.* 30mm
across with 5 pale yellow petals and an orange-yellow 'eye'. Pink-flowered
forms occur near houses (suggesting hybridisation with garden primulas),
but sometimes also far from any habitation.

pin

Most primulas have flowers of two types: 'pin-eyed', with a pinhead-like stigma visible in the
throat of the flower and the shorter stamens hidden below, and 'thrum-eyed', in which this is
reversed and the long stamens are visible but the short stigma is concealed. Their pollen also
differs. Cross-pollination between flowers of the two different types produces far more seed than
pollination by a flower of the same type.

thrum

Bird's-eye Primrose
Primula farinosa

Nationally Scarce and
listed as Vulnerable.
A beautiful primula,
confined to the N
Pennines and adjacent
areas of N England,
where very locally
common in wet,
spring-fed grassy
places with plenty of

GROWTH Perennial.
HEIGHT To 15cm.
FLOWERS May–Jun (Jul).
STATUS Native.
ALTITUDE 0–570m, but
mostly 200–400m.

bare ground, mostly on limestone.
DETAIL Flowers 7–15mm across, several together
at the tip of the mealy-white stem. Leaves short
(2–6cm), mealy-white below. **SIMILAR SPECIES**
Scottish Primrose *P. scotica*, a Scottish endemic,
is confined to Orkney and the coast of N
Scotland. Shorter, with darker purple flowers and
no gaps between the petals, it flowers May–Jun
and again, after a break, Jul–Aug.

Cowslip *Primula veris*

Locally common. Species-rich grassland, usually on calcium-rich soils. Hedgebanks, ancient earthworks, churchyards, scrub and woodland rides. Often sown on the verges of new trunk roads, but seldom now found in grassland or meadows unless managed for conservation.
❀ Note clusters of up to 30 subtly sweet-scented flowers, hanging more or less to one side. **DETAIL** Leaves grouped at the base of the stem, 5–15cm long, *narrowing abruptly around halfway towards the base,* evergreen, wrinkled. Flowers cup-shaped, 10–15mm across, base of petals with a *fine notch* and distinct orange spot. *Stem with short straight hairs.*

GROWTH Perennial.
HEIGHT To 30cm.
FLOWERS Apr–May.
STATUS Native.
ALTITUDE 0–750m.

Oxlip *Primula elatior*

Nationally Scarce and listed as Near Threatened. Locally common in ancient woodland on chalky boulder clay (rarely old hedgerows) in N Essex, W Suffolk, Cambs, Beds and Hunts, and a few woods on the Herts-Bucks border.
❀ *Flowers larger, paler yellow*, and *opening more widely than Cowslip.* Stems *shaggy hairy.* Leaves similar but usually *hairier, especially towards the base.* **DETAIL** Flowers mostly all facing the same way, 15–20mm across. Lacks notch at base of petals and calyx lobes more pointed, with dark green midribs (not uniformly green). Seed capsule at least as long as calyx. **SIMILAR SPECIES False Oxlip** *P. × polyantha*, the hybrid between Primrose and Cowslip, occurs sporadically. It has larger, richer yellow flowers that face in various directions, a notch at the base of the petals (as Cowslip), and leaves that taper to the base (as Primrose).

GROWTH Perennial.
HEIGHT 10–30cm.
FLOWERS Apr–May.
STATUS Native.
ALTITUDE Lowland.

Brookweed
Samolus valerandi

Very locally common. Bare, wet ground, mostly brackish or over chalk and limestone: flushed sea cliffs, dune slacks, spring-fed fens, and the edges of streams, ditches and ponds. ❀ The *daisy-like leaves* form a basal rosette and are also arranged alternately on the lower stem; the flower-spike elongates as the season advances, with small, bell-shaped, white flowers at the tip and pea-like capsules below. **DETAIL** Hairless. Leaves spoon-shaped, 1–8cm long. Flowers 2–4mm across, with 5 petals, sepals and stamens and a single stigma; each flower stalk has a tiny leaf-like bract. **SIMILAR SPECIES** An enigmatic, hard-to-place species that may recall a crucifer (member of the cabbage family), but has 5 petals rather than 4.

GROWTH Perennial.
HEIGHT To 45cm, but often rather shorter.
FLOWERS Jun–Sep; most plants die after flowering.
STATUS Native.
ALTITUDE Lowland.

Sea Milkwort *Glaux maritima*

Common around all coasts and on the banks of tidal rivers, in short turf and damp bare places: dune slacks, damp sand, shingle and mud, spray-drenched cliffs and crevices and the drier parts of saltmarshes. Very rare inland around salt springs in Worcs and Staffs. ❀ Small and low-growing, sometimes found in dense patches, sometimes straggling through other vegetation. Stems and leaves fleshy, flowers small and pinkish. **DETAIL** Hairless. Leaves opposite, stalkless, 4–12mm long. Flowers solitary, stalkless, *c.* 5mm across, calyx 5-lobed, pale pink with a reddish centre (sometimes whitish); petals absent. 1 stigma and 5 stamens. **SEE ALSO** Sea Sandwort (p. 165).

GROWTH Perennial.
HEIGHT Shoots 10–30cm.
FLOWERS May–Aug.
STATUS Native.
ALTITUDE Lowland.

Bog Pimpernel
Anagallis tenella

Locally common. Short, permanently wet vegetation around springs and flushes with areas of bare mud or peat: fens, dune-slacks, damp pastures and bogs. ❀ Small, with slender, creeping, stems flat to the ground and pairs of rounded leaves at regular intervals, resembling a necklace. The exquisite flowers grow singly on slender upright stalks and often only open in the sun. DETAIL Hairless. Stems rooting at the nodes, often reddish. Leaves 5mm long, evergreen. Flowers bell-shaped, 10–14mm across, the 5 white petals have fine red veins and the 5 stamens have hairy, bottle-brush-like filaments.

GROWTH Perennial.
HEIGHT Stems to 20cm.
FLOWERS May–Sep.
STATUS Native.
ALTITUDE 0–610m.

Scarlet Pimpernel *Anagallis arvensis*

A common weed of arable fields and gardens, also dunes, shingle banks, cliffs and heathland. ❀ Low-growing, flowers red with a reddish-purple base (occasionally white, pink or blue), and only opening fully in sunshine. DETAIL Hairless. Stems square, sprawling, to 40cm long. Leaves opposite, 10–15mm long, unstalked. Flowers on stalks up 3.5cm long, solitary, 10–15mm across, with 5 petals, 5 sepals and 5 stamens.

GROWTH Usually annual.
HEIGHT To 15cm.
FLOWERS Apr–Oct.
STATUS Native.
ALTITUDE 0–320m.

Chaffweed
Centunculus minimus

Near Threatened. Very local. Bare, damp, sandy ground, usually on acid soils and often near the sea: dunes, heathland tracks, forest rides. ❀ *Tiny*, the *pinkish, globular seed capsules* are more obvious than the flowers. DETAIL Hairless. Leaves oval, 3–5mm long, near stalkless, bordered black below; *upper leaves alternate*. Flowers solitary, in the leaf axils, less than 2mm across, the 5 white to pink corolla lobes much shorter than the 5 narrowly triangular calyx lobes. SEE ALSO Allseed (p. 94).

GROWTH Annual.
HEIGHT 2–5cm (8cm).
FLOWERS Jun–Jul (Aug).
STATUS Native.
ALTITUDE Lowland.

GROWTH Perennial.
HEIGHT Stems to 40cm.
FLOWERS May–Sep.
STATUS Native.
ALTITUDE 0–820m.

Yellow Pimpernel
Lysimachia nemorum

Fairly common in deciduous woodland and other moist, shady places, usually on mildly acid soils, in the N also by moorland streams and on cliffs. ✿ Delicate, low-growing or sprawling with solitary, star-like flowers. **DETAIL** Leaves opposite, oval, 1–4cm long, *more or less pointed*, hairless, bright shiny green. Flowers on slender stalks arising from the leaf axils, 10–15mm across, corolla divided almost to the base into 5 petals, sepals green, *narrowly triangular or strap-shaped, pointed*. **SIMILAR SPECIES** Buttercups have several flowers per stem and, in most species, divided leaves. **SEE ALSO** Yellow Star of Bethlehem (p. 331).

Creeping Jenny *Lysimachia nummularia*

GROWTH Perennial.
HEIGHT Stems to 60cm.
FLOWERS Jun–Aug. Seed is rarely set but stem fragments can grow into new plants.
STATUS Native.
ALTITUDE 0–570m.

Locally fairly common. Damp, shady places in woods, wet grassland and near water, often on clay soils; also grown in gardens and frequently escaping. ✿ Rather like Yellow Pimpernel, with yellow, cup-shaped flowers, but has much more obviously *prostrate, creeping stems*. **DETAIL** As Yellow Pimpernel but *leaves broader and more rounded, with blunt tips*, flower stalks stouter, flowers larger, 15–25mm across, minutely spotted with numerous yellow and orange glands (especially near margins, and forming a *fine fringe to the edge*), sepals much broader – triangular-oval; leaves and sepals with a scatter of tiny, inconspicuous black gland-dots.

Yellow Loosestrife

Lysimachia vulgaris

Fairly common. Marshes, the margins of rivers, streams, ditches and ponds, on wet, peaty soils. Spreads via rhizomes to form extensive patches. ❀ *Tall*, with clusters of yellow flowers. **Detail** Softly-hairy, leaves stalkless, opposite or in whorls of 3–4 around stem, oval-lanceolate, 3–12cm long, with tiny orange gland-dots along edges and scattered on both sides. Flowers 15–20mm across, the 5 petals have tiny yellow gland-dots and hairless margins, sepals (calyx-teeth) glandular-hairy with conspicuous narrow golden fringes.

GROWTH Perennial.
HEIGHT 60–150cm.
FLOWERS Jul–Aug.
STATUS Native.
ALTITUDE Lowland.

Dotted Loosestrife

Lysimachia punctata

Often thrown out with garden rubbish and naturalised in damp places on rough ground, woodland edges and verges. ❀ Similar to Yellow Loosestrife but more often found away from water, flowers larger, with an *orange-brown centre, petals glandular-hairy along margins* and *sepals uniformly green*. **Detail** Stem more distinctly ridged and more densely hairy than Yellow Loosestrife. Leaves short-stalked (4–15mm), lacking gland-dots on upper surface.

GROWTH Perennial.
HEIGHT To 120cm.
FLOWERS Jul–Oct.
STATUS Introduced to cultivation from E Europe by 1658. Recorded from the wild from 1853 and still increasing.
ALTITUDE To 460m.

Tufted Loosestrife

Lysimachia thyrsiflora

Nationally Scarce, confined as a native to the central lowlands of Scotland and a few sites in Yorkshire. Shallow water in fens, ditches, lakes etc. ❀ Dense sprays of small flowers grow from the base of the mid-stem leaves. **Detail** Nearly hairless. Leaves and narrow petals with many fine black gland-dots, the long-stalked anthers are orange-yellow. A shy flowerer.

GROWTH Perennial.
HEIGHT To 70cm.
FLOWERS Jun–Jul.
STATUS Native.
ALTITUDE To 310m.

Dwarf Cornel *Cornus suecica*

Near Threatened. Locally common on upland moors, on acid peat amidst Heather, Bilberry, etc. ✿ Low-growing, with Dogwood-like leaves and clusters of tiny purple flowers set off by 4 whitish bracts, 5–8mm long. Fruit a bright red berry, 5–10mm across. DETAIL Grows from a rhizome. Stems 4-angled, variably hairy. Leaves opposite, oval, pointed, 1–3cm long, hairy above (hairs flat to the surface, attached in the centre), with 3–5 conspicuous veins, greyish and hairless below; more or less stalkless. Flowers 2–4mm across, with 4 sepals, petals and stamens, in clusters of 8–25.

FAMILY ERICACEAE: HEATHERS, WINTERGREENS ETC.

Usually described as 'saprophytic', in other words living on decaying organic matter, but recent research shows that it is actually 'epiparasitic', using fungi of the genus *Tricholoma* to extract nutrients from living trees.

Yellow Birdsnest *Hypopitys monotropa*

Scarce and local, and listed as Endangered. Bare woodland floors under Beech and Hazel on chalky soils and pines on acid soils, also damp dune slacks with Creeping Willow. ✿ Lacks chlorophyll and is *uniformly dull, pale yellow*. Flower spikes nodding, becoming erect in fruit. DETAIL Stems usually hairy above. Leaves small and scale-like. Flowers hairless on outside, variably hairy inside, sepals 4–5, petals 4–5, 8–13mm long. SEE ALSO Broomrapes (pp. 244–245), Birdsnest Orchid (p. 315).

Common Wintergreen *Pyrola minor*

Local and uncommon.
Damp woodland and,
in the N, damp places
on heaths, rocky ledges
and, rarely, dunes.
❁ The long-stalked
spikes of *delicate,
small, globular
pinkish-white flowers*
are distinctive but
surprisingly hard to
spot. DETAIL Leaves
evergreen, all near base
of stem, 2.5–6cm long,
more or less rounded,
with wavy, finely
toothed edges; stalks
shorter than blades.
Flowers 4–7mm
across. Petals 5, *style
1–2mm long, enclosed
by the petals and
not normally visible*,
stigma 5-lobed. SIMILAR
SPECIES Intermediate

GROWTH Perennial.
HEIGHT 10–20cm (30cm).
FLOWERS Jun–Aug.
STATUS Native.
ALTITUDE 0–1130m.

Wintergreen *P. media* has larger flowers with a longer style (4–6mm) that
projects slightly. Common in the far N, rare to S.

Round-leaved Wintergreen
Pyrola rotundifolia

Nationally Scarce.
Very local in fens,
dunes slacks and
associated plantations,
and old quarries
flushed with chalky
water. In the N also
pine woodland, damp
moorland gullies and
cliff ledges in the
mountains.
❁ Long-stalked spikes
of *open, saucer-shaped
whitish flowers*,
8–12mm across, with
a distinctive *subtly
S-shaped style*.
DETAIL Leaves as
Common Wintergreen
but rounder and longer
stalked. Style 4–10mm
long.

GROWTH Perennial.
HEIGHT 10–30cm.
FLOWERS Jun–Sep.
STATUS Native.
ALTITUDE 0–760m.

Lady's Bedstraw *Galium verum*

GROWTH Perennial.
HEIGHT Stems 15–60cm
(100cm).
FLOWERS Jun–Aug (Sep).
STATUS Native.
ALTITUDE 0–780m.

Honey-scented when fresh
but smells of new-mown
hay when dry. Formerly
believed to discourage
fleas and was incorporated
into straw mattresses,
especially for the beds of
women about to give birth,
hence its name.

Common. Short grassland on well-drained, poor soils, especially over chalk, limestone and sands, including verges, cliffs, dunes and machair. ✿ Low growing and sprawling, with *frothy masses of tiny, bright yellow, 4-petalled flowers*. DETAIL Stems smooth, subtly 4-angled, hairless to sparsely hairy. Leaves in whorls of 6–12, strap-shaped, with down-rolled margins and a tiny fine point at the tip, slightly rough above, hairy below, to 25mm long. Flowers 2–4mm across. Fruit 2 fused nutlets, hairless, smooth to minutely wrinkled.

Crosswort *Cruciata laevipes*

GROWTH Perennial.
HEIGHT 15–60cm.
FLOWERS Late Apr–Jun.
STATUS Native.
ALTITUDE 0–570m.

Common. Rough, ungrazed grassland on moist but well-drained soils: verges, hedgebanks, scrub, woodland rides. ✿ *Densely but softly hairy*, with yellowish-green leaves in whorls of 4 and *clusters of small, fragrant, bright yellow flowers at the base of the leaves*. DETAIL Patch-forming, stems spreading then erect. Leaves 10–25mm, oval, 3-veined. Flowers 2–3mm across. Fruit hairless, smooth, ripening to black.

Cleavers *Galium aparine*

A familiar weed of gardens, arable land, hedgebanks and waste ground, favouring fertile soils, also fens, scree and shingle beaches. ❁ Sprawling or scrambling, the bristly leaves and stems cling to clothing and fur. Flowers tiny. DETAIL Stem with *hooked, downward-pointing bristles on angles.* Leaves 10–60mm long, in whorls of 6–8: bristly-hairy, *edged with backward-pointing prickles, and with a fine point at the tip* (to 1.5mm). Flowers just 1–2mm across, in stalked clusters in the leaf axils. Fruit 2 fused nutlets *with numerous hooked bristles.*

GROWTH Annual.
HEIGHT Shoots to 300cm.
FLOWERS Apr–Sep.
STATUS Native.
ALTITUDE 0–460m.

Hedge Bedstraw
Galium album

Common. Rough grassland on verges, hedgebanks, churchyards, scrub, woodland edges, avoiding wet or very acid soils. ❁ Sprawling to erect, with *large sprays of small white flowers.* DETAIL Stems well branched, *smooth*, square, variably downy. Leaves in whorls of 5–10, strap-shaped, 5–30mm long, with fine, more or less forward pointing prickles on the edges and *a short, fine point at the tip.* Flowers 3–4mm across, *petals tapering to a fine point.* Fruit 2 fused nutlets, hairless, *minutely wrinkled.*

GROWTH Perennial.
HEIGHT Stems to 100cm (150cm).
FLOWERS Jun–Sep.
STATUS Native.
ALTITUDE 0–845m.

Northern Bedstraw
Galium boreale

Local in damp, grassy and rocky places in the hills, often near water, especially on base-rich soils where protected from grazing, also dunes. ❁ The only white-flowered bedstraw with *3-veined leaves in whorls of 4 around the stem.* DETAIL Stems 4-angled, smooth, hairy (at least towards base). Leaves dark green, 15–40mm long, narrowly elliptical, blunt, hairy to hairless. Flowers 2–5mm across. Fruit 2 fused nutlets with many tiny hooked bristles.

GROWTH Perennial.
HEIGHT 20–45cm.
FLOWERS Jul–Sep.
STATUS Native.
ALTITUDE 0–1065m.

Heath Bedstraw
Galium saxatile

Locally abundant. Grassy places on poor, acid soils: heathland, woodland rides, unimproved hill pastures, etc. ✿ Forms dense *mats*, often with many non-flowering shoots. DETAIL Hairless. Stems *smooth*, 4-angled. Leaves in whorls of 5–8, *short* (5–11mm long), *with a fine point at the tip and forward-pointing prickles on the edges.* Flowers *c.* 3mm across, petals blunt or with a tiny point at the tip. Fruit 2 fused nutlets, *finely warted.*

GROWTH Perennial.
HEIGHT 10–30cm.
FLOWERS May–Aug.
STATUS Native.
ALTITUDE 0–1220m.

Limestone Bedstraw *Galium sterneri*

Locally common. Short grassland over limestone and other basic rocks, also scree, limestone pavement and other rocky places. Mat-forming, with abundant non-flowering shoots. ✿ Hairless, with smooth, 4-angled stems and curved, backward-pointing prickles on the leaf edges (at least at base). *Note range.* DETAIL Leaves in whorls of 5–8, strap-shaped, 3–10mm long, with a fine point at tip. Flowers 2–3mm across, petals blunt or with a tiny point at the tip. Fruit finely warted.
SIMILAR SPECIES Slender Bedstraw *G. pumilum* is very similar, but has blunter warts on the fruits. Rare, only S of a line from the Severn to the Thames.

GROWTH Perennial.
HEIGHT 10–20cm (30cm).
FLOWERS Jun–Jul.
STATUS Native.
ALTITUDE To 975m.

Squinancywort
Asperula cynanchica

Very locally abundant in short, dry, species-rich turf over chalk and limestone (mostly now reserves and road verges), also calcareous dunes. ✿ Ground-hugging, with clusters of tiny, 4-petalled, *pinkish flowers. Note habitat.*
DETAIL Stems sprawling, 4-angled, often with short, rough hairs.
Leaves in whorls of 4, hairless, pointed or with a fine point at tip, up to 10(20)mm long, *usually unequal in length, with 2 long and 2 short.* Flowers 3–4mm across, petals about as long as corolla tube, inner surface very pale pink, flushed and lined deep pink; outer surface white to pale lilac-pink. Fruit 2 fused nutlets, finely warted. SEE ALSO Field Madder (p. 194).

GROWTH Perennial.
HEIGHT 5–20cm (50cm).
FLOWERS Jun–Aug (Sep).
STATUS Native.
ALTITUDE 0–305m.

Marsh Bedstraw *Galium palustre*

Very common in a wide variety of marshy places, usually on neutral to mildly acid soils (generally commoner than Fen Bedstraw, and often in wetter places). ❀ Scrambling, but not especially rough to the touch, with open clusters of small white flowers; *leaves rounded, without a bristle at the tip.* DETAIL Hairless. Stem 4-angled, with tiny prickles on angles (rarely smooth), and often purplish. Leaves 5–35mm long, in whorls of 4–8, edged with sparse, weak prickles. Flowers 3–4.5mm across. Fruit a pair of nutlets, slightly wrinkled.

GROWTH Perennial.
HEIGHT Stems to 100cm.
FLOWERS Jun–Aug.
STATUS Native.
ALTITUDE 0–825m.

Fen Bedstraw *Galium uliginosum*

Locally common. Marshes and tall-herb fens, usually where the ground water is alkaline. ❀ Much like Marsh Bedstraw but *rough to the touch and leaves with an obvious bristle-point at the tip* (at least 0.5mm long). DETAIL Hairless. Stems scrambling, often purplish, *very rough,* with down-turned prickles on the 4 angles. Leaves 7–20mm long, in whorls of 5–8, fringed with strong curved prickles. Flowers 2.5–3mm across. Fruit a pair of nutlets, finely warted.

GROWTH Perennial.
HEIGHT Stems 10–60cm.
FLOWERS (Jun) Jul–Aug.
STATUS Native.
ALTITUDE 0–750m.

Woodruff *Galium odoratum*

Locally common. Deciduous woodland (especially ancient woods), scrub and shady hedge banks, usually on fairly damp, neutral to calcareous soils. ❀ Creeping and patch-forming, with numerous *erect,* unbranched stems tipped by umbel-like clusters of white flowers. The dark green, slightly glossy leaves are hay- or vanilla-scented when bruised. DETAIL Leaves in whorls of 6–8(9), long-oval, up to 40mm long, pointed, hairless, with fine forward-pointing prickles along the margins. Stems 4-angled, smooth, hairy below each whorl. Corolla 4–6mm long, *cut halfway to base, forming 4 petals.* Fruit a pair of nutlets with *abundant hooked bristles.*

GROWTH Perennial.
HEIGHT 15–30cm (45cm).
FLOWERS (Apr) May–Jun.
STATUS Native.
ALTITUDE 0–640m.

Field Madder
Sherardia arvensis

GROWTH Annual.
HEIGHT 10–30cm (40cm).
FLOWERS May–Oct.
STATUS Native.
ALTITUDE To 365m.

A common arable weed, especially on light soil; often in stubble. Also dry, disturbed ground on poor grassland, dunes, and sheltered rocky places, especially near the coast. ❀ Low growing, *prickly-hairy*, with clusters of tiny, *pale pinkish-mauve flowers*. DETAIL Stems with stiff, downward-pointing hairs on angles. Leaves in whorls of 4–6, pointed, 5–18mm long, with prickles on edges. Flowers in dense, stalked clusters of 4–10 with a *ruff of leaf-like bracts immediately below*. Calyx 0.5–1mm (minute in *Galium* bedstraws), cut into 4–6 sepals, corolla 4–5mm long, with a long, slender tube with 4 spreading petals at the tip. Fruit 2 fused nutlets, 4mm long, each with the remains of the calyx on top.

Wild Madder *Rubia peregrina*

GROWTH Perennial.
HEIGHT Stems to 150cm.
FLOWERS Jun–Aug.
STATUS Native.
ALTITUDE Lowland.

Rather local. Scrubby, often rocky, places near the sea, also a few localities on limestone inland. ❀ Scrambling evergreen with *strong, curved prickles on the stem angles and edges of the shiny, leathery, dark green leaves, and on the midrib below*. Open clusters of small yellowish-green 5-petalled flowers are followed by *black berries*. DETAIL Hairless. Stems woody near base, sharply 4-angled. Leaves 10–60mm long, with 1 vein, in whorls of 4–6. Flowers 4–6mm across, calyx minute, corolla with a very short tube. Fruit *c.* 5mm across, with 1 seed.

Common Centaury
Centaurium erythraea

Locally common. Dry, sparsely-vegetated places: grassland, grassy heaths, dunes and woodland rides. ❂ Upright, often well branched, with dense heads of *bright pink flowers* 10–12mm across that only open fully in bright sunshine. On the coast, however, may be *extremely short*, with crowded heads of tiny flowers, *c.* 6–7mm across. DETAIL *Basal rosette usually present when flowering*, leaves oval, blunt, stalkless, to 5 × 2cm. *Stem leaves more or less oval with a pointed tip and 3(5) veins on underside.* Flowers almost stalkless (stalk separating base of calyx from bracts below 0–1mm long; bracts often tiny). Corolla lobes 4.5–6mm long, calyx usually less than 0.75 times as long as corolla tube. *Stigmas narrowly rounded to nearly conical at tip.*

FLOWERS Jun–Oct.
STATUS Native.
ALTITUDE 0–435m.
GROWTH Biennial (annual).
HEIGHT 10–40cm (2–50cm).
FLOWERS Jun–Oct.

Seaside Centaury
Centaurium littorale

Nationally Scarce. Rather local. Dunes, the upper saltmarsh and other sparse calcareous turf by the sea. ❂ Very like Common Centaury but flowers more bluish-pink, leaves sometimes more leathery. DETAIL Rosette leaves usually present when flowering, these and stem leaves *strap-shaped, near parallel-sided*, to 2 × 0.5cm, with a *blunt tip* and *1 (0–3) vein below*. Flowers near stalkless (as Common Centaury). Corolla lobes 5–6.5mm long, calyx usually more than 0.75 times as long as corolla tube. *Stigmas broadly rounded to nearly flat at tip.*

GROWTH Biennial.
HEIGHT To 25cm.
FLOWERS Jul–Aug
(Jun–Sep).
STATUS Native.
ALTITUDE Lowland.

Lesser Centaury
Centaurium pulchellum

Local. Sparse, dry grassland: grassy heaths, woodland rides, coastal dunes and the upper saltmarsh. ❂ Short to very short: larger plants with *widely-branching, rather open flower heads*, the smallest unbranched, sometimes with just 1 flower. Flowers very small, 4–8mm across, deep pink, with an *obvious stalk, 1–4 mm long, separating the calyx from the bracts below*. DETAIL *Annual, usually without a basal rosette at flowering*. Lower leaves more or less oval, blunt, to 2 × 1cm, with *1–3 weak veins*, becoming narrower and often more pointed above. Corolla lobes 2–4mm.

GROWTH Annual.
HEIGHT 2–15cm (20cm).
FLOWERS Jun–Sep.
STATUS Native.
ALTITUDE Lowland.

Marsh Gentian
Gentiana pneumonanthe

Nationally Scarce. Very local on damp heathland. ❀ Flowers large, trumpet-shaped, bright blue, striped with green on the outside and spotted with silvery-green inside. DETAIL Stem leafy (basal leaves usually withered by flowering), leaves 15–40mm long, strap-shaped, with 1 vein. Calyx with 5 lobes. Corolla 25–40mm long, cut into 5 large triangular lobes at the tip, alternating with 5 small lobes. Stamens 5.

GROWTH Perennial.
HEIGHT 10–40cm.
FLOWERS Mid Jul–Sep.
STATUS Native.
ALTITUDE Lowland.

Spring Gentian
Gentiana verna

Nationally Rare. Confined to limestone grassland in Upper Teesdale and the Burren, W Ireland. ❀ The brilliant, gentian-blue flowers are distinctive. DETAIL Leaves oval, 18–20mm long, mostly in a basal rosette. Flowers solitary, 17–31mm across, with corolla 15–25mm long, cut into 5 large and 5 small lobes.

GROWTH Perennial.
HEIGHT 2–8cm (0.5cm).
FLOWERS Apr–Jun.
STATUS Native (Sch 8).
ALTITUDE To 730m.

Yellow-wort
Blackstonia perfoliata

Locally common in dry, open grassland over chalk and limestone and on calcareous dunes. ❀ Foliage distinctly grey-green, with *pairs of triangular-oval leaves fused around the stem*. The flowers only open in sunshine. DETAIL Hairless. Basal rosette often withering by flowering time. Flowers 10–15mm across, calyx cut almost to the base into 6–8 narrow sepals, corolla cut into 6–8 petals, each 5–10mm long (longer than tube). Stigmas 2, style 1.

GROWTH Annual (biennial).
HEIGHT 15–50cm.
FLOWERS Jun–Sep (Oct).
STATUS Native.
ALTITUDE Lowland.

Autumn Gentian
Gentianella amarella

Locally common in short, dry turf over chalk and limestone, also calcareous dunes and machair. ❀ Short and usually well-branched, with *dark green foliage, often washed purple*. Flowers dull purple, with a ring of white hairs in the throat; they only open fully in sunshine. **DETAIL** Leaves 10–30mm long. Calyx cut into 4–5 sepals, with the *2 outer sepals not more than twice as wide as the 2 inner*. Corolla 12–18mm long (to 22mm in Ireland), *less than twice as long as calyx*, tube more or less cylindrical, divided at the tip into 4–5 petals, each 4–7mm long. Stigmas 2. Flowers may be creamy-white, washed purplish-red on the outside, in N Britain. **SIMILAR SPECIES Early Gentian** *G. anglica* is found in similar habitats in S England, but is much scarcer (Sch 8). Flowers Apr–Jun.

GROWTH Biennial (annual).
HEIGHT 3–30cm.
FLOWERS Mid Jul–Oct.
STATUS Native.
ALTITUDE To 750m.

Chiltern Gentian
Gentianella germanica

Nationally Scarce and listed as Vulnerable. Very locally common from N Hants through the Chilterns to Herts and Beds, in short, dry, grassland on chalk, often where sheltered in open scrub or near woodland. ❀ Much like Autumn Gentian but flowers brighter – bright bluish-purple – and larger, with more obvious transverse wrinkles across the calyx.
DETAIL Corolla 25–35mm, *more than twice as long as calyx*, broadening towards the mouth, petals 6–11mm long.

GROWTH Biennial (annual).
HEIGHT 3–40cm.
FLOWERS Aug–early Oct.
STATUS Native.
ALTITUDE Lowland.

Field Gentian
Gentianella campestris

Local; has disappeared from most of S Britain and listed as Vulnerable. Unimproved, species-rich grassland on neutral to mildly acidic soils, including pastures, verges, dunes and machair. ❀ Very like Autumn Gentian but flowers slightly paler and bluer – bluish-purple to whitish – with the *2 outer sepals much broader than the 2 inner sepals*. Flowers always with 4 petals. **DETAIL** Corolla 15–25cm long, petals 6–11mm long.

GROWTH Annual or biennial.
HEIGHT To 30cm.
FLOWERS Jul–early Oct.
STATUS Native.
ALTITUDE To 915m.

Yellow Centaury
Cicendia filiformis

Nationally Scarce, listed as Vulnerable. Local on bare sandy and peaty ground that is wet in winter and kept open by grazing and other disturbance: heathland (often alongside tracks), heavily poached pastures, woodland rides, dune slacks and cliffs. ❁ Tiny and hard to spot, with slender, upright stems. Flowers 2–4mm across, only opening in full sun in the morning. DETAIL Stems with 1–2 pairs of opposite, strap-shaped leaves, each 2–6mm long. Flowers long-stalked, terminal, calyx with 4 triangular lobes, corolla cut into 4 petals, stigma 1.

GROWTH Annual.
HEIGHT 2–10cm (18cm).
FLOWERS Late Jun–Sep.
STATUS Native.
ALTITUDE Lowland.

FAMILY APOCYNACEAE: PERIWINKLES

Greater Periwinkle *Vinca major*

A frequent garden throw-out. Verges, woodland, waste ground – wherever garden rubbish can be dumped. ❁ Sprawling, with fine but tough, woody shoots, shiny, dark green evergreen leaves and solitary purplish-blue flowers. Petals distinctively twisted in bud. DETAIL Stems rooting at tips. Leaves 2.5–9cm long, oval, opposite, the *margins with tiny hairs*, stalks 6–12mm. Flowers 3–5cm across, *calyx teeth narrow, hairy*, corolla divided into 5 petals that lie flat; stamens 5, stigma 1.

GROWTH Perennial.
HEIGHT Stems to 200cm.
FLOWERS Feb–May
(Jan–Jun).
STATUS Introduced
to cultivation from the
Mediterranean by 1597. First
recorded in the wild by 1650.
ALTITUDE Lowland.

Lesser Periwinkle
Vinca minor

A fairly frequent garden throw-out. Verges, banks, woodland and waste places. ❁ As Greater Periwinkle but leaves smaller and narrower and flowers sky blue and smaller, 25–30mm across; more often found in 'natural' sites and sometimes considered native. DETAIL *Leaves without a fringe of hairs*, 1.5–4.5cm long, lanceolate, stalks 3–4mm. *Calyx teeth hairless.*

GROWTH Perennial.
HEIGHT Stems to 100cm.
FLOWERS Apr–May
(Jan–Jul).
STATUS Introduced to
cultivation from Europe by 995
and long-naturalised.
ALTITUDE To 380m.

Viper's Bugloss
Echium vulgare

Common. Grassland and bare, disturbed ground on dry soils on sand, chalk and limestone: verges, old pits, heaths, field margins, waste ground, dunes and sandy shingle. ❁ Tall, forming sheets of bright blue when *en masse*, the whole plant is roughly hairy. DETAIL Basal leaves in a rosette, stalked, strap-shaped, to 15cm, dying off as the flowering stem develops. Stem with red-based bristles and stalkless, strap-shaped leaves. Buds purplish-pink, flowers bright blue, trumpet-shaped, 10–19mm long; calyx divided almost to base into 5 narrow teeth, corolla cut into 5 unequal lobes, with 4–5 projecting purplish stamens and a projecting forked style.

GROWTH Biennial.
HEIGHT To 100cm.
FLOWERS Jun–Sep.
STATUS Native.
ALTITUDE Lowland.

The English name reflects both the rough, bristly feel ('bugloss' derives from the Greek for 'ox-tongued') and the supposed resemblance of various parts of the plant to a snake; indeed, the fruits, said to resemble an Adder's head, were used as a cure for snake bite, although the plant is actually poisonous.

Common Fiddleneck *Amsinckia micrantha*

GROWTH Annual.
HEIGHT To 70cm.
FLOWERS Apr–Aug.
STATUS Introduced from N America and in cultivation by 1836, but not recorded from the wild until 1910; still spreading.
ALTITUDE Lowland.

Locally common weed of arable crops and rough ground, mostly on light, sandy soils. ❁ Roughly hairy due to abundant bristly white hairs; has *coiled spikes of yellow flowers*. DETAIL Leaves strap-shaped, stalkless, to 8cm. Flowers in spikes that lengthen as new flowers are formed, pale yellow, 3–5mm long, calyx divided almost to base into 5 teeth (calyx 5–6mm long in fruit), corolla divided into 5 petals, with 5 stamens and 1 style within the corolla tube.

Common Gromwell *Lithospermum officinale*

GROWTH Perennial.
HEIGHT To 80cm (100cm).
FLOWERS Late May–Jul.
STATUS Native.
ALTITUDE Lowland.

Fairly local. Rough, scrubby grassland, verges and wood borders, mostly on basic soils. ✱ Tall, with narrow, spear-shaped, stalkless leaves and *disproportionately small creamy flowers*. **DETAIL** Hairy, with tiny swollen bases to the hairs. Stems often grouped, sometimes branched. Leaves alternate, 5–8cm long, dark green. Flowers in loose spikes in leaf axils, appearing crowded when first flowering but elongating in fruit. Flowers 3–6mm across, sepals 5, corolla cut into 5 petals, with the corolla tube concealing the 5 stamens and single style. *Seeds whitish, smooth*, shiny. **SIMILAR SPECIES Field Gromwell** *L. arvense* is a declining annual weed, introduced in the Bronze Age. Shorter (to 50cm), with larger white flowers (5–9mm across) and *pale brown, minutely-warted* seeds.

Lungwort *Pulmonaria officinalis*

Commonly grown in gardens and sometimes naturalised in woodland, scrub and shady places on rough and waste ground. ✱ Early-flowering, *flowers reddish* to *bluish-violet* when fresh, becoming *bluer with age*. **DETAIL** Roughly hairy. Leaves usually *spotted white*, basal leaves *long-stalked*, oval, pointed, narrowing abruptly to a *rounded* or *heart-shaped base*; stem leaves stalkless. Flowers in clusters at the tip of sprawling, unbranched stems, 9–11mm across; calyx with 5 teeth and *many glandular hairs*, corolla cut into 5 petals, stamens

GROWTH Perennial.
HEIGHT 20–30cm.
FLOWERS Mar–May.
STATUS Introduced to gardens by 1597 and recorded in the wild by 1793 (Europe).
ALTITUDE To 385m.

5, style 1 (either long or short, as in Primrose, see p. 182). **SIMILAR SPECIES Narrow-leaved Lungwort** *P. longifolia*, native to woods around the Solent and Poole Harbour, has narrower leaves that *taper into the stalk*, smaller, bluer flowers (5–6mm across), a more bell-shaped calyx with longer, narrower teeth (half length of tube) and *few or no glandular hairs*.

Green Alkanet
Pentaglottis sempervirens

Common in gardens but very vigorous and frequent as a garden throw-out in light shade on verges, woodland margins and waste ground, mostly near habitation. ❀ Bristly-hairy, with pointed-oval leaves and piercing blue flowers with a white eye. DETAIL Leaves and stems with bristle-like white hairs, those on the leaves often with white blisters at the base. Basal leaves to 30cm, long-stalked. Flowers 8–10mm across, calyx deeply cut into 5 narrow teeth, corolla cut into 5 petals, with 5 stamens and single style enclosed in the short corolla tube. SEE ALSO Bristly Oxtongue (p. 268) has similarly blistered leaves.

GROWTH Perennial.
HEIGHT 30–60cm (100cm).
FLOWERS Mar–Jul.
STATUS Introduced to gardens prior to 1700 and recorded from the wild by 1724 (SW Europe).
ALTITUDE To 380m.

Bugloss
Anchusa arvensis

A common weed of arable fields, usually on light soils, also on dunes, heaths and waste ground on disturbed, sandy ground near the sea. ❀ Very bristly, with strap-shaped, scalloped, wavy-edged leaves and coiled spikes of small blue flowers with a white eye. DETAIL Sprawling to erect, stem and leaves with stiff hairs that have swollen bases. Leaves to 15cm, more or less stalkless, strap-shaped; basal rosette usually dead by flowering time. Flowers 4–6mm across, calyx cut almost to the base into 5 teeth, corolla cut into 5 slightly unequal petals, with 5 stamens and single style enclosed in the curved, 4–7mm long corolla tube.

GROWTH Annual.
HEIGHT 15–50cm.
FLOWERS Jun–Sep.
STATUS Ancient introduction.
ALTITUDE Lowland.

Borage *Borago officinalis*

Common in gardens and occasionally spreading to verges and waste ground. ❀ *Roughly hairy*, with blue flowers, the *petals spreading widely* to reveal the stamens. DETAIL Basal leaves long-stalked, oval, to 15cm long, with a heart-shaped base, grey-green and prominently wrinkled; stem leaves stalkless, strap-shaped, wavy-edged. Flowers 20–25mm across. Calyx very hairy, cut into 5 narrow teeth, corolla with 5 lobes.

GROWTH Annual.
HEIGHT To 60cm.
FLOWERS Jun–Sep.
STATUS In cultivation since at least 1200 and first recorded in the wild in 1777 (Mediterranean).
ALTITUDE Lowland.

Common Comfrey *Symphytum officinale*

Local. The banks of rivers, streams and ditches, damp verges and marshes. ❀ Comfreys are roughly hairy, with coiled flower spikes. Common Comfrey is large and upright, with pale green leaves and *creamy-white flowers*, less often reddish, purple or 'peppermint-striped'. Stem leaves stalkless, running down the stem at the base into *broad wings* that *extend beyond the next leaf-base*. DETAIL Basal leaves 15–30cm long, tapering into the stalk. Calyx about half length of corolla, deeply cut into 5 long, narrow teeth. Corolla 8–20mm long, tubular, divided into 5 short teeth at the tip, with 5 stamens and 1 long style. Seeds smooth, shiny.

GROWTH Perennial.
HEIGHT To 150cm.
FLOWERS Apr–Jun.
STATUS Native.
ALTITUDE To 320m.

calyx & stigma

Tuberous Comfrey *Symphytum tuberosum*

Local. The banks of rivers, streams and ditches and wet woodland, tolerant of some shade. Also introduced to verges and waste ground. ❀ Upright, medium-sized comfrey with *pale yellow flowers*. *Upper stem leaves stalkless*, with wings only running a *short way down the stem*. DETAIL Rather roughly hairy. No leafy runners. Stems purplish or with purple bases to the hairs. Leaves tapering into the stalk; lower leaves stalked, mid stem leaves the longest (to 17cm), stalkless. Calyx very deeply cut into long, narrow, finely pointed teeth. Seeds dull, with minute warts.

GROWTH Perennial.
HEIGHT 30–60cm.
FLOWERS May–Jul.
STATUS Native in N England and Scotland, introduced elsewhere.
ALTITUDE To 335m.

calyx & stigma

Russian Comfrey *Symphytum × uplandicum*

The hybrid between Common Comfrey and Rough Comfrey *S. asperum*. The *commonest comfrey in most places*, on verges, woodland margins and waste ground. ❀ Large and upright, with large, stiff, dark green leaves, pink to dark purple buds and pink, pinkish-blue, purple or violet flowers; stem leaves stalkless, with *narrow* wings running a *short way* down the stem (rarely beyond the next leaf-base). DETAIL Roughly hairy. Basal leaves 8–18cm long. Calyx deeply cut into long, narrow, pointed teeth. Seeds dull, with minute warts.

GROWTH Perennial.
HEIGHT To 100cm (150cm).
FLOWERS May–Jun (Aug).
STATUS Introduced as a forage plant in 1870 and known from the wild by 1884.
ALTITUDE 0–470m.

White Comfrey *Symphytum orientale*

Fairly common on verges and waste ground. ❀ Tall, upright comfrey, flowers *pure white* (often discoloured by brown blotches), *stems unwinged*, and calyx *shallowly cut* into 5 short teeth. DETAIL Very hairy, but only slightly rough to the touch. Stem leaves to 20cm long, with rounded or heart-shaped bases, stalked or stalkless but without wings running down stem. Seeds dull, with minute warts.

GROWTH Perennial.
HEIGHT To 70cm (150cm).
FLOWERS Mar–May.
STATUS Introduced to cultivation by 1952, and known from the wild by 1849 (NW Turkey, Russia and Caucasus).
ALTITUDE Lowland.

Creeping Comfrey *Symphytum grandiflorum*

Common in gardens, parks and churchyards, and an occasional garden throw-out on shady verges etc. ❀ Low-growing and *patch-forming*, spreading aggressively via sprawling leafy runners that root at the nodes. Leaves *small*, flowers pinkish-red in bud, *creamy-yellow* when open. DETAIL Stems bristly, flowering stems unbranched. Leaves rather roughly hairy, 4–8cm long, pointed-oval, with a *heart-shaped* base. Stem leaves mostly stalked, *without* wings running down stem. Calyx cut into 5 narrow, *blunt* teeth.

GROWTH Perennial.
HEIGHT Stems to 40cm.
FLOWERS Mar–May.
STATUS Introduced to cultivation in the late 19th century and recorded from the wild from 1898 (Caucasus).
ALTITUDE To 400m.

Hound's Tongue *Cynoglossum officinale*

GROWTH Biennial.
HEIGHT 30–60cm (90cm).
FLOWERS May–Jul (Aug).
STATUS Native.
ALTITUDE To 400m.

Near Threatened. Very locally common. Disturbed ground on light, sandy and chalky soils and shingle, on scrubby grassland, woodland fringes and field margins; unpalatable to grazing animals and often stands proud on short, Rabbit-cropped turf. ❀ Greyish-green, softly downy to the touch, with coiled spikes of drooping, *maroon-red flowers*, the plant is said to smell of mice. **DETAIL** Leaves oval to strap-shaped, 10–25cm long, the lower stalked but dying off towards summer, upper leaves stalkless. Flowers short-stalked, 6–10mm across, calyx cut almost to the base into 5 oblong sepals, corolla funnel-shaped, cut into 5 petals, with 5 stamens and 1 style concealed inside the corolla tube. Fruits with 4 large seeds, 5–6mm across, with hooked spines.

Phacelia

Phacelia tanacetifolia

GROWTH Annual.
HEIGHT To 70cm (100cm).
FLOWERS Aug–Sep.
STATUS Introduced
from California to
gardens in 1832
and first
recorded
in the wild
in 1885.
ALTITUDE
Lowland.

Often sown as green manure or in wildflower margins beside arable fields, and also found as a casual on waste ground etc., originating from wild bird seed and pheasant feed. ❀ Distinctive *coiled spikes of blue or pale mauve flowers* with *prominent projecting stamens*. **DETAIL** Slightly hairy. Leaves in a basal rosette and on the stem, 4–15cm long, cut into 3–5 pairs of leaflets (or lobes). Flowers 6–10mm across, calyx cut almost to base into 5 teeth, corolla bell-shaped, with 5 projecting stamens and 1 style, forked towards the tip.

Field Forget-me-not
Myosotis arvensis

Common. Disturbed dry ground, including arable fields. ❀ *Flowers small* and *blue* with a yellow 'eye'. Calyx with numerous *erect, hooked hairs*, hairs on stem *erect* on lower stem but *flattened* towards flower spike. DETAIL Flowers to 3mm across; in fruit flower stalk *1.2–2 times as long as calyx*. Var. *sylvestris*, sometimes found in shady places, is larger, with flowers up to 5mm across, and can be mistaken for Wood Forget-me-not, but is more greyish-green, with saucer-shaped rather than flat flowers (the corolla tube is *shorter* than the calyx). In fruit, *tips of sepal teeth pressed together*, concealing the ripe seeds (erect, exposing the seeds, in Wood Forget-me-not).

GROWTH Annual (perennial).
HEIGHT Stems 15–40cm.
FLOWERS Apr–Sep.
STATUS Ancient introduction.
ALTITUDE 0–610m.

Early Forget-me-not *Myosotis ramosissima*

Locally common. Bare ground on poor, dry soils, especially sand, chalk and limestone: grassy heaths, dunes, field margins, walls and waste ground.

GROWTH Annual.
HEIGHT 2–15cm (25cm).
FLOWERS Apr–Jun.
STATUS Native.
ALTITUDE 0–430m.

❀ As Field Forget-me-not (hairs on calyx *erect* and *hooked*, hairs on stem *erect*, becoming *flattened* towards flowers), but flowers typically *smaller*, stems *much finer* (c. 1mm across), and in fruit the flower stalk is *no longer than the calyx*. DETAIL Flowers 2–3mm across, corolla bright blue (occasionally pink, rarely white), saucer-shaped, with the corolla tube *shorter* than the calyx teeth.

Changing Forget-me-not *Myosotis discolor*

Scattered in a wide range of grassy habitats, both dry and damp, where the vegetation is relatively short and often rather sparse.

GROWTH Annual.
HEIGHT 8–35cm.
FLOWERS May–Jun.
STATUS Native.
ALTITUDE 0–610m.

❀ As Field and Early Forget-me-nots (*hooked, erect* hairs on calyx, *erect* hairs on lower stem, becoming *flattened* towards flowers), but the tiny flowers are *yellow or cream* as they open, *changing to pink or blue* as they age. DETAIL Stems erect. Flowers to 2mm across, the corolla tube lengthening with age until *obviously longer than sepal teeth*. Flower stalk always shorter than the calyx.

Tufted Forget-me-not
Myosotis laxa

Fairly common. Wet grassland and the edges of ponds, streams and rivers, often where disturbed by livestock etc. ✿ As Water Forget-me-not but flowers slightly smaller and hairs on lower stem *flattened throughout*. DETAIL Stems erect to somewhat sprawling, tufted, *without runners*. Flowers to 5mm across, calyx with flattened hairs and *proportionally narrow teeth* (calyx often cut more than halfway to base, with the base of the teeth shorter than the sides). Style shorter than calyx tube. Flowers sometimes very pale (cf. Pale Forget-me-not).

GROWTH Annual to biennial.
HEIGHT Stems to 40cm.
FLOWERS May–Oct.
STATUS Native.
ALTITUDE To 550m.

Pale Forget-me-not
Myosotis stolonifera

Nationally Scarce. Rather local, around springs and flushes in the hills and small streams, pools and ditches in the valleys. ✿ As Tufted Forget-me-not, with hairs on calyx and stem flattened throughout, but *flowers small and whitish*, leaves short, and *produces abundant runners* in late summer, resulting in dense floating mats. DETAIL Stems erect. *Leaves rarely more than 3 times as long as wide* (usually more than 4 times longer in Tufted). Flowers up to 5mm across. Calyx always cut more than halfway to base.

GROWTH Perennial.
HEIGHT To 20cm (30cm).
FLOWERS Jun–Aug.
STATUS Native.
ALTITUDE 130–820m.

Creeping Forget-me-not *Myosotis secunda*

Common, usually on acid, peaty soils, in damp places by rivers and streams, marshy pastures, and around springs and flushes on moors and bogs. ✿ Much as Tufted Forget-me-not but produces rooting runners as the season advances and thus *patch-forming*. Hairs on stem flattened towards flowers, but *long and erect towards base* (also a good distinction from Pale Forget-me-not). Flowers always pale. DETAIL Stems erect to somewhat sprawling. Flowers to 6(8)mm across, calyx tube with hairs flattened, *calyx teeth narrow*, with the base shorter than the sides (calyx cut at least halfway to base). *Fruit stalks long, 2.5–5 times length of calyx.*

GROWTH Annual to perennial.
HEIGHT Stems to 50cm.
FLOWERS May–Aug.
STATUS Native.
ALTITUDE 0–805m.

Water Forget-me-not
Myosotis scorpioides

Common by ditches and slow-flowing streams and rivers, and in marshes; avoids very acid soils. Usually terrestrial, but may grow submerged or as a floating raft. ❀ Small, sky-blue flowers with a yellow 'eye'. DETAIL Patch-forming, growing from rhizomes and runners; stems erect to sprawling. Hairs on stems often sparse, *flattened* towards flowers, *erect* towards base. Flowers to 8(12)mm across. *Hairs on calyx flattened, calyx teeth short* (calyx cut less than halfway to base, base of teeth as long as sides). Style longer than calyx tube, sometimes projecting beyond the teeth. Fruit stalk 1–2 times length of calyx.

GROWTH Perennial.
HEIGHT 15–30cm (stems to 70cm).
FLOWERS May–Sep.
STATUS Native.
ALTITUDE 0–600m.

Wood Forget-me-not
Myosotis sylvatica

Locally common. Light shade in woodland and damp hedgebanks, also rocky places in the uplands; a *common garden escape* in a wide variety of habitats. ❀ Flowers relatively *large and bright*, hairs on stem *erect throughout*. DETAIL Stems erect to somewhat sprawling, tufted, with some non-flowering basal shoots but no runners. Calyx densely covered by curled and hooked hairs. Flowers to 8mm across, flat. In fruit, stalks 1.5–2 times length of calyx, *calyx teeth spreading*, exposing the ripe seeds. Garden escapes usually have brighter flowers and are often pinkish.

GROWTH Perennial (garden escapes usually annual).
HEIGHT 15–50cm.
FLOWERS Apr–Jun.
STATUS Native.
ALTITUDE 0–485m.

Oysterplant
Mertensia maritima

Nationally Scarce and listed as Near Threatened. Very local on shingle, rocky or occasionally sandy beaches in Scotland and N Ireland. Seed is dispersed by the sea. ❀ *Mat-forming*, with *grey-green foliage* and blue flowers (pink in bud). DETAIL Stems prostrate, green or purple. Leaves 0.5–6cm long, oval, hairless, the lower stalked. Calyx *c.* 6mm across, cut to the base into 5 sepals, corolla split into 5 petals.

GROWTH Perennial.
HEIGHT Stems 30–60cm.
FLOWERS Jun–Aug.
STATUS Native.
ALTITUDE Lowland.

Dodder
Cuscuta epithymum

Local on heathland, chalk downland and dune grassland. Listed as Vulnerable. ❁ A parasite that attaches itself to a variety of plants, but especially gorse, Heather and Wild Thyme. The tangles of red, thread-like stems and clusters of tiny pale pink flowers are unique. **DETAIL** Stems *c.* 0.5mm in diameter, with minute scale-like leaves. Flowers in stalkless clusters, corolla 3–4mm across, cut into 5 lobes, styles 2, stamens 5. **SIMILAR SPECIES Greater Dodder** *C. europaea* is found scattered in S England, mostly parasitic on nettles. Slightly larger, its flowers have blunt (not pointed) sepals and the stamens do not project from the flower.

GROWTH Annual.
HEIGHT Stems to 100cm (perhaps much more).
FLOWERS Jun–Sep.
STATUS Native.
ALTITUDE Lowland.

Dodder seeds germinate on or near the soil's surface and produce a small, swollen, root-like anchor that lasts just 2–5 days, and a shoot, which must reach a host plant within a few days. Once a host is found, the Dodder's shoots attach themselves with sucker-like absorptive pads and start to extract nutrients. Dodder can go on to cover the host, sometimes 'leapfrogging' to additional plants, forming mats at ground level or extensive sheets draped over large gorse bushes. Although an 'annual', when growing on perennial hosts Dodder can induce the formation of galls (growths of host tissue) where parasitic tissue is able to overwinter. The following spring, new Dodder plants develop from these galls.

Field Bindweed *Convolvulus arvensis*

Common. Gardens and other cultivated ground, roadsides and grassland. ❁ A *small-flowered bindweed*, usually low-growing and trailing, with trumpet-shaped white, pink, or striped pink and white flowers. **DETAIL** Stems more or less hairless. Leaves 2–6cm long, alternate, hairless to hairy. Flowers 10–30mm across, calyx deeply cut into 5 rounded lobes, corolla shallowly 5-lobed, style 1, tipped by 2 stigmas. *Epicalyx absent* – merely 2 small, narrow bracts on stem some way below flower.

GROWTH Perennial.
HEIGHT Stems to 75cm (200cm).
FLOWERS Jun–Sep.
STATUS Native.
ALTITUDE Lowland.

Hedge Bindweed
Calystegia sepium

Common. Gardens, hedgerows, scrub, woodland, tall herb fen, waste ground. ❂ A rampant climber, with trumpet-shaped white flowers (less often pink with 5 white stripes). The sepals are enclosed by 2 pouch-like bracteoles, but these *do not (or hardly) overlap* and the *sepals remain visible*. **DETAIL** Leaves to 15cm, hairless to sparsely hairy. Calyx deeply cut into 5 pointed sepals, corolla shallowly 5-lobed, style 1, tipped by 2 stigmas. *Epicalyx of 2 bracteoles, 10–18mm wide when flattened.* Subspecies *sepium*, with hairless stems and flowers 3–5(5.5)cm across, is common; subspecies *roseata*, with sparsely hairy stems and flowers 4–5.5cm across and always pink with 5 white stripes, is scarce and mostly found near the sea in the W.

GROWTH Perennial.
HEIGHT Stems to 200cm (300cm).
FLOWERS Jul–Sep.
STATUS Native.
ALTITUDE 0–365m.

Large Bindweed
Calystegia silvatica

Common. Gardens, hedgerows, waste ground, usually near houses. ❂ As Hedge Bindweed, but *flowers larger* and *bracteoles pouched and strongly overlapping, hiding the sepals*. **DETAIL** Hairless. *Bracteoles 18–45mm wide when flattened.* Flowers 5–9cm across, white, occasionally pink-striped on outside. **SIMILAR SPECIES Hairy Bindweed** *C. pulchra* is a scarce garden escape. It has similarly overlapping bracteoles but pink or pink-striped flowers, with the flower stalks more or less hairy and usually narrowly winged towards the tip.

GROWTH Perennial.
HEIGHT Stems to 300cm (500cm).
FLOWERS Jul–Sep.
STATUS Introduced to cultivation in 1815 and first recorded in the wild in 1863.
ALTITUDE 0–350m.

Sea Bindweed
Calystegia soldanella

Fairly common. The strand-line on sandy and occasionally shingle beaches, also dunes, both on bare ground and amongst grassy swards. ❂ *Prostrate rather than climbing*, with *fleshy, kidney-shaped leaves* and trumpet-shaped pink flowers with 5 white stripes and a yellowish centre. **DETAIL** Hairless. Flowers 30–55mm across. Calyx enclosed by rounded bracteoles that overlap slightly but are shorter than the sepals.

GROWTH Perennial.
HEIGHT Stems to 100cm.
FLOWERS Jun–Aug.
STATUS Native.
ALTITUDE Lowland.

Black Nightshade *Solanum nigrum*

GROWTH Annual.
HEIGHT Stems to 70cm.
FLOWERS Jul–Oct.
STATUS Native and introduction.
ALTITUDE Lowland.

A common weed of arable fields and other disturbed ground. ❀ White flowers are followed by *shiny black berries.* DETAIL Variably hairy. Stems erect to sprawling, often blackish. Leaves 2.5–5cm long, variably toothed. Flowers (3)5–10mm across, sepals blunt. Berries 6–10mm across. Subspecies *nigrum*, the common form, is sparsely hairy. Subspecies *schultesii* is scarce and sporadic, and is glandular-hairy.

Green Nightshade *Solanum physalifolium*

GROWTH Annual.
HEIGHT Stems 25–80cm.
FLOWERS May–Jul.
STATUS Introduced from S America and first recorded in the wild in 1949. Still spreading.
ALTITUDE Lowland.

Fairly common arable weed, especially in Sugar Beet. ❀ Much like Black Nightshade but densely glandular-hairy, foliage more yellowish-green, and fruits ripening to green or purplish-brown, *partly enclosed by sepals, which have more pointed tips.* DETAIL Stems always green. Leaves 3–7cm long. Flowers 4–8mm (3–10mm).

Deadly Nightshade *Atropa belladonna*

Rather local. Open woodland, scrub and hedgerows on dry, usually chalky, soils; also a casual around towns.
❀ Rather shrubby, with nodding, bell-shaped flowers growing singly in the leaf axils. Fruit a shiny black berry 15–20mm across.
DETAIL Stems upright, purplish towards base, glandular-hairy. Leaves 8–20cm long, variably sticky-hairy. Calyx deeply cut into 5 lobes. Corolla 24–30mm long, greenish- or brownish-purple, cut at the tip into 5 shallow lobes.

All parts of the plant are poisonous. The leaves have been mistaken for comfrey and used to make tea, but it is the berries that are notorious due to their supposed attraction to young children – in which just 2 berries can produce symptoms. Despite the name, however, cases of fatal poisoning seem to be very rare and Deadly Nightshade is just one of many potentially lethal plants.

GROWTH Perennial.
HEIGHT To 150cm (200cm).
FLOWERS Jun–Aug.
STATUS Native (but some populations are probably ancient introductions).
ALTITUDE Lowland.

Bittersweet *Solanum dulcamara*

GROWTH Perennial.
HEIGHT Stems 50–300cm (700cm).
FLOWERS May–Sep.
STATUS Native.
ALTITUDE Lowland.

Common. Woodland, scrub and hedgerows, but especially damp places beside water and in fens. ❀ A distinctive scrambler, with small flowers, the 5 purple petals setting off the cone of yellow stamens. Aka Woody Nightshade.
DETAIL Hairless to hairy. Stems woody towards base. Leaves alternate, 4–11cm long, at least some with 2(4) lobes or leaflets near the base. Flowers *c.* 10mm across. Fruits 8–12mm, ripening through green and yellow to *bright red*. Var. *marinum* of shingle beaches is prostrate, with fleshier leaves.

Henbane
Hyoscyamus niger

GROWTH Biennial.
HEIGHT To 80cm.
FLOWERS Jun–Aug (Sep).
STATUS Introduced in the Bronze Age.
ALTITUDE Lowland.

Listed as Vulnerable. Very locally common. Dry, disturbed chalky or sandy soils: field margins, Rabbit warrens, waste ground, also sand and shingle by the sea. ❀ Unpleasantly sticky-hairy and foul-smelling, with solitary, sinister-looking flowers in vertical rows up the stem. DETAIL Stems woody at base, glandular-hairy towards tip. Basal leaves 15–20cm, irregularly lobed, stem leaves smaller, stalkless, clasping. Calyx cut into 5 pointed teeth at the tip, corolla 2–3cm, cut into 5 rounded lobes. Fruit a capsule enclosed within the swollen base of the persistent calyx.

Thorn Apple
Datura stramonium

GROWTH Annual.
HEIGHT 30–100cm (150cm).
FLOWERS Jul–Oct.
STATUS In cultivation by 1597 and first recorded in the wild in 1777.
ALTITUDE Lowland.

Local and often erratic. Arable fields, gardens, waste ground. ❀ Very distinctive, with white (rarely purple), trumpet-shaped flowers, jagged leaves and spiny green fruits. DETAIL Leaves shiny dark green, 5–20cm, irregularly and coarsely toothed. Calyx 3–5cm, with 5 teeth, each up to 10mm long, corolla 5–10cm, shallowly-lobed. Capsule 3.5–7cm, with slender spines up to 15mm long, splitting open via 4 valves. SIMILAR SPECIES **Apple of Peru** *Nicandra physalodes* is up to 80cm tall, with bell-shaped, blue or violet flowers 25–40mm across and net-veined, heart-shaped sepals. Fruit a dry brown berry enclosed by the enlarged dry sepals. An increasing alien.

Foxglove
Digitalis purpurea

GROWTH Biennial or short-lived perennial.
HEIGHT To 200cm.
FLOWERS Jun–Jul (Sep).
STATUS Native.
ALTITUDE 0–880m.

Common on acid soils, especially following soil disturbance (or fires): hedgebanks, woodland clearings, clear-fell, heaths, moors, sea cliffs.
❀ The tall spires of pinkish-purple or white, pouch-like flowers, spotted darker inside the throat, are distinctive.
DETAIL Densely hairy. Stems unbranched. Leaves 10–30cm long, oval, pointed, wrinkled, the lower forming an overwintering basal rosette. Calyx short, cut into 5 teeth. Corolla 40–55mm, with 5 unequal lobes at the tip. Stamens 4. **SEE ALSO** Ploughman's Spikenard (p. 274), whose leaves are very similar.

Fairy Foxglove
Erinus alpinus

GROWTH Short-lived perennial.
HEIGHT Stems 5–20cm.
FLOWERS May–Aug.
STATUS Introduced to gardens by 1739 and recorded from the wild by 1867 (S Europe).
ALTITUDE To 350m.

Local. Sunny, old walls and other stony places. ❀ Sprawling, most leaves in an untidy rosette, the flowering shoots with loose clusters of showy, 5-petalled pinkish-purple (sometimes white) flowers, 6–9mm across.
DETAIL Stems hairy, sprawling. Leaves alternate, to 2cm, strap-shaped, shallowly toothed, variably hairy. Calyx split into 5 strap-shaped lobes. Corolla tube 3–7mm, split into 5 notched petals, stamens 4.

Common Field Speedwell *Veronica persica*

An abundant weed of disturbed ground: arable fields, gardens and waste ground. ❀ Speedwells have bluish flowers with 4 petals, the lowest a little smaller and narrower than the rest and often white, 2 prominent stamens and a projecting stigma.
DETAIL Sparsely hairy. Stems sprawling. Leaves oval, coarsely-toothed, 10–30mm, rather short-stalked. *Flowers solitary, on long stalks in the leaf axils*, relatively large (*c.* 8–12mm across), bright blue. Calyx lobes 5–6mm long, oval, pointed. *Capsule 2-lobed, the obviously flattened, sharp-edged lobes diverging at 90° or more*, with short, non-glandular hairs as well as erect glandular hairs.

GROWTH Annual.
HEIGHT Stems 10–30cm (50cm).
FLOWERS Jan–Dec.
STATUS Introduced from SW Asia, the first record from the wild was in 1826.
ALTITUDE 0–350m.

Green Field Speedwell *Veronica agrestis*

Local and declining. Gardens, allotments and arable fields on well-drained, often acidic soils. ❀ Much like Common Field Speedwell but *flowers usually smaller and paler; best identified by the capsule*. **DETAIL** Leaves mid green. Flowers 3–8mm across, pale blue or pale lilac to whitish. *Capsule with 2 rounded, pea-shaped lobes lying roughly parallel with a narrow notch between their tips, covered with erect glandular hairs only.*

GROWTH Annual.
HEIGHT Stems to 30cm.
FLOWERS Mar–Nov.
STATUS Ancient introduction.
ALTITUDE 0–455m.

Grey Field Speedwell *Veronica polita*

Local and declining. Arable fields, gardens and allotments, on well-drained soils. As Green Field Speedwell but *leaves duller, dark grey-green, and flowers bright dark blue*; conclusively identified by the capsules. ❀ **DETAIL** Flowers 4–8mm across. Capsule with 2 *rounded, pea-shaped lobes* (as Green Field Speedwell) but *many short non-glandular hairs as well as erect glandular hairs.*

GROWTH Annual.
HEIGHT Stems to 30cm.
FLOWERS Mar–Nov.
STATUS Introduced from Europe, the first record from the wild was in 1777.
ALTITUDE 0–350m.

Slender Speedwell
Veronica filiformis

Fairly common. Mown grass (lawns, churchyards, playing fields – spread via grass clippings), and sometimes other grassy places. ❀ Mat-forming, with creeping stems that *root at the nodes, rounded to kidney-shaped leaves* and *solitary, relatively large* bright blue flowers on *long stalks* from the lead axils. DETAIL Stems minutely hairy. Leaves relatively small (4–10 mm across), sparsely hairy, short-stalked. Flowers 8–15mm across, lilac-blue with a whitish lower petal, stalks 2–3 times as long as leaves.

GROWTH Perennial.
HEIGHT Stems to 50cm.
FLOWERS Late Mar–Jun.
STATUS Introduced to gardens by the early 19th century, escaping and spreading rapidly in the 20th century (Turkey and the Caucasus).
ALTITUDE To 450m.

Ivy-leaved Speedwell *Veronica hederifolia*

Common. Gardens, arable land, waste ground and pavements cracks, also shady places such as woodland rides and hedgebanks. ❀ *Flowers rather small, solitary, on short stalks in the leaf axils; leaves wider than long with 3–5 lobes and obvious veins.* DETAIL Sparsely hairy. Stems more or less sprawling, sometimes against trees and walls. Leaves circular to kidney-shaped, to 12mm, stalked. Flowers 4–9mm across, calyx lobes oval with a heart-shaped base, seed capsule hairless. There are 2 subspecies.

GROWTH Annual.
HEIGHT Stems to 60cm.
FLOWERS Apr–May.
STATUS Ancient introduction.
ALTITUDE 0–380m.

V. h. hederifolia Flowers mostly 6mm or more across, whitish to blue, anthers blue, 0.7–1.2mm long; capsule stalks mostly 2–4× as long as calyx, which expands greatly after flowering; terminal lobe of leaf usually wider than long.
V. h. lucorum Flowers mostly 6mm or less across, whitish to pale lilac-blue, anthers whitish to pale blue, 0.4–0.8mm long; capsule stalks mostly 3.5–7× as long as calyx, which expands slightly; terminal lobe of leaf usually longer than wide.

Wall Speedwell
Veronica arvensis

Common. Disturbed ground and sparse grassland, especially on dry soils, including paths, tracksides, walls, arable fields and dunes. ❀ Upright, with *tiny, almost stalkless, bright blue flowers* that only open fully in bright sun. DETAIL Hairy. Stems sprawling to erect, glandular-hairy. Leaves oval-triangular, variably toothed or lobed, to 15mm. Flowers 2–3mm across, *clustered at the tip of the stem amongst stalkless, leaf-like bracts.*

GROWTH Annual.
HEIGHT 1–20cm (30cm).
FLOWERS Mar–Oct.
STATUS Native.
ALTITUDE 0–820m.

Pink Water Speedwell
Veronica catenata

Fairly common. Waterside mud and still or slow-flowing water up to 100cm deep; favours alkaline soils and tolerates slightly brackish conditions (e.g. coastal grazing marshes). ❁ Upright and rather fleshy, with small pink flowers in *stalked clusters*, mostly *in pairs at each node*. DETAIL Leaves strap-shaped with a few shallow teeth, stalkless, clasping. *Flower stalks shorter than the spear-shaped bracts.* Flowers 5–6mm across, usually (but not always) pink; *dark veins on petals ending well short of edge.* Capsules wider than long, deeply notched, held more or less at 90° to the stem, splitting into 4 valves.

GROWTH Annual to perennial.
HEIGHT To 30cm.
FLOWERS Jun–Sep.
STATUS Native.
ALTITUDE Lowland.

Blue Water Speedwell
Veronica anagallis-aquatica

Locally common. The margins of rivers, ditches, ponds etc., growing in or beside the water or as a fully-submerged, non-flowering aquatic. ❁ As Pink Water Speedwell but flowers *pale blue* and *flower stalks at least as long as the tiny strap-shaped bracts.* DETAIL Sends out rooting runners. Leaves broader than Blue Water Speedwell, lower leaves sometimes short-stalked. *Veins on petals almost reaching edges.* Capsules more or less circular, either not notched or only shallowly notched, held erect or spreading at 45°. SIMILAR SPECIES The hybrid between the water speedwells, V. × lackschewitzii, is common in some areas, and commoner than either parent in Hampshire. Sterile or nearly sterile, the flower spikes continue to elongate through the season (not ceasing growth as the fruit ripens).

GROWTH Annual to perennial.
HEIGHT To 50cm.
FLOWERS Jun–Aug.
STATUS Native.
ALTITUDE 0–380m.

Marsh Speedwell
Veronica scutellata

Rather local. Marshes, wet grassland, bogs, wet heaths, upland flushes, and the margins of ponds and lakes. Commonest on acid soils, and in the N and W. ❁ Flowers in loose, few-flowered spikes, *always only 1 spike per leaf node.* DETAIL Usually hairless. Stems slender, sprawling, sometimes scrambling up through other vegetation; grazed plants may be mat-forming. Leaves strap-shaped, pointed, 20–40mm long, with a few, scattered teeth, stalkless, often purplish, especially below. Flower stalks twice length of tiny, strap-shaped bracts. *Flowers 5–8mm across, whitish to pale pink or lilac.* Capsule kidney-shaped, splitting into 2 valves.

GROWTH Perennial.
HEIGHT Stems to 60cm.
FLOWERS Jun–Aug.
STATUS Native.
ALTITUDE 0–780m.

Brooklime
Veronica beccabunga

Common. Short, open vegetation in a wide range of wet habitats, from shallow water on the margins of rivers and streams to marshes and wet woodland rides; avoids very acid soils.
❁ Low-growing and patch-forming. Hairless, stems and leaves distinctly fleshy, *all leaves short-stalked.* Flowers in slender spikes, *in pairs from the leaf axils.*

GROWTH Perennial.
HEIGHT Stems 20–60cm.
FLOWERS May–Sep.
STATUS Native.
ALTITUDE 0–845m.

DETAIL Stems sprawling, rooting at the nodes. Leaves opposite, long-oval, blunt, 25–60mm long, with shallow blunt teeth. Flowers 7–8mm across, deep blue. Capsule 2–4mm, as long as wide, splitting into 4 valves; stalks at *c.* 90° to stem in fruit, short to longer than bracts.

Thyme-leaved Speedwell *Veronica serpyllifolia*

Common in a wide variety of open, moist places, from lawns to damp rock ledges, but especially characteristic of compacted ground along the sides of paths, tracks and woodland rides.
❁ Patch-forming, with *shiny, oval, untoothed leaves* and *short, erect spikes of small, pale flowers.*
DETAIL More or less hairless (upper stem sometimes hairy). Stems prostrate, rooting at the nodes. Leaves 5–15mm long, very short-stalked. Flowers 5–8mm across, whitish to pale blue with darker veins (6–10mm and bright blue in mountain populations). Flower stalks longer than calyx. Capsule wider than long, shorter than calyx; style shorter than capsule.

GROWTH Perennial.
HEIGHT To 30cm.
FLOWERS May–Oct.
STATUS Native.
ALTITUDE 0–1160m.

Spiked Speedwell *Veronica spicata*

Nationally Scarce. Rare and very local in dry grassland on limestone in a few places in the W, also in Breckland. ❁ Spikes of deep blue flowers distinctive. DETAIL Hairy. Leaves in pairs, 15–30mm; lowest narrowly oval, widest in the middle, shallowly and bluntly toothed, stalked; upper narrower, unstalked. Calyx lobes blunt.
SIMILAR SPECIES Garden Speedwell *V. longifolia* has similar spikes of blue flowers, but is hairless or only minutely hairy, with leaves in clusters of 2–4, broadest at the base, sharply-toothed, pointed. Calyx lobes pointed.

GROWTH Perennial.
HEIGHT 8–30cm (80cm).
FLOWERS Jun–Sep.
STATUS Native (Sch 8).
ALTITUDE To 400m.

Germander Speedwell *Veronica chamaedrys*

Very common in grassy places, and prominent on springtime verges. ❀ The bright blue flowers with a white 'eye', in loose, long-stalked clusters growing from the axils of the upper leaves, are very conspicuous.

GROWTH Perennial.
HEIGHT To 20cm (stems to 50cm).
FLOWERS Mar–Jul.
STATUS Native.
ALTITUDE 0–820m.

DETAIL Stems sprawling, rooting at the nodes, becoming erect towards the tip, with 2 *opposite rows of long white hairs* (may be hairy all round in shaded or young plants). Leaves hairy, 10–25mm, oval with a heart-shaped base, bluntly toothed, *stalkless or short-stalked* (to 5mm). Flowers 8–12mm across. The buds are often galled by a tiny midge, forming a small, cotton-wool like ball.

Wood Speedwell *Veronica montana*

Locally common in damp deciduous woodland, scrub and shady hedge-banks. ❀ Much like Germander Speedwell but *all leaves stalked* (stalks 5–15mm).

GROWTH Perennial.
HEIGHT To 10cm (40cm).
FLOWERS Apr–Jul.
STATUS Native.
ALTITUDE To 435m.

DETAIL Hairy. Stems sprawling, becoming erect at the tip, *hairy all round*. Leaves oval, coarsely-toothed, 20–30mm. Flowers in loose, long-stalked clusters growing from the axils of the upper leaves, 8–10mm across, *pale lilac-blue*.

Heath Speedwell
Veronica officinalis

Locally common in open woodland, grassland and heathland, on well-drained, often acid, soils. ❀ Stems prostrate, sometimes *mat forming*, with *erect* spikes of pale lilac flowers.

GROWTH Perennial.
HEIGHT Stems to 40cm.
FLOWERS May–Aug.
STATUS Native.
ALTITUDE 0–880m.

DETAIL Hairy. Leaves *oval*, very shallowly toothed, 10–20mm long, at least the lower short-stalked. Flowers in dense spikes growing from the leaf axils, each flower 5–9mm across, on a *short* stalk (c. 2mm, shorter than bract and calyx).

Sharp-leaved Fluellen *Kickxia elatine*

Local and often uncommon arable weed, favouring sandy and especially chalky soils. ✸ *Sprawling*, with spurred, 2-lipped flowers; *upper lip violet*, lower lip with a large yellow boss. The flowers are suspended on long stalks from the leaf axils and often partly hidden below the grey-green leaves, which have pointed, triangular bottom corners. **DETAIL** Hairy (but few glandular hairs). Stems usually well-branched towards base. Leaves 1.5–3cm long, long-stalked. Calyx cut into 5 pointed triangular lobes. Corolla 7–12mm, of which about half is the spur, stamens 4; flower stalk only hairy immediately below flower.

GROWTH Annual.
HEIGHT Stems 20–50cm.
FLOWERS Jul–Oct.
STATUS Ancient introduction.
ALTITUDE Lowland.

Round-leaved Fluellen *Kickxia spuria*

Local and often uncommon arable weed, favouring chalky soils, including chalky boulder clays. ✸ Very like Sharp-leaved Fluellen, but more robust and *densely sticky-hairy*, with *rounded leaves*. Upper lip more brownish-purplish. **DETAIL** Leaves 2.5–6cm. Corolla 8–15mm, spur curved, flower stalks hairy throughout.

GROWTH Annual.
HEIGHT Stems 20–50cm.
FLOWERS Jul–Oct.
STATUS Ancient introduction.
ALTITUDE Lowland.

Cornish Moneywort *Sibthorpia europaea*

Nationally Scarce. Local in damp, shady places on acid soils, especially mossy carpets by sheltered streams and ditches. ✸ Small, *shaggy-hairy, mat-forming evergreen*, with long-stalked leaves, 5–20mm across, cut into shallow lobes, and tiny whitish or pinkish flowers. **DETAIL** Stems root at the nodes. Leaves alternate. Flowers solitary, in leaf axils. Calyx cut into 5 strap-shaped lobes, corolla 1–2.5mm across, 5-lobed. Stamens 4. **SEE ALSO** Marsh Pennywort (p. 311).

GROWTH Perennial.
HEIGHT Stems to 40cm.
FLOWERS Late Jun–Oct.
STATUS Native.
ALTITUDE 0–515m.

Common Toadflax
Linaria vulgaris

GROWTH Perennial.
HEIGHT 30–80cm.
FLOWERS Late Jun–Oct.
STATUS Native.
ALTITUDE 0–360m.

Very common. Roadside banks, verges, and other rough, grassy places. ✿ Distinctive, showy, yellow, snapdragon-like flowers. DETAIL Hairless, or glandular-hairy above. Leaves 3–8cm long, very narrow, strap-shaped, grey-green. Corolla 18–35mm, including a more or less straight spur of 6–13mm, 2-lipped, yellow with a large orange-yellow boss on the lower lip; stamens 4.

Purple Toadflax
Linaria purpurea

A common garden escape. Walls, pavement cracks, rough ground, often around houses. ✿ Tall, slender spikes of small, purplish (occasionally pink), snapdragon-like flowers distinctive.

GROWTH Perennial.
HEIGHT To 100cm.
FLOWERS Jun–Sep.
STATUS Introduced to gardens from Italy. Recorded in the wild around 1830 and still spreading.
ALTITUDE Lowland.

DETAIL Hairless. Stem usually purplish, leaves strap-shaped, 4–6mm wide, dark blue-green, sometimes purplish. Corolla 7–15mm long, including a curved, pointed spur (3–6mm long, at least half the length of the rest of the corolla), purplish or mauve with slightly darker purplish veins, boss on lower lip the same colour or white; flowers pink in some cultivars.

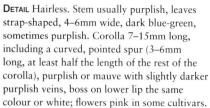

Pale Toadflax
Linaria repens

GROWTH Perennial.
HEIGHT To 80cm.
FLOWERS Jun–Sep.
STATUS Ancient introduction.
ALTITUDE 0–415m.

Locally common. Railway tracks, waste ground and rough grassy places, especially on chalky soils. ✿ Slender spikes of whitish to pale mauve flowers, marked with bluish-violet veins. The *short spur* and *orange-yellow patch on the boss on the lower lip* separate it from pale forms of Purple Toadflax. DETAIL Stems usually green, often glandular-hairy near tip (otherwise hairless), leaves to 2.5mm wide, grey-green. Corolla 8–15mm, including a short, straight, blunt spur (1–5mm long, less than half the length of the rest of the corolla).

Ivy-leaved Toadflax
Cymbalaria muralis

Common. Old walls, pavement cracks, railway ballast, usually near houses, also shingle beaches and occasionally other dry, rocky places. ❀ Straggling, with small, usually 5-lobed, glossy, ivy-like leaves. Flowers lilac (occasionally white) with a yellow spot. **Detail** Hairless (or sparsely hairy on calyx and young shoots). Stems sometimes rooting at lower nodes, often purple. Leaves 1–4cm, long-stalked with palmate veins. Flowers solitary in leaf axils, corolla 2-lipped, 9–15mm long, including the short, curved spur.

GROWTH Perennial.
HEIGHT Stems to 60cm.
FLOWERS May–Sep
(Apr–Dec).
STATUS Introduced to gardens from S Europe. Recorded in the wild by 1640.
ALTITUDE 0–570m.

Small Toadflax
Chaenorhinum minus

Rather local on dry, often chalky soils: railway lines, forestry rides, waste ground; also an uncommon and declining arable weed.

GROWTH Annual.
HEIGHT 8–25cm.
FLOWERS May–Oct.
STATUS Ancient introduction.
ALTITUDE 0–425m.

❀ Small but often well branched, with blunt, strap-shaped, dull grey-green leaves and small pale purple flowers with a small pale yellow boss at the mouth of the lower lip. **Detail** Sticky hairy (rarely hairless). Leaves alternate, 10–25mm. Flowers solitary, on long stalks from the leaf axils. Corolla 2-lipped, 6–9mm long including a narrow conical spur at the base.

Weasel's Snout
Misopates orontium

Aka Lesser Snapdragon. A scarce arable weed of light soils. Listed as Vulnerable.

GROWTH Annual.
HEIGHT 20–50cm.
FLOWERS Jul–Oct.
STATUS Ancient introduction.
ALTITUDE Lowland.

❀ Small, bright pink snapdragon-like flowers distinctive. **Detail** Stems usually glandular-hairy towards tip. Leaves 30–50mm, narrow, strap-shaped, sparsely glandular-hairy. Calyx lobes strap-shaped, unequal, longer than corolla tube. Corolla 10–17mm, 2-lipped, with broad, rounded pouch at base of tube and the mouth closed by a boss on the lower lip; stamens 4. Fruit a sticky-hairy capsule, said to resemble a Weasel's snout.

Ribwort Plantain *Plantago lanceolata*

Abundant. Almost any sort of grassland, pavement cracks, waste ground, etc., avoiding only very acid soils. ❁ Plantains have their leaves all at the base of a leafless flower stalk. In this species the leaves are *strap-shaped to long-oval, tapering gradually to the stalk, with 3–5 bold, parallel veins,* and usually held more or less erect. Flowers tiny, in a dense spike. **Detail** Leaves 5–30cm, pointed, hairy to more or less hairless, sometimes weakly toothed. *Flower stalk strongly ribbed,* silky-hairy. Flowers spike up to 4(8)cm long, individual flowers *c.* 4mm across, with 4 sepals, chaffy-white with a brown midrib, corolla divided at tip into 4 lobes, colour as sepals; 4 long stamens with yellowish anthers and white filaments, and 1 long, brush-like stigma. Each flower lies behind a tiny greenish bract.

GROWTH Perennial.
HEIGHT 10–50cm.
FLOWERS Apr–Oct.
STATUS Native.
ALTITUDE 0–790m.

Greater Plantain *Plantago major*

GROWTH Perennial or annual.
HEIGHT 10–40cm.
FLOWERS Jun–Oct.
STATUS Native.
ALTITUDE 0–700m.

Abundant. The sides of paths and tracks, pavement cracks, gardens, bare and waste ground, arable fields; tolerant of trampling. ❁ *Leaves broadly oval, long-stalked.* Flower spike a 'rat's tail', with *short stamens* and small, dull purple anthers. **Detail** Leaves 7–15cm, more or less hairless, with 3–9 prominent veins. Flower spike 10–20cm. Sepals green with a white border, corolla lobes brownish-white.

Buck's-horn Plantain
Plantago coronopus

Locally abundant on bare, sandy
and gravelly ground, in short
turf and on rocks. Especially
common on the coast, and very
tolerant of trampling (can be
abundant on rough car parks
and paths). ❁ Forms ground-
hugging rosettes 1–25cm across,
*leaves usually with several teeth
or triangular lobes on each side*
(the 'tines' of the buck's antlers).
Detail Leaves to 12cm, variably
hairy and sometimes slightly

GROWTH Annual to perennial.
HEIGHT To 20cm.
FLOWERS May–Jul (Oct).
STATUS Native.
ALTITUDE 0–455m (790m).

fleshy. Flower spike 2–4(7)cm long, floral bracts green, *tapering
to a long, fine point*; sepals translucent white with a green midrib; corolla
lobes pale brown *without a conspicuous midrib*; stamens yellow.

Sea Plantain
Plantago maritima

Common. Saltmarshes, cliffs and
short turf by the sea, also inland
in the uplands in damp grassy
and rocky places and sometimes
on road verges. ❁ Has *narrow,
fleshy leaves* in upright tufts.
Detail Tufted, growing from a
woody rootstock. Leaves 2–22cm
× 3–8mm, half-cylindrical or
channelled in cross-section,

GROWTH Perennial.
HEIGHT To 30cm.
FLOWERS Jun–Sep.
STATUS Native.
ALTITUDE 0–790m.

pointed, sometimes slightly toothed, obscurely veined, usually hairless.
Flower spikes 2–7(10)cm long, bracts green, arrow-shaped; sepals
translucent white with a green midrib, *corolla lobes translucent pink with
a brown midrib*; stamens yellow. See Also Sea Arrowgrass (p. 327), which
has narrower, more deeply channelled leaves.

Hoary Plantain *Plantago media*

Locally common. Dry grassland, usually over
chalk and limestone but also heavy clays,
including verges, churchyards, downland.
❁ Leaves oval, *hairy*, rather bluish-
green, *tapered to a short, winged
stalk* and conspicuously 'pleated'
with 5–9 prominent veins.
Flowers scented, with *obvious
long white stamens*, the filaments
variably tinged pinkish-purple.
Detail Leaves 5–8cm, weakly

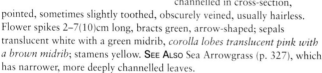

GROWTH Perennial.
HEIGHT To 40cm.
FLOWERS May–Aug.
STATUS Native.
ALTITUDE 0–540m.

toothed. Flower spike 2–6(12)cm. Sepals green with a
white border, corolla lobes translucent white.

GROWTH Perennial.
HEIGHT To 150cm.
FLOWERS
Jun–Sep.
STATUS
Native.
ALTITUDE
Lowland.

Water Figwort
Scrophularia auriculata

Fairly common. Wet places beside water, also marshes and wet woodland. ✿ Figworts are tall, with small, dark, inconspicuous globular flowers, the entrance overhung by the upper lip. DETAIL Leaves and stem hairless, flower spike glandular-hairy. Stem 4-angled, *'winged'*. Leaves 6–12cm long, opposite, usually with a *blunt tip* and rounded to heart-shaped base, *margins typically bluntly-toothed; leaf stalk winged,* often with a *small pair of leaflets at the base.* Sepals rounded, with a *pale border c. 0.5–1mm wide.* Corolla greenish to purplish-brown, *c.* 10mm long, divided at the mouth into 5 lobes, 2 forming the upper lip. Stamens 4, plus a fifth dark maroon, disc-shaped, sterile stamen at the top of the flower, the *staminode.*

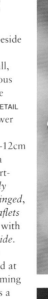

main stem in cross-section, showing 'wings'

Green Figwort

Water Figwort

Commmon Figwort

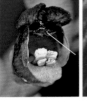

Green Figwort   Water Figwort
the staminode is arrowed

GROWTH Perennial.
HEIGHT To 150cm.
FLOWERS Jul–Sep.
STATUS Native.
The first record was not until 1840 and may have been introduced by wildfowl. Still spreading.
ALTITUDE Lowland.

Green Figwort
Scrophularia umbrosa

Local. Damp, fertile places beside rivers and streams and in wet woodland. ✿ Much like Water Figwort and easily overlooked, but flower spike with *numerous small leafy bracts,* especially towards base, thus not so conspicuously bare, and *stems rather more broadly 'winged',* especially towards base. DETAIL Leaves pointed to blunt at the tip, rounded at base or *tapering* to the stalk, margins typically with fine, slightly pointed teeth; stalk obviously winged, but *without leaflets.* Calyx teeth with similar broad pale margins. Staminode obviously *rather broader than tall* and *more or less 2-lobed.*

Common Figwort
Scrophularia nodosa

Fairly common. Damp, fertile places in woods and alongside ditches and rivers, sometimes on drier waste ground.
✹ As Water Figwort but stems 4-angled, *not winged* (or only very slightly so), *leaf stalks not winged,* and *leaves pointed*, with *saw-toothed edges*. DETAIL Stems hairless to hairy. Leaves hairless (rarely hairy), 6–13cm long with a rounded to heart-shaped base; leaf stalks short. Sepals 5, oval, blunt, green with a very *narrow pale border* (<0.3mm wide). Corolla greenish to purplish-brown. Staminode disc-shaped, sometimes notched at the tip.

GROWTH Perennial.
HEIGHT 40–100cm.
FLOWERS Jun–Sep.
STATUS Native.
ALTITUDE To 480m.

Balm-leaved Figwort
Scrophularia scorodonia

Nationally Scarce. Local. Note restricted range in SW England; rarely naturalised elsewhere. Rough, scrubby grassland along hedges, field margins and cliff-tops, and old quarries and other waste places. ✹ *Stem and leaves hairy,* with distinctive *deeply wrinkled, sage-like leaves*. DETAIL Stem bluntly to sharply 4-angled. Leaves 5–10cm long, either pointed or blunt at the tip, *heart-shaped at base*; margins scalloped to double-toothed. Flowers dull purple, structure as Common Figwort but staminode disc-shaped.

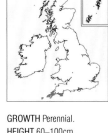

GROWTH Perennial.
HEIGHT 60–100cm.
FLOWERS Jun–Aug
(May–Oct).
STATUS Introduced? First recorded in 1712 (Cornwall) and may have been introduced accidentally via ports.
ALTITUDE Lowland.

Yellow Figwort
Scrophularia vernalis

Very local in shady places under hedges, in woodland clearings and on waste and rough ground. ✹ Whole plant *glandular-hairy*, very slightly sticky, with a faintly unpleasant smell. Leaves *nettle-like*, flowers *greenish-yellow*; unlike other figworts they have *5 equal corolla lobes forming a pouch*, with the 5 stamens all similar and *projecting from its mouth*. DETAIL Leaves 4–15cm long with a heart-shaped base, strongly double-toothed, the lower long-stalked.

GROWTH Biennial
(perennial).
HEIGHT To 50cm (80cm).
FLOWERS Apr–Jun.
STATUS Introduced from Europe. Recorded in the wild from 1633 and still spreading.
ALTITUDE Lowland.

Great Mullein
Verbascum thapsus

Common. Scrub, verges, railway banks, old quarries and other rough, grassy places, usually on dry soils. Sometimes a garden escape.

❁ Tall and striking, with the flower spike usually unbranched (or with small branches at the base) and the *foliage densely covered with greyish-white, felt-like hairs that do not wear off.*
DETAIL Lower leaves 15–45cm, narrowing to a short, winged stalk; middle and upper stem leaves stalkless, the wings *extending down the stem.*
Flowers 15–30mm across, the 3 upper stamens with orange, kidney-shaped anthers at 90° to the filaments, which have abundant white hairs, the 2 lower stamens with anthers upright or at 45° and *filaments more or less hairless.* Stigma knob-like.
SEE ALSO Evening-primrose (p. 113).

GROWTH Biennial.
HEIGHT to 200cm.
FLOWERS Jun–Aug.
STATUS Native.
ALTITUDE 0–570m.

Dark Mullein *Verbascum nigrum*

Locally common. Verges, field margins and other grassy places, usually on well-drained, often chalky, soils. Sometimes a garden escape.
❁ Separated from other mulleins by the clear dark green foliage, with the lower leaves long-stalked, wrinkled and heart-shaped at the base, and dense reddish-purple hairs on the filaments of the anthers. **DETAIL** Basal leaves to 20cm, slightly hairy above, more densely so below, where grey-green.

GROWTH Biennial or short-lived perennial.
HEIGHT To 120cm.
FLOWERS Jun–Sep.
STATUS Native.
ALTITUDE 0–335m.

Flowers 12–20mm across, blotched with reddish-purple at the base. Anthers orange-yellow, all placed at 90° to the filaments.

Hoary Mullein
Verbascum pulverulentum

Locally common; note range. Verges, field borders, coastal shingle. ❁ Flower spike usually well-branched, resembling a candelabra. Foliage with a marked greyish-white bloom when young, due to a dense covering of felted hairs, eventually becoming patchily green as the hairs wear off.
DETAIL Basal leaves short-stalked or stalkless, 10–30cm. Flowers 18–25mm across. Anthers orange-scarlet, all at 90° to the filaments, which have abundant white hairs. **SIMILAR SPECIES White Mullein** *V. lychnitis* usually has white flowers, but they can be pale yellow. The leaves are usually only sparsely mealy above, even when fresh.

GROWTH Biennial.
HEIGHT To 150cm.
FLOWERS Jul–Aug.
STATUS Native.
ALTITUDE Lowland.

FAMILY LAMIACEAE: DEAD-NETTLES, MINTS ETC. ('LABIATES')

A distinctive family, most species having a square stem and flowers with a distinct lower lip, grouped into clusters at the base of the leaves and bracts. Many are strongly scented, and the family includes herbs such as Basil and Mint.

Looking at the flower in more detail, the 5 sepals are fused into a calyx tube with 5 teeth at the mouth, and these teeth are sometimes divided between an upper lip (3 teeth) and lower lip (2 teeth). The 5 petals are fused into a corolla, which is split into 3–5 lobes at the mouth, generally grouped into an upper lip (usually 2-lobed) and a lower lip (usually 3-lobed), but in some species the lobes are almost identical.

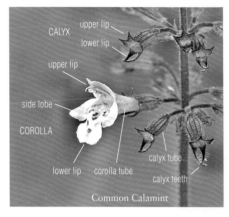

Common Calamint

White Dead-nettle *Lamium album*

Common. Gardens, hedgebanks, verges and waste ground. ❀ White flowers, in dense, well-spaced whorls, distinctive. The leaves resemble a nettle but lack stinging hairs. DETAIL Grow from rhizomes and may be patch-forming. Hairy. Stems often purplish. Leaves 3–7cm long. Calyx 9–15mm, with 5 narrow teeth. Corolla 18–25mm, upper lip long, hooded, enclosing the 4 stamens; lower lip marked with greenish-gold, side lobes small, with 2–3 teeth; corolla tube slightly pouched near base.

GROWTH Perennial.
HEIGHT 20–60cm.
FLOWERS Mar–Dec.
STATUS Ancient introduction.
ALTITUDE 0–345m.

Spotted Dead-nettle *Lamium maculatum*

Occasional but increasing garden escape. Verges, rough and waste ground, usually near houses. ❀ Whitish blotch along centre of leaves, flowers pinkish-purple. DETAIL Hairy. Stems grow from rhizomes and may be patch-forming. Leaves 2–5cm, usually heart-shaped, toothed. Flower structure as White Dead Nettle but corolla 20–35mm with just 1 tooth on side lobes of lower lip. Rarely, white patch on leaves absent or flowers white.

GROWTH Perennial.
HEIGHT To 30cm (60cm).
FLOWERS Mar–Oct.
STATUS Introduced to gardens from Italy in 1683. Recorded in the wild from c. 1730.
ALTITUDE To 490m.

GROWTH Annual.
HEIGHT 10–25cm (40cm).
FLOWERS Mar–Oct (all year).
STATUS Ancient introduction.
ALTITUDE 0–610m.

Red Dead-nettle *Lamium purpureum*

Abundant weed of disturbed ground: gardens, arable fields, etc. ☻ Softly downy, with whorls of reddish-purple flowers. *Leaves all stalked*, with small, blunt teeth. DETAIL Stems sparsely hairy to hairless. Leaves and bracts 1–5cm long, teeth <2mm long; uppermost bracts stalkless. Calyx 5–7.5mm, with 5 narrow teeth. Corolla 10–18(20)mm long, upper lip enclosing the 4 stamens, lower lip with small side lobes, corolla tube straight, with a ring of hairs inside.

GROWTH Annual.
HEIGHT 10–25cm.
FLOWERS Mar–Oct.
STATUS Ancient introduction.
ALTITUDE 0–320m.

Cut-leaved Dead-nettle
Lamium hybridum

Common but easily overlooked. Arable fields, gardens, waste ground and other disturbed places, on dry, fertile soils. ☻ Very like Red Dead-nettle but often more slender, less hairy, and *leaves more deeply toothed*. DETAIL As Red Dead-nettle but leaves with many teeth over 2mm long. Flowers often much shorter and not opening, corolla with few or no hairs inside.

GROWTH Annual.
HEIGHT 5–25cm.
FLOWERS Apr–Aug.
STATUS Ancient introduction.
ALTITUDE 0–455m.

Henbit Dead-nettle *Lamium amplexicaule*

Local in small numbers. Arable fields and other disturbed ground on light soils, also walls and pavement cracks. ☻ *Upper leaves and bracts stalkless, forming a collar below the whorls of pinkish-purple flowers.* Only 1–2 flowers open at once, with the *long, straight corolla tubes held up at 45°*, surrounded by beautiful rose-purple buds. DETAIL Downy. Stems branched from base. Leaves 0.7–2.5cm, lower leaves with stalks 3–5cm long. Calyx 5–7mm, densely covered with *erect white hairs, calyx teeth erect to converging in fruit*. Corollas 14–20mm, including the 10–14mm-long tube, which is hairless inside; lower lip <3mm long. Some flowers are very short (no longer than calyx) and may not open. SIMILAR SPECIES **Northern Dead-nettle** *L. confertum* is confined to Scotland and parts of Ireland, mostly near the coast. More robust, calyx 8–12mm long, with hairs flattened and teeth spreading in fruit, lower lip of corolla more than 3mm long.

Hedge Woundwort
Stachys sylvatica

Common. Woodland, hedge-
bottoms and other shady places
on moist, fertile soils; a garden
weed. ❀ Flowers in whorls,
reddish-purple blotched white.
Leaves nettle-like, *all stalked*,
with a *strong, unpleasant smell
when bruised.* DETAIL Stems
erect, bristly-hairy, with glandular
hairs on upper part, growing from
a creeping rhizome. Leaves softly
hairy, 4–9cm, stalks 4–12cm long. Calyx

GROWTH Perennial.
HEIGHT 30–80cm (100cm).
FLOWERS Jun–Sep.
STATUS Native.
ALTITUDE 0–500m.

with 5 long-triangular teeth. Corollas 13–18mm, upper lip hooded,
enclosing the 4 stamens, lower lip with 3 rounded lobes. SIMILAR SPECIES
Hybrid Woundwort *S.* × *ambigua*, the cross with Marsh Woundwort,
is widespread, and often found without 1 or both parents. Sterile, it
has shorter leaf stalks (10–20% of combined length of leaf and stalk in
middle and upper stem leaves; in Hedge Woundwort 25–50%).

Marsh Woundwort
Stachys palustris

Locally common. Damp places
near water, also wet grassland
and marshes; occasionally an
arable weed. ❀ Much like Hedge
Woundwort but flowers pinkish-
purple and *leaves on central and
upper stem stalkless, narrowly
spear-shaped and only faintly
scented when bruised.*
DETAIL Roughly hairy. Stems erect,
with glandular hairs on upper parts,
growing from strongly creeping rhizomes.

GROWTH Perennial.
HEIGHT 30–80cm (100cm).
FLOWERS Jun–Sep.
STATUS Native.
ALTITUDE 0–540m.

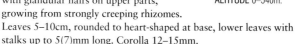

Leaves 5–10cm, rounded to heart-shaped at base, lower leaves with
stalks up to 5(7)mm long. Corolla 12–15mm.

Field Woundwort
Stachys arvensis

Near Threatened. An uncommon arable
weed on neutral to acid soils. ❀ Sparse
whorls of *small*, dull pinkish flowers with a
disproportionately large calyx.
DETAIL Hairy. Stems erect to
sprawling. Leaves 1.5–4cm,
stalked. Calyx with 5 narrow,
pointed teeth, corollas small,
6–8mm, pale pink streaked
with purple, upper lip hooded,
enclosing the 4 stamens, lower
lip with 3 rounded lobes.

GROWTH Annual.
HEIGHT To 25cm.
FLOWERS Apr–Oct.
STATUS Ancient introduction.
ALTITUDE To 380m.

Black Horehound
Ballota nigra

Common. Hedge-bottoms, verges, walls, often on rich soils. ❁ Upright, with tight whorls of bright pinkish-purple flowers, each with a conspicuous trumpet-shaped calyx. Foliage with an unpleasant small when bruised, often turning purplish-back.
DETAIL Roughly hairy. Leaves 2–5cm long, stalked. Calyx *c.* 10mm long, 10-veined, expanding towards the tip into 5 short, triangular teeth (no longer than wide). Corolla 9–15mm long, upper lip hooded, enclosing the 4 stamens, lower lip 3-lobed.

GROWTH Perennial.
HEIGHT 40–80cm (100cm).
FLOWERS Jun–Sep (Oct).
STATUS Introduced in the Iron Age.
ALTITUDE Mostly lowland, but to 480m.

White Horehound
Marrubium vulgare

Nationally Scarce. Very local. Rough, grassy slopes over chalk and limestone near the sea and grassy and heathy places in Breckland; otherwise a much-decreased relict of cultivation. ❁ Foliage *prominently wrinkled*, thyme-scented, with a *variably dense covering of felted white hairs*. Flowers in tight, well-spaced whorls. **DETAIL** Stem white-felted, often branched. Leaves 1.5–4.5cm, roughly circular, bluntly-toothed, upper short-stalked, lower leaves long-stalked. *Calyx with 10 small teeth, hooked at the tip* and spreading in fruit; corolla 15mm, white, upper lip narrow, 2-lobed, lower lip 3-lobed (central lobe much the largest); stamens 4, all *within corolla tube.*

GROWTH Perennial.
HEIGHT 15–60cm.
FLOWERS Jun–Jul (Oct).
STATUS Native and introduced.
ALTITUDE Lowland.

Catmint *Nepeta cataria*

Listed as Vulnerable. Uncommon and local in rough or scrubby grassland on chalk, especially verges, roadside banks and hedge-bottoms. ❁ Erect and often well branched, *softly grey-downy, mint-scented* when bruised. Flowers crowded into dense heads at the tip of the stems. **DETAIL** Leaves 3–7cm. Flowers in dense whorls in an oblong head. Calyx with 15 ribs and 5 narrow-triangular teeth; corolla 7–12mm, tube shorter than calyx, *upper lip more or less flat,* lower lip 3-lobed; stamens 4, shorter than upper lip, anthers crimson-purple.
SIMILAR SPECIES Garden Catmint *N.* × *faassenii* has blue flowers.

GROWTH Perennial.
HEIGHT 40–100cm.
FLOWERS Jul–Sep.
STATUS Ancient introduction.
ALTITUDE Lowland.

Common Hemp-nettle *Galeopsis tetrahit*

Common. Moist, lightly shaded places: woodland rides and clearings, hedgebanks, fens; also an arable weed. ❀ *Roughly bristly*, with *relatively small, white to pinkish flowers, variably marked with fine purplish scribbles*, opening a few at a time in each whorl. DETAIL Stems usually widely branching, slightly swollen below nodes. Leaves 2.5–10cm, coarsely-toothed. Calyx 12–14mm long, with 5 widely-spreading spine-tipped teeth. Corolla 13–20(25) mm long, rarely more than 1.5 times total length of calyx, upper lip hooded, enclosing the 4 stamens, lower lip 3-lobed with conical spurs at the base of the side lobes, *central lobe flat and, at most, only slightly notched, the dark markings falling well short of the tip.*

GROWTH Annual.
HEIGHT 10–100cm.
FLOWERS Jul–Sep.
STATUS Native.
ALTITUDE 0–570m.

Bifid Hemp-nettle *Galeopsis bifida*

Very like Common Hemp-nettle and easily overlooked – they are found in similar habitats, often together, but Bifid is more of an arable weed. ❀ Flower colour very variable (sometimes pale yellow), but the central lobe of the lower lip is always *distinctly notched* and *folded downwards along the margins*, while the *dark markings extend almost to the edge* – sometimes the whole lip may be dark.
DETAIL Corolla 13–16mm.

GROWTH Annual.
HEIGHT 10–100cm.
FLOWERS Jul–Sep.
STATUS Native.
ALTITUDE To 460m.

Large-flowered Hemp-nettle *Galeopsis speciosa*

Listed as Vulnerable. Becoming very scarce to S, commoner to N. A declining arable weed, favouring root crops on peaty soils and other disturbed ground. ❀ As Common Hemp-nettle but flowers clearly larger, with a *bold purple blotch on the lower lip.* DETAIL As Common Hemp Nettle but stems more strongly swollen below the nodes, where there are yellowish-tipped glandular hairs (especially on upper stem). Calyx 12–17mm; corolla 27–35mm (rarely only 22mm), *about twice total length of calyx.*

GROWTH Annual.
HEIGHT To 100cm.
FLOWERS Jul–Sep.
STATUS Ancient introduction.
ALTITUDE 0–445m.

Yellow Archangel
Lamiastrum galeobdolon

Locally common. Damp, deciduous woodland and shady verges, usually on heavy soils, also limestone pavement. ❂ Distinctive whorls of *large, yellow flowers marked with reddish-brown streaks*. DETAIL More or less hairy. Patch-forming, with long, leafy runners. Leaves 3–7cm long, stalked. Calyx with 5 long-triangular teeth. Corolla *c*. 20mm long, upper lip hooded, enclosing the 4 stamens, lower lip 3-lobed. SIMILAR SPECIES The cultivated form, subspecies *argentatum*, is an increasingly common garden throw-out, with silvery patches on the leaves. It comes into flower earlier and may form extensive patches.

GROWTH Perennial.
HEIGHT 20–30cm (60cm).
FLOWERS May–Jun.
STATUS Native.
ALTITUDE 0–425m.

argentatum

Bastard Balm
Melittis melissophyllum

Nationally Scarce, listed as Vulnerable. Uncommon. Light shade on damp, base-rich soils: woods (especially following coppicing), scrub and hedgebanks. ❂ *Flowers large and conspicuous, creamy or pink with a rosy-purple blotch on the lower lip*. DETAIL Shaggy hairy. Leaves 5–9cm long, oval, stalked, bluntly-toothed, smelling of mock-orange. *Calyx bell-like, 2-lipped*, upper lip with 2–3 small teeth, lower with 2 deeper lobes. Corolla 25–40mm long, upper lip flat or slightly hooded, lower lip longer, 3-lobed; stamens 4, shorter than upper lip. SEE ALSO Large-flowered Hemp-nettle (p. 231).

GROWTH Perennial.
HEIGHT 20–50cm (70cm).
FLOWERS May–Jun (Apr–Jul).
STATUS Native.
ALTITUDE Lowland.

Balm *Melissa officinalis*

Long-established escape from cultivation on hedgebanks, verges and waste ground, usually near houses. ❂ *Leaves strongly lemon-scented when crushed*, flowers small and rather inconspicuous, whitish or pinkish (buds rich yellow). DETAIL Variably hairy. Leaves 3–7cm, wrinkled, lower leaves long-stalked. *Calyx 2-lipped*, upper lip with 3 short, triangular teeth, lower lip with 2 long, narrow teeth. Corolla 8–15mm, *tube curved upwards*, upper lip 2-lobed, flat or slightly hooded, lower lip 3-lobed; stamens 4, shorter than upper lip.

GROWTH Perennial.
HEIGHT 30–60cm (100cm).
FLOWERS Aug–Sep.
STATUS Introduced to cultivation from S Europe by 995 and recorded in the wild from 1763. Increasing.
ALTITUDE Lowland.

Skullcap
Scutellaria galericulata

Locally fairly common. Marshes,
fens, damp ground by rivers,
streams and ponds, wet woodland,
dune slacks and, in Scotland,
boulder beaches. ✿ *Flowers
bright blue, in pairs at the base
of the leaves.* DETAIL Hairy. Stems
upright or sprawling. Leaves
2–7cm, short-stalked or stalkless,
wrinkled, heart-shaped at base
and tapering to a blunt tip, with
subtly-toothed edges. Calyx

GROWTH Perennial.
HEIGHT 20–50cm.
FLOWERS Jun–Sep.
STATUS Native.
ALTITUDE To 365m.

2-lipped, *upper lip with an erect flange on its back.* Corolla
10–20mm, *strongly curved upwards from the base*, upper lip
hooded, lower lip flattened, vaguely lobed; stamens 4.

Lesser Skullcap
Scutellaria minor

Locally fairly common. Bogs, wet
heaths and damp woodland on
acid soils. ✿ Small and slender,
with *small, pinkish-purple flowers
in pairs at the base of the leaves.*
DETAIL Sparsely hairy to hairless.
Leaves 1–3cm long, sometimes
with 1–4 teeth near base, short-
stalked. *Calyx with 2
untoothed lips, upper lip
with a small, erect flange
on its back.* Corolla

GROWTH Perennial.
HEIGHT 10–20cm (25cm)
FLOWERS Jul–Sep.
STATUS Native.
ALTITUDE 0–440m.

6–10mm, straight, upper lip short, hooded, lower
lip flattened,   vaguely lobed; stamens 4.

Gypsywort
Lycopus europaeus

Common. The banks
of streams, rivers,
ponds and lakes, fens,
wet woodland and dune
slacks. ✿ Upright, with
*dense whorls of tiny white
flowers, finely spotted purple. Leaf
margins cut into large, jagged teeth
(becoming narrow, pointed lobes
at base of lower leaves).*
DETAIL Patch forming, more or less
hairy. Leaves 5–10cm,

GROWTH Perennial.
HEIGHT To 60cm (100cm).
FLOWERS Jul–early Sep.
STATUS Native.
ALTITUDE 0–485m.

stalkless. Calyx with 5 teeth; corolla 3–5mm, with *4
almost equal lobes* (uppermost lobe slightly broader,
notched); *stamens 2, projecting from corolla.*

Wild Clary *Salvia verbenaca*

Locally common in S and E England, scarcer elsewhere. Dry, sunny grassland, usually on calcareous soils: verges, banks, churchyards and dunes. ❀ Rosettes of deeply wrinkled, blue-green, rather sage-like (and faintly sage-scented) leaves produce erect spikes, with *whorls of arched, violet-blue flowers*, often with 2 white spots on lower lip. DETAIL Finely hairy, with upper stem glandular-hairy. Basal leaves 4–12cm, usually lobed, long-stalked; stem leaves few, stalkless. Calyx *c.* 7mm long, sticky-hairy, longest hairs white, without glands, 2-lipped, upper lip with 3 minute teeth at tip, lower lip with 2 long teeth. Corollas of 2 sizes – larger 10–17mm, upper lip 2-lobed, strongly hooded, more or less arched, lower lip 3-lobed; stamens 2, shorter than upper lip, style short. Smaller corollas 6–12mm, not opening and more or less hidden in calyx.

GROWTH Perennial.
HEIGHT 30–80cm.
FLOWERS May–Aug.
STATUS Native.
ALTITUDE Lowland.

Meadow Clary

Salvia pratensis

Nationally Scarce. Near Threatened. Rare on unimproved grassland on dry soils over chalk and limestone. ❀ Flowers larger, brighter and more striking than Wild Clary. DETAIL As Wild Clary but basal leaves 7–15cm long, toothed (sometimes strongly double-toothed) but not lobed. Longest hairs on calyx brownish, gland-tipped. Corolla 15–30mm in bisexual flowers, *c.* 10mm in female flowers, with many glandular hairs. Style long, projecting.

GROWTH Perennial.
HEIGHT 30–80cm (100cm).
FLOWERS Jun–Jul.
STATUS Native (Sch 8).
ALTITUDE Lowland.

Betony *Betonica officinalis*

Locally common. Rough grassland, including verges, grassy heaths and woodland rides, mostly on slightly acid to chalky soils. ❀ Tufted, with a basal rosette of leaves. Flowering stems slender, with few pairs of leaves and short spikes composed of whorls of *reddish-purple flowers*. DETAIL Sparsely hairy. Rosette leaves long-stalked, 3–7cm long; 2–4 pairs of stem leaves, the upper stalkless; lowest bracts like the leaves. Calyx 7–9mm with 5 bristle-pointed teeth, not obviously hairy. Corollas 12–18mm, upper lip hooded, enclosing the 4 stamens, lower lip 3-lobed.

GROWTH Perennial.
HEIGHT 10–60cm (75cm).
FLOWERS Late Jun–Sep.
STATUS Native.
ALTITUDE 0–480m.

rosette leaf

GROWTH Perennial.
HEIGHT 10–30cm.
FLOWERS Apr–Jun.
STATUS Native.
ALTITUDE 0–760m.

Bugle *Ajuga reptans*

Common. Damp deciduous woodland and other shady places, also unimproved grassland, on neutral to acid soils. ❀ Patch-forming, with rosettes of *glossy dark green or purplish leaves* and short, erect, leafy stems bearing *whorls of powder blue flowers marked with white*. DETAIL Produces long, leafy rooting runners. Stems more or less hairy, but on 2 opposite faces only. Basal leaves forming a rosette, sparsely hairy, 4–8cm long; stem leaves few, the upper unstalked, grading into upper bracts, which are shorter than the flowers and usually purple. Calyx 5–6mm long, with 5 teeth. Corolla *c.* 12mm long, with tiny upper lip, 3-lobed lower lip, and ring of hairs inside tube; stamens 4.

Ground Ivy *Glechoma hederacea*

Very common. Shady places in woods, scrub and hedge-bottoms, also in grassland, where perhaps always a relict of former cover? ❀ Creeping and patch-forming, with *dull dark green, kidney-shaped leaves* and erect leafy stems bearing several *whorls of 2–4 bright blue flowers*; in full sun foliage often a rich bronzy-red. DETAIL More or less softly hairy. *Sends out slender, rooting runners up to 100cm long.* Leaves 1–4cm wide, those on runners having stalks up to 10cm long. Calyx with 5 slightly unequal teeth. Corolla 15–20mm in bisexual flowers (smaller in female flowers), tube shorter than calyx, upper lip flat, lower lip 3-lobed; stamens 4.

GROWTH Perennial.
HEIGHT 10–20cm (30cm).
FLOWERS Mar–Jun.
STATUS Native.
ALTITUDE 0–570m.

Selfheal
Prunella vulgaris

Common. Grassland, woodland rides, lawns, tracksides and waste ground on neutral to calcareous, often moist, soils. ✱ Patch-forming, with creeping runners and short erect flowering stems bearing *crowded oblong heads* of small bluish-violet flowers at the tip, surrounded by more or less purplish bracts. DETAIL Sparsely hairy. Leaves 1–5cm, widest at rounded or wedge-shaped base, dull, not or only slightly toothed, lower stalked, upper nearly stalkless. Calyx purplish, upper lip flattened, square-tipped, with 3 tiny teeth, lower lip shorter, cut into 2 narrow-triangular teeth. Corolla 10–15mm long, upper lip rounded, strongly hooded, enclosing the 4 stamens, lower lip 3-lobed.

GROWTH Perennial.
HEIGHT To 20cm (30cm).
FLOWERS Jun–Sep.
STATUS Native.
ALTITUDE 0–780m.

Wood Sage
Teucrium scorodonia

Common. Heathland, woodland rides, hedge-bottoms, dunes, shingle and limestone pavement, on acid to mildly alkaline soils. ✱ Clump or patch-forming, with *strongly wrinkled, sage-like (but hardly aromatic) leaves* and loosely-branched, leafless spikes of *small, pale greenish-cream flowers* all facing 1 side. DETAIL Downy. Leaves 3–7cm long, stalked. Flowers in opposite pairs with very small bracts. Calyx with 5 unequal teeth, the uppermost rounded, much broader than other teeth. *Corolla with upper lip absent,* lower lip 5-lobed, folded downwards 5–6mm long; 4 prominent stamens.

GROWTH Perennial.
HEIGHT 15–50cm.
FLOWERS Jul–Sep.
STATUS Native.
ALTITUDE 0–550m.

Wild Marjoram
Origanum vulgare

Locally common. Rough
grassland and bare ground
on poor, dry, calcareous soils:
verges, hedgebanks, scrub and
woodland rides. ❁ Upright, with
branched clusters of small, reddish-
purple (sometimes white) flowers. Foliage
pleasantly aromatic – the cultivated form
is the herb Oregano. DETAIL Sparsely hairy.

GROWTH Perennial.
HEIGHT 30–50cm (80cm).
FLOWERS Jul–Sep.
STATUS Native.
ALTITUDE 0–410m.

Leaves 1.5–5cm long, more or less
untoothed, stalked. Lower bracts leaf-
like, upper bracts small, purple. Calyx
tube hairy inside, with 5 short teeth. Corolla 4–7mm, upper
lip flat, shallowly 2-lobed, lower lip 3-lobed; stamens 4, longer
than flower in bisexual flowers (much shorter in female flowers).

Common Calamint
Clinopodium ascendens

Local. Hedgebanks, verges, and rough,
scrubby grassland, on dry, chalky or sandy
soils. Foliage distinctly mint-scented.
❁ Flowers pinkish-white with small purple
spots on lower lip, in several short-stalked
clusters in each whorl. DETAIL Hairy. Leaves
1.5–4cm, short-stalked. Calyx 5–8mm,
upper lip with 3 teeth, lower with 2 slightly
narrower, longer teeth (2–3.5mm, with
long hairs on margins over 0.2mm);
hairs in throat usually not projecting
from tube. Corolla 10–16mm, upper lip
flat, notched, lower lip short, 3-lobed;
stamens 4, shorter than upper lip. Style 1,
topped by 2 unequal stigmas.

GROWTH Perennial.
HEIGHT 30–60cm.
FLOWERS
Jul–Sep.
STATUS
Native.
ALTITUDE
To 380m.

Wild Basil *Clinopodium vulgare*

Locally common in S and E, scarcer elsewhere.
Rough grassland, hedgebanks, scrub, woodland
rides and dunes, on dry, calcareous soils.
❁ Flowers pinkish-purple, in dense whorls
around the stem, but *only a few open at any
one time.* DETAIL Hairy. Leaves 1.5–5cm long,
subtly-toothed, stalked (very like Marjoram
but more than 3 teeth per side and side veins
loop round to join at tip); weakly scented.
Calyx subtly S-curved, 7–9.5mm, upper lip
with 3 teeth, lower with 2 narrower, more
deeply-cut teeth. Corolla 12–22mm, upper
lip flat, notched, lower lip short, 3-lobed;
stamens 4, shorter than upper lip. Style 1,
tipped by 2 unequal stigmas.

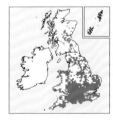

GROWTH Perennial.
HEIGHT 10–40cm
(75cm).
FLOWERS
Jul–Sep.
STATUS Native.
ALTITUDE
0–395m.

pouch

Basil Thyme
Clinopodium acinos

Listed as Vulnerable. Rather local on disturbed ground and short, sparse grassland on dry, chalky (sometimes sandy) soils, often on forestry rides and old railways. ❁ Low-growing, sometimes prostrate, with short, upright flower spikes and whorls of 4–6 *small violet or white flowers.* DETAIL Hairy. Leaves weakly scented, 5–15mm long, oval, shallowly-toothed. Calyx 4.5–7mm, *subtly S-curved, with a swollen, pouched base*; upper lip with 3 teeth, lower with 2 slightly longer and narrower teeth. Corolla 7–10mm, upper lip flat, notched, lower lip 3-lobed; stamens 4, shorter than upper lip. Style 1, topped by 2 unequal stigmas.

GROWTH Annual (perennial).
HEIGHT To 10cm (25cm).
FLOWERS May–Sep.
STATUS Native.
ALTITUDE Lowland.

Wild Thyme *Thymus polytrichus*

Common. Short turf and bare ground on dry soils over chalk, limestone, sands and gravels: mostly on chalk downland in the SE, elsewhere on upland grassland, heaths, cliff-tops, dunes and rocky places. ❁ Forms *dense mats*. Foliage *only faintly thyme-scented*. *Short*, erect stems with *dense, rounded heads* of pinkish-purple flowers. DETAIL Evergreen. Stems prostrate, rooting at nodes, reddish. Leaves hairless or sparsely hairy, to 8mm long. Flower stems more or less square in cross-section, *2 opposite faces hairy, sometimes densely so, alternating with 2 hairless to sparsely hairy faces; no long hairs on angles*. Calyx c. 4mm long, upper lip with 3 short teeth, lower with 2 longer, finer teeth. Corolla c. 7mm, upper lip slightly notched, lower 3-lobed. Stamens 4, longer than corolla.

GROWTH Perennial.
HEIGHT To 10cm.
FLOWERS May–Sep.
STATUS Native.
ALTITUDE 0–1125m.

Large Thyme *Thymus pulegioides*

Locally common. Short grassland on chalky soils, less often sands and gravels. ❁ Similar to Wild Thyme but stems often more erect, not forming dense mats, *flower heads more elongated* and *foliage strongly thyme-scented.* DETAIL As Wild Thyme but leaves 6–12mm, usually hairless. Flower stems more sharply 4-angled, with *long hairs on the angles* (note that 2 faces are typically broad, the other 2 narrow, thus hairs may appear as 2 parallel lines on opposite sides of the stem); the 2 narrow faces are sometimes finely downy.

GROWTH Perennial.
HEIGHT Shoots to 20cm (25cm).
FLOWERS Jul-Aug.
STATUS Native.
ALTITUDE Lowland.

GENUS *MENTHA:* WILD & GARDEN MINTS

Mints are a difficult group, due to wide variation and hybridisation, but most are easily recognised as a mint – the bruised foliage is strongly scented. The flowers are small, the calyx having 5 teeth and the corolla 4 more or less equal lobes; there are 4 stamens. Four species of mint are native, but several others, including hybrids, are widely cultivated and may be found as garden throw-outs – mints easily establish from root fragments and can be very persistent.

Spear Mint M. *spicata* is the commonest garden mint and fairly frequent as a garden throw-out. Flower heads terminal (as Water Mint) but *narrower (5–15mm wide), leaves more or less stalkless*; fertile, it has protruding stamens. Spearmint-scented. **Apple Mint** M. × *villosa*, the hybrid between Spear Mint and Round-leaved Mint M. *suaveolens* is also a common throw-out, with terminal flower heads 5–15mm wide and stalkless leaves, but these are usually rounded-oval, wrinkled and hairy. Stamens not protruding. **Pepper Mint** M. × *piperita*, the hybrid between Water Mint and Spear Mint, is also much like Water Mint, with a terminal flower head and stalked leaves, but the leaves and calyx tube are hairless or only sparsely hairy. Stamens not protruding. Peppermint scented.

Water Mint
Mentha aquatica

Common – by far the commonest mint. Wet places near water (may grow partially or wholly submerged), marshes, wet woods and dune slacks. ❀ Creeping and patch-forming, with a sweet, peppermint scent. Flowers lilac, with projecting stamens tipped by red anthers, in *dense, rounded heads, 12–25mm wide, at the tip of the stem*; often also smaller flower heads lower on the stem.

GROWTH Perennial.
HEIGHT 15–60cm (90cm).
FLOWERS Jul–Oct.
STATUS Native.
ALTITUDE 0–455m.

Leaves stalked. DETAIL Grows from creeping rhizomes. Variably hairy, plant sometimes washed purplish. Leaves oval, 2–6cm long, with shallow, blunt teeth; upper bracts small. Calyx hairy, teeth narrow, pointed, 3–4.5mm long. Corolla *c.* 6mm.

Corn Mint
Mentha arvensis

Locally common. Arable fields, wet grassland, woodland rides and waste ground. ❀ As Water Mint but *flowers in whorls in the leaf axils, terminal flower head absent.* Scent sweet and fruity, plant never purplish. DETAIL Hairy. Calyx 1.5–2.5mm long, teeth triangular, small (<0.5mm), exterior hairy, but no hairs in throat; corolla *c.* 6mm. *Stamens project from flower.*

GROWTH Perennial (annual).
HEIGHT To 60cm.
FLOWERS Jul–Sep.
STATUS Native.
ALTITUDE 0–435m.

SIMILAR SPECIES Whorled Mint M. × *verticillata*, the sterile hybrid with Water Mint, is widespread. Very similar, but calyx 2.5–3.5mm, teeth narrow, 1–1.5mm long, *stamens usually enclosed within flower*.

Monkeyflower *Mimulus guttatus*

GROWTH Perennial.
HEIGHT 20–50cm (80cm).
FLOWERS Jul–Sep.
STATUS Introduced to gardens from N America by 1812 and known from the wild by 1824.
ALTITUDE Lowland.

Locally common. Wet ground by streams, rivers and ponds, marshes and wet meadows. ❀ Creeping and patch-forming, with pale green foliage and large yellow flowers, often with red spots in the throat.
DETAIL Hairless below, bracts, flower stalks and flowers densely hairy, with many glandular hairs. Shoots often root at the nodes. Leaves opposite, 2–10cm, oval, toothed. Calyx 5-lobed, corolla 2.5–5cm long with a small, 2-lobed upper lip and 3-lobed lower lip (the latter with 2 bosses almost blocking the throat). Stamens 4, style 1, stigma 2-lobed.

Hybrid Monkeyflower *Mimulus × robertsii*

GROWTH Perennial.
HEIGHT To 60cm.
FLOWERS Jun–Sep.
STATUS Recorded from 1872.
ALTITUDE To 610m.

The hybrid between Monkeyflower and Blood-drop-emlets, *common in the uplands*. Habitat as Monkeyflower. ❀ *Flowers variably marked with orange, red or purplish-brown blotches on both the throat and lower lip.* **DETAIL** Sterile, producing little or no seed. Flower variably hairy, with the *bosses on lower lip of the flower small, not closing the throat.* **SIMILAR SPECIES Blood-drop-emlets** M. *luteus* is more or less hairless, with only sparse glandular hairs in the flowers. Dark blotches on flowers often large, sometimes on lower lip only. Fertile. Very scarce, *mostly in lowlands.*

FAMILY OROBANCHACEAE: BROOMRAPES & ALLIES

A group of plants that are either total or partial parasites. Broomrapes *Orobanche* and toothworts *Lathraea* are total parasites. Growing underground attached to the roots of other plants, they only appear above ground to flower. Their leaves are reduced to small scales and lack chlorophyll, and they cannot manufacture their own food. Cow-wheats *Melampyrum*, eyebrights *Euphrasia*, bartsias *Odontites* and *Parentucellia*, yellow rattles *Rhinanthus* and louseworts *Pedicularis* are partial parasites (sometimes termed 'root-hemiparasites'). They have green leaves and photosynthesise, but also extract nutrients from the roots of other plants.

Common Broomrape

Eyebrights
Euphrasia spp.

A genus of 18 closely related species, many of which are rarities. Although easily recognisable as a group, identification of the individual species is very difficult and made even more so by hybridisation: no less than 71 wild hybrids have been described. Eyebrights are locally abundant on a wide variety of sunny, permanent or semi-permanent grassland.
❀ Variable in stature, from tiny, compact plants in chalk grassland and cliff-top turf to long, straggling individuals in fenland vegetation. Leaves small and well-toothed, flowers small – fingernail-sized – but always distinctive: white to purple, veined darker and with a yellow blotch on the lower lip. All eyebrights are root-hemiparasites. DETAIL Stems usually well-branched. Leaves mostly opposite, sometimes flushed purple. Calyx split into 4 teeth. Corolla 2-lipped, lower lip split into 3 notched lobes.

GROWTH Annual.
HEIGHT 2–45cm.
FLOWERS May–Sep.
STATUS Native.
ALTITUDE 0–1215m.

Common Cow-wheat *Melampyrum pratense*

Locally common. Woodland, scrub, heaths and moors on poor, dry soils, usually acidic but sometimes over chalk and limestone. A root-hemiparasite.
❀ Patch-forming, with dark green or purplish, spear-shaped leaves and 2-lipped, tubular yellow flowers in pairs, both facing the same way, at the base of leaf-like bracts. DETAIL More or less hairless. Stems often purplish. Leaves opposite, more or less stalkless, strap-shaped, 1.5–8cm, grading towards tip of stem into green, leaf-like bracts, usually with 1–2 pairs of teeth near base. Calyx with 4 long, pointed teeth. Corolla yellow with purple marks near the mouth, 10–18mm long (much longer than calyx); upper lip forming a hood, lower lip with 3 small teeth and raised bosses partly closing the mouth.

GROWTH Annual.
HEIGHT 8–40cm (60cm).
FLOWERS May–Sep.
STATUS Native.
ALTITUDE 0–960m.

Yellow Rattle
Rhinanthus minor

Locally abundant. Unimproved, rough grassland, including fens and meadows. A root-hemiparasite. ❀ Erect, with yellow flowers and an *inflated calyx that becomes brown and bladder-like in fruit*, the seeds rattling around inside. DETAIL Variably but sparsely hairy. Leaves dark green, often tinged purplish, 2–6cm long, opposite, *stemless*, strap-shaped, expanding to a heart-shaped base, *prominently veined with coarse, blunt teeth*. Calyx 4-toothed, sometimes tinged reddish-purple, variably hairy. Corolla 12–15mm, upper lip flattened, hooded, with *2 tiny violet teeth on either side of the tip*, lower lip 3-lobed.

GROWTH Annual.
HEIGHT To 50cm.
FLOWERS May–Aug.
STATUS Native.
ALTITUDE 0–1075m.

Yellow Bartsia
Parentucellia viscosa

Local. Rough grassy places on damp, sandy soils, often on disturbed ground and mostly near the sea: dune slacks, grassy heaths, verges; sometimes sown with grass seed. A root-hemiparasite. ❀ Conspicuously *sticky-hairy*, with a leafy spike of bright yellow flowers. DETAIL Stems usually unbranched. Leaves opposite, 1.5–4cm. Calyx ribbed, with 4 teeth. Corolla 16–24mm, upper lip short, hooded, lower 3-lobed.

GROWTH Annual.
HEIGHT 10–50cm.
FLOWERS
Jun–Oct.
STATUS Native.
ALTITUDE
Lowland.

Red Bartsia
Odontites vernus

Locally common. Rough grassy places, including verges and field margins; in N and W Scotland also gravelly and rocky seashores and saltmarsh margins. A root-hemiparasite. ❀ Usually well-branched, with slender, erect stems and long, curved, leafy, 1-sided spikes of small, pinkish-purple flowers. DETAIL Downy. Stems often washed purple. Leaves opposite, unstalked, strap-shaped, 1–4cm long with 1–5 blunt teeth each side. Calyx 5–8mm long, split into 4 triangular teeth. Corolla 8–10mm, upper lip hooded, lower lip 3-lobed.

GROWTH Annual.
HEIGHT To 30cm (50cm).
FLOWERS Jun–Sep.
STATUS Native.
ALTITUDE 0–540m.

GROWTH Perennial (rarely biennial).
HEIGHT 10–20cm.
FLOWERS Apr–Jul.
STATUS Native.
ALTITUDE 0–915m.

Lousewort *Pedicularis sylvatica*

Locally common. Short grassland on damp (but not wet), acid soils, on heaths, moors and other rough grassy places. A root-hemiparasite. ✿ Usually *low-growing* and somewhat sprawling, *multi-stemmed*, with intricately cut, crimped leaves and leafy spikes of 2-lipped, pinkish-purple flowers, the upper lip erect and hooded at the tip, with a terminal tooth and *single tiny teeth to either side*, the lower lip flat and broadly 3-lobed. DETAIL Stems variably hairy, often purple, woody at base. Leaves mostly alternate, to 3cm, the margins strongly down-rolled, dark green, often washed reddish-purple. Calyx pouch-like, ribbed, split at the mouth into 4 short, toothed, leaf-like lobes, *usually more or less hairless, but may be hairy in the W*; the calyx becomes inflated and bladder-like in fruit. Corolla 2–2.5cm. *Capsule not longer than calyx*, seeds usually partly winged.

Marsh Lousewort *Pedicularis palustris*

GROWTH Annual to biennial.
HEIGHT To 60cm.
FLOWERS May–Sep.
STATUS Native.
ALTITUDE 0–670m.

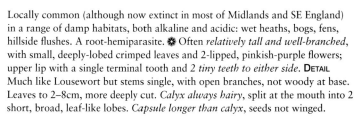

Locally common (although now extinct in most of Midlands and SE England) in a range of damp habitats, both alkaline and acidic: wet heaths, bogs, fens, hillside flushes. A root-hemiparasite. ✿ Often *relatively tall and well-branched*, with small, deeply-lobed crimped leaves and 2-lipped, pinkish-purple flowers; upper lip with a single terminal tooth and 2 *tiny teeth to either side*. DETAIL Much like Lousewort but stems single, with open branches, not woody at base. Leaves to 2–8cm, more deeply cut. *Calyx always hairy*, split at the mouth into 2 short, broad, leaf-like lobes. *Capsule longer than calyx*, seeds not winged.

Common Broomrape
Orobanche minor

GROWTH Usually annual.
HEIGHT 10–30cm (60cm).
FLOWERS May–Jul (Sep).
STATUS Native.
ALTITUDE Lowland.

Locally common and by far the commonest broomrape. Parasitic on a wide variety of plants, but especially

members of the pea and daisy families. May appear in gardens, municipal plantings, supermarket car parks etc. ❀ Stigmas usually purple. DETAIL Stems reddish-brown, variably washed purple. Corolla 10–18mm, creamy-yellow, usually strongly washed purple, lower lip cut into *3 equal, blunt lobes, without prominent bosses*. Stamens with filaments hairless or sparsely hairy at base, stigmas with the *2 lobes clearly separate*. Occasional variants are overall rather yellow, with yellow stigmas. Var. *maritima* is found on the S coast, usually on Wild Carrot. It often has darker purple stems with a bulbous base, the lower lip of the corolla has *large yellow bosses*, with the central lobe the largest, and the *stigma lobes are partly fused*.

Ivy Broomrape
Orobanche hederae

Locally common. Rocky woodland, quarries, cliffs etc. Parasitic on ivy, especially Atlantic Ivy. ❀ Often has a distinct *'nose cone' of unopened buds*, stem with flowers *from ground level*. Stigmas usually dull yellow, sometimes purple. DETAIL Stems brownish-purple. Corolla dull cream, washed and veined purplish, 10–22mm long, subtly pinched-in just behind the mouth (uniformly parallel-sided in Common), lower lip with more or less pointed lobes. Stamens with filaments usually hairless.

GROWTH Annual or perennial.
HEIGHT 10–60cm (105cm).
FLOWERS Jun–Oct.
STATUS Native.
ALTITUDE Lowland.

Knapweed Broomrape
Orobanche elatior

Very locally common. Rough grassland on chalk, limestone and other lime-rich soils. Parasitic on Greater Knapweed. DETAIL Stems yellowish. Corolla 18–25mm, dull yellow, subtly washed pink and finely veined reddish-brown. Stamens with filaments usually hairy to tip. Stigma yellow.

GROWTH Perennial.
HEIGHT 25–75cm.
FLOWERS Mid Jun–Jul (Aug).
STATUS Native.
ALTITUDE Lowland.

Yarrow Broomrape
Orobanche purpurea

Nationally rare, listed as
Vulnerable. Very local on rough
cliff-top grassland and verges.
Parasitic on Yarrow. ✿ *Purplish-
blue flowers distinctive.* Although
Common Broomrape sometimes
has purplish flowers, Yarrow
Broomrape is unique amongst
British Broomrapes in having
*2 small, pointed, strap-shaped
bracteoles at the base of each
calyx, in addition to the 4 calyx*

GROWTH Annual or short-
lived perennial.
HEIGHT 15–45cm.
FLOWERS Late May–Jul.
STATUS Native.
ALTITUDE Lowland.

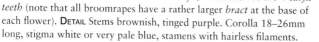

teeth (note that all broomrapes have a rather larger *bract* at the base of
each flower). DETAIL Stems brownish, tinged purple. Corolla 18–26mm
long, stigma white or very pale blue, stamens with hairless filaments.

Thyme Broomrape
Orobanche alba

Nationally Scarce. Very local on
rocky turf by the sea, mostly in
W Scotland (especially the Inner
Hebrides) and N and W Ireland,
also inland on limestone. Parasitic
on Wild Thyme. ✿ *Distinctively
reddish.* Flowers sweetly scented.
DETAIL Stems purplish-red. Corolla
15–20(25)mm, pale cream, usually
strongly washed dark red. Stigmas
reddish, the 2 lobes partly joined.
Stamens with hairy filaments.

GROWTH Annual.
HEIGHT 8–25cm (35cm).
FLOWERS Jun–Aug.
STATUS Native.
ALTITUDE 0–490m.

Greater Broomrape
Orobanche rapum-genistae

Nationally Scarce, listed as
Near Threatened. Very local
in scrubby places. Parasitic on
woody members of the pea family,
principally Common Gorse and
Broom. ✿ *Robust, often tall, and
overall reddish-brown* (Knapweed
Broomrape can be as tall or taller,
but is yellowish-brown). The
flowers have a strong, unpleasant
aroma. DETAIL Stems yellowish,
strongly washed purplish-red.
Corolla 20–25mm, pale yellow to
reddish, deeper red inside. Stigmas
yellow. Stamens with the filaments
glandular-hairy towards tip, basal
third hairless.

GROWTH Perennial.
HEIGHT 20–90cm.
FLOWERS May–early Aug.
STATUS Native.
ALTITUDE Lowland.

GROWTH Perennial.
HEIGHT 8–30cm.
FLOWERS Apr–May.
STATUS Native.
ALTITUDE To 350m.

Toothwort
Lathraea squamaria

Parasitic on the roots of a variety of trees, especially Hazel. Rather local. Deciduous woodland and hedgerows, often near the banks of streams and rivers; may have a preference for chalk and limestone. ❀ The pallid, ghostly flowers are unique, although easily overlooked; they die down rapidly after flowering. **Detail** Stems white or pale pink with a few scale-like bracts. Flowers in a 1-sided spike, each with an oval to diamond-shaped bract. Calyx sticky-hairy, cut towards the tip into 4

triangular lobes; corolla slightly longer (14–20mm), rose- or purplish-pink, tubular, divided at the tip into 2 lips; stigma projecting, creamy. See Also Yellow Birdsnest (p. 188), Birdsnest Orchid (p. 315).

Purple Toothwort *Lathraea clandestina*

GROWTH Perennial.
HEIGHT Around 5cm.
FLOWERS Mar–May.
STATUS Introduced from SW Europe to gardens *c.* 1888. First reported from the wild in 1908. Sometimes planted.
ALTITUDE Lowland.

Parasitic on the roots of a variety of trees, especially willows, poplars and alders. Very scattered in damp, shady places in woodland and hedgerows, especially near the banks of streams and rivers. ❀ Very distinctive. **Detail** No aerial stem, the flowers arising in clusters on short stalks directly from the underground rhizome. Calyx hairless; corolla 40–50mm, purplish-violet (white in var. *pallidiflora*).

Common Butterwort
Pinguicula vulgaris

Locally common (although extinct in most of central and S England). Bare, permanently wet ground, low in nutrients, in bogs, fens, flushed rocks and even alongside forestry rides. ✿ Rosettes of greenish-yellow leaves produce 1–3 slender, leafless stems, each carrying a violet-blue flower with a white throat. DETAIL Whole plant with abundant very short glandular hairs. Leaves 2–8cm long. Calyx and corolla 5-lobed, the corolla with a broad, flat, 3-lobed lower lip; 14–22(25) mm long, including the pointed, 4–7(10)mm-long spur. Fruit a capsule. Overwinters as a bud.

GROWTH Perennial.
HEIGHT 5–18cm.
FLOWERS May–Jul.
STATUS Native.
ALTITUDE 0–970m.

SIMILAR SPECIES **Large-flowered Butterwort** *P. grandiflora* is native to SW Ireland and introduced to a few sites in England and Wales. Flowers obviously larger (25–35mm, including spur 10–14mm), with a broad lip.

Butterworts are insectivorous, enabling them to thrive in nutrient-poor habitats. Their leaves have a distinctive 'greasy' feel due to a coating of slime produced by many tiny glandular hairs. Insects become trapped in this slime and the leaf then excretes additional enzymes to digest their bodies. The resultant nutrient-rich soup is absorbed by the leaf.

Pale Butterwort
Pinguicula lusitanica

Locally frequent. Bare, permanently wet, peaty ground on bogs (especially alongside drains or cuttings), moorland flushes and wet heaths, only on acid soils. ✿ A small, often tiny, version of Common Butterwort, easy to overlook, with *whitish flowers*. DETAIL *Prominently sticky-hairy.* Leaves 1–2.5cm long, translucent dull pale green with reddish-purple veins. Corolla 7–11mm long, including 2–4mm spur that is angled downwards; very pale lilac with a dull, pale yellow throat, washed and veined reddish-purple around the throat and spur. Overwinters as a rosette.

GROWTH Perennial.
HEIGHT 3–10cm (15cm).
FLOWERS May–Aug (Oct).
STATUS Native.
ALTITUDE 0–490m.

GROWTH Perennial.
HEIGHT 10–30cm (stems to 150cm).
FLOWERS May–Jun.
STATUS Native.
ALTITUDE 0–1005m.

Bogbean
Menyanthes trifoliata

Locally common away from SE England. Shallow water on the margins of lakes, ponds and slow-flowing rivers, also wet, swampy places and dune slacks. Sometimes introduced as an ornamental. ❁ The upright spikes of star-shaped white flowers with 'hairy' petals are unique.
DETAIL Hairless. Stems sprawling and may spread over large areas. Leaves alternate, 3-lobed, leaflets 3–9cm long, held up on 7–20cm long stalks. Flowers white, 15–20mm across, calyx cut into 5 lobes, corolla divided into 5 petals, flushed pink on the outside when fresh; inner surface with abundant long white hairs. Stamens 5, style 1. The flowers are of two types: 'pin' and 'thrum' (see p. 182). Fruit an oval capsule.

FAMILY VERBENACEAE: VERVAIN

Vervain
Verbena officinalis

GROWTH Perennial.
HEIGHT 30–75cm (140cm).
FLOWERS Jun–Sep.
STATUS Introduced in the Stone Age.
ALTITUDE Lowland.

Locally fairly common. Bare, disturbed ground and rough grassy places on dry, often chalky, soils and at the foot of S-facing slopes: verges, scrub, woodland rides, coastal cliffs, old quarries. ❁ Flowers small, opening 1–2 at a time on long, spreading, slender spikes; leaves distinctive.
DETAIL Stems tufted, 4-angled, hairy. Leaves opposite, the lower stalked, sparsely hairy, 2–8cm long, deeply lobed, tapering into the stalk. Calyx with 5 short teeth. Corolla pinkish-lilac, 3–5mm across, split at the mouth into 5 lobes that are divided between a 2-lobed upper lip and 3-lobed lower lip. Stamens 4, stigma 1, knob-like. Fruits with the style on top, splitting into 4 nutlets.

Clustered Bellflower

Campanula glomerata

Locally common. Rough grassland, dunes, scrub and open woodland on chalk and limestone. An occasional garden escape. ❀ Erect, but often rather short, with *clusters of stalkless, violet-blue or purplish-blue, bell-shaped flowers.* **DETAIL** Hairy. Basal leaves long-stalked, oval-oblong, more or less pointed, often heart-shaped at base, 2–4cm long, with fine, blunt teeth; lower stem leaves narrower and short-stalked, upper stem leaves clasping. Calyx teeth triangular, held erect. Corolla 12–25mm, cut 0.25–0.5 to base into 5 spreading lobes. Fruit a capsule shedding seeds through pores near the base. **SEE ALSO** Gentians (p. 197).

GROWTH Perennial.
HEIGHT 3–30cm (80cm).
FLOWERS Jun–Oct.
STATUS Native.
ALTITUDE 0–355m.

Harebell

Campanula rotundifolia

Locally common in sunny spots on poor, dry soils, both acid and alkaline: verges, unimproved grassland, dunes and rocky places. ❀ The commonest bellflower, stems somewhat sprawling, *upper leaves very narrow.* Flowers nodding on thread-like stems; bell-like, with *very fine, narrow sepals,* they are pale to bright blue and very variable in size. **DETAIL** Sparsely hairy to more or less hairless. *Basal leaves oval to circular,* 4–15mm across, usually with a heart-shaped base and sometimes toothed; very long-stalked, they wither early. Stem leaves 15–25mm × 1–3mm, pointed, more or less stalkless. Calyx teeth <1mm wide, more or less spreading, corolla 5–20(30)mm long, cut 0.25–0.3 to base. Fruit a capsule shedding seeds through pores near the base.

GROWTH Perennial.
HEIGHT 15–30cm (50cm).
FLOWERS Jul–Sep (Nov).
STATUS Native.
ALTITUDE 0–1210m.

GROWTH Perennial.
HEIGHT 50–80cm (100cm).
FLOWERS Jul–Sep.
STATUS Native.
ALTITUDE 0–320m.

Nettle-leaved Bellflower
Campanula trachelium

Local. Woodland, hedgebanks and scrubby grassland on dry, base-rich soils. An occasional garden escape in a wider range of habitats. ❀ Erect, with spikes of striking bluish-mauve flowers. *Leaves heart-shaped or square-cut at base, not tapering into the stalk.* DETAIL *Stems tufted, sharply 4-angled,* sparsely rough-hairy. Leaves roughly hairy, *coarsely-toothed,* lower leaves 6–10cm, stalks up to 10cm long, stem leaves short-stalked. Flowers short-stalked, in clusters of 1–4 at intervals along a leafy spike. *Calyx teeth narrow-triangular, more or less erect.* Corolla 25–35mm, lobed 0.25–0.4 to base, hairy. Fruit a capsule, shedding seeds through pores near the base. SIMILAR SPECIES **Creeping Bellflower** *C. rapunculoides* is a declining garden escape. *Creeping and patch-forming,* with drooping, bell- or funnel-shaped, violet-blue flowers 20–30mm long. *Calyx teeth usually spread at 90° or angled backwards.*

Giant Bellflower
Campanula latifolia

GROWTH Perennial.
HEIGHT 50–120cm.
FLOWERS Jul–Aug (Sep).
STATUS Native.
ALTITUDE 0–390m.

Local. Woodland, hedgerows and other shady places on moist, fertile, often base-rich soils. Also an occasional but increasing garden escape in a wider range of habitats. ❀ A tall, very striking bellflower, flowers purplish-blue to pale blue or sometimes whitish. *Lower leaves tapering into the stalk.* DETAIL Stems tufted, erect, softly hairy to nearly hairless, faintly and bluntly ridged. Leaves hairy, 6–10cm, *finely toothed,* lower leaves with *a winged stalk,* upper leaves stalkless. Flowers short-stalked, solitary, at intervals along a leafy spike. Calyx teeth narrowly triangular, erect or spreading at *c.* 45°. Corolla 35–55mm, lobed 0.3–0.5 to base, hairy. Fruit a capsule shedding seeds through pores near the base.

Venus's Looking-glass
Legousia hybrida

An uncommon and declining arable weed, mostly on chalky soils. ❀ Flowers in loose clusters, purple to pale lilac, 5–10mm across, *sepals rather longer than petals*. The flowers only open in sunshine. DETAIL Stems erect to sprawling, roughly hairy. Leaves roughly hairy to hairless, to 8cm long, stalkless, narrowly strap-shaped, *wavy-edged, clasping the stem*. Flowers stalkless, calyx teeth narrow-triangular, variably pointed; corolla 4–10mm long, cut halfway to the base into 5 petals. Stigma 3-lobed. *Fruit long and narrow (15–30mm)*, 3-ribbed in cross-section, with the sepals persisting at the tip.

GROWTH Annual.
HEIGHT Stems 5–30cm.
FLOWERS May–Aug.
STATUS Ancient introduction.
ALTITUDE Lowland.

Peach-leaved Bellflower
Campanula persicifolia

A garden escape on verges, waste ground and woods. ❀ Erect, with *narrow, pointed, strap-shaped leaves* and large flowers, the petals spreading widely. DETAIL Tufted, stems hairless. Leaves 5–12cm, weakly toothed. Calyx teeth narrow-triangular, held at 45°–90° to the flower stalk. Corolla 25–50mm, divided less than a quarter of the way to the base. Style split into 3 stigmas, each more than half as long as the style. Fruit a capsule, seed shed through pores in upper half.

GROWTH Perennial.
HEIGHT 30–80cm.
FLOWERS Jun–Aug.
STATUS Long grown in gardens. Recorded in the wild from c. 1900.
ALTITUDE Lowland.

Trailing Bellflower
Campanula poscharskyana

A fairly frequent garden escape on walls, pavements and waste ground, usually near houses. ❀ More or less sprawling, with slate-blue flowers, the *petals widely spread into a star shape*. DETAIL Stems hairy. Leaves roughly hairy, long-stalked, 3–5cm across, finely toothed, *circular with a heart-shaped base*. Calyx teeth narrowly triangular, held erect or at 45°. Corolla 15–25mm long, *at least as wide as long, cut 0.5–0.75 to the base*. Capsule with pores near base. SIMILAR SPECIES **Adria Bellflower** C. *portenschlagiana* is similarly popular in gardens and also a frequent garden escape, mostly in the S and W. Stems and leaves not so hairy (may be hairless), leaves smaller, flowers violet-blue, *funnel-shaped (much longer than wide)*, the corolla cut 0.25–0.4 to base.

GROWTH Perennial.
HEIGHT Stems to 30cm.
FLOWERS Jun–Oct.
STATUS Introduced to gardens from SE Europe in 1931. First recorded in the wild in 1957.
ALTITUDE Lowland.

Sheep's-bit

Jasione montana

Locally common. Sunny, grassy, heathy and rocky places on dry, acid soils, especially near the sea, including cliffs and dunes. ❀ Tiny, sky-blue to deep blue flowers are gathered into *flattened, circular flower clusters* 5–35mm across, with a *collar of green bracts just below*. DETAIL Hairy. Stems sprawling to nearly erect. Leaves 1–5cm, strap-shaped, wavy-edged, basal leaves narrowing to a short stalk, stem leaves shorter, stalkless. Calyx cup-shaped, split into 5 long, fine, pointed sepals. Corolla 5mm long, *straight in bud*, split almost to the base into 5 strap-shaped petals; stigmas 5, short, anthers creamy, *style club-shaped*, longer than petals, *notched at the tip*, forming 2 blue stigmas. SEE ALSO Devil's-bit Scabious (p. 294).

GROWTH Biennial.
HEIGHT 5–30cm (50cm).
FLOWERS May–Aug.
STATUS Native.
ALTITUDE 0–955m.

Round-headed Rampion *Phyteuma orbiculare*

Nationally Scarce, but very locally common on species-rich chalk grassland. ❀ Upright, often less than 15cm tall, with *1–2cm wide globular clusters* of tiny, deep blue flowers that are characteristically *curved in bud*. DETAIL *Hairless* to very sparsely hairy. *Basal leaves long-stalked*, 2–4cm, long-oval, shallowly-toothed, rounded at the base; lower stem leaves strap-shaped, tapering to a stalk, upper stem leaves narrow, stalkless. Flowers with a tiny bract, but no collar of bracts below the flower head. Calyx cup-shaped, with 5 triangular teeth. Corolla 5–8mm long, split almost to the base into 5 narrow petals; stamens 5, anthers creamy, *style curved, split at the tip into 2–3 deep blue stigmas*.

GROWTH Perennial.
HEIGHT 5–50cm.
FLOWERS Late Jun–Aug.
STATUS Native.
ALTITUDE Lowland.

Ivy-leaved Bellflower

Wahlenbergia hederacea

Near Threatened. Local in damp places on acid soils: moorland, heathy pastures, woodland rides, and by streams and flushes. ❀ Low-growing, with *small, pale blue flowers; creeps through the grass with thread-like stems*. DETAIL Hairless. Leaves alternate, stalked, 5–10mm long, *angular* (resembling ivy). Flowers solitary, on 1–4cm long stalks from the leaf axils. Calyx teeth short, very narrow. Corolla 6–10mm long, cut at the tip into 5 lobes, style shorter than petals, split at the tip into 3 stigmas.

GROWTH Perennial.
HEIGHT Stems to 30cm.
FLOWERS Jul–Aug.
STATUS Native.
ALTITUDE 0–520m.

FAMILY ASTERACEAE: DAISIES

Probably the largest family of flowering plants, with at least 23,000 species worldwide, only rivalled by the orchids, Orchidaceae. Around 100 species are native or ancient introductions to the British Isles, *plus* 25 recent introductions that are at least fairly widespread, *plus* several hundred microspecies of dandelions and hawkweeds. Many are very common and conspicuous on roadsides and in grassland, especially in mid to late summer. Some are easy to identify, but there are some slightly tricky groups: the yellow-flowered, dandelion-like catsears, hawksbits, hawksbeards and hawkweeds can be especially daunting, but it is often easier to identify them (by carefully checking a few characters) than to remember their names (see box on p. 263).

The old name for the family is Compositae, and they are often known as 'composites'. This name reflects the composite nature of the flowers. Each 'flower' (referred to here as 'flower head', technically a *capitula*) is made up of numerous small florets.

Each floret is an individual flower, with calyx, corolla, ovary, stigma and stamen, although some of these may be modified or absent. In particular, the calyx may be absent, represented by a ring of scales or teeth, or modified into a plume of hairs (technically a *pappus*), the 'parachute' that helps to disperse the seed. The corolla may be tubular, with 5 small teeth at the tip (reflecting the origin of the corolla as 5 fused petals). Alternatively, the corolla may have a very short tube, extended on one side into a strap-shaped ligule – the ray. In some species the florets are all tubular, in others they are all strap-shaped, but in Daisy and many others the central florets are tubular and known as disc-florets, while the outer florets are strap-shaped and known as ray-florets. There are 5 stamens, with the anthers fused into a cylinder around the single style, which usually forks at the tip into 2 stigmas. In some species the disc-florets are bisexual, with both stigma and stamens, while the ray-florets are either functionally female, with the male parts much reduced, or sterile.

The florets are attached to a receptacle – a flat or domed expanded tip to the stem. There may be small scales mixed in with the flowers, and these may be an important identification feature. Surrounding the receptacle is a variable number of bracts (technically *phyllaries*), which often resemble sepals. The seeds (technically *achenes*, that is 1-seeded fruits), may have a downy plume and sometimes also a beak.

Dandelion

ray-florets only

stigma

Daisy

ray-floret

disc-florets

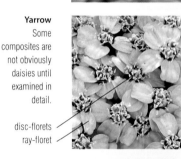

Yarrow
Some composites are not obviously daisies until examined in detail.

disc-florets
ray-floret

Catsear

ray-florets

bracts

Dandelion
A 'clock' showing the mature seeds.

hairy plume

receptacle

seed

beak

Greater Burdock *Arctium lappa*

GROWTH Biennial.
HEIGHT To 150cm (200cm).
FLOWERS Jul–Sep.
STATUS Ancient introduction.
ALTITUDE Lowland.

The burrs cling to fur and clothing, and were the inspiration for 'Velcro'.

Fairly common. Rough grassy places, especially on moist, disturbed soils: verges, the banks of rivers and streams, field borders, woodland rides, waste ground. ❁ Tall and robust with a well-branched stem and very large, rounded leaves. *Flower heads a globular mass of narrow, hook-tipped bracts – the burrs – surrounding a relatively small cluster of reddish-purple florets.* **DETAIL** Basal leaves up to 40cm × 30cm, rough to the touch, heart-shaped at base, stalked – *all leaf stalks solid*; upper leaves stalkless. *Flower heads large, 24–47mm across, hairless to sparsely hairy, in flat-topped clusters* (stalks of outer flowers longer than inner to bring them all to roughly the same height); *stalks of terminal flowers in main clusters at least 2.5cm long.* Seed with a plume of rough, yellowish hairs.

Lesser Burdock

Arctium minus

GROWTH Biennial.
HEIGHT 60–130cm.
FLOWERS Jul–Sep.
STATUS Native.
ALTITUDE Lowland.

Common. Habitats similar to Greater but grows on a wider range of soils. ❁ As Greater Burdock but *stalks of lower leaves hollow (at least near their base), flower heads average smaller, 15–32mm across,* usually in well-branched, spike-like clusters, with the *flower heads at various levels; stalks of terminal flower heads typically 0–2cm (but may be longer).* **SIMILAR SPECIES Wood Burdock** *A. nemorosum.* Scarcer than Lesser in the S (and absent SW England), but commoner in the N, favouring old woods and hedgerows; distribution poorly understood due to problems with identification – many botanists do not try to separate them. *Terminal flower heads always stalkless,* arrangement as Lesser, but *size near Greater* (flower heads average 27–40mm across).

Melancholy Thistle

Cirsium heterophyllum

Locally common in the uplands. Verges, meadows, streamsides etc. on damp soil. ❁ *Tall and statuesque,* with *solitary flower heads* 20–35mm across on *cottony, unbranched stems, near leafless towards tip.* DETAIL Leaves shiny green and hairless above, *silvery-white below* due to dense, felted hairs. Basal leaves 15–40cm, variably lobed, with prickles along the edges, long-stalked; stem leaves few, heart-shaped at base, clasping the stem. Bracts tipped purple. Corolla *c.* 27mm long, reddish-purple. Seeds with a plume of *feathery hairs.*

GROWTH Perennial.
HEIGHT 45–120cm.
FLOWERS Late Jun–Aug.
STATUS Native.
ALTITUDE To 760m.

Meadow Thistle

Cirsium dissectum

Rather local. Fens, wet meadows and the margins of bogs, on neutral to acid, peaty soils. ❁ *No harsh spines. Flower heads usually solitary,* 12–30mm across, on cottony, unbranched stems that are *almost leafless towards the tip.* DETAIL Basal leaves 6–15cm long, long-stalked, variably toothed, with weak spines along margins, *whitish below* due to dense, cottony hairs, and dull green and hairy above; stem leaves few, stalkless, more or less clasping the stem. Bracts tipped purple. Corolla *c.* 20mm long, reddish-purple. Seeds with a plume of *feathery hairs.* SIMILAR SPECIES **Tuberous Thistle** *C. tuberosum*, rare on chalk grassland in Wessex and S Wales, is very similar but leaves deeply lobed, green both sides.

GROWTH Perennial.
HEIGHT 15–50cm (80cm).
FLOWERS Late May–Aug.
STATUS Native.
ALTITUDE 0–595m.

Dwarf Thistle *Cirsium acaule*

Common. Short, unimproved grassland over chalk and limestone. ❁ Distinctive, with a *rosette of spiny leaves tight to the ground.* Flower heads 12–25mm wide, *usually single* (but may be 2–3 together), and *mostly stalkless* (but sometimes on leafy, *unwinged stalks* to 10cm long). DETAIL Leaves 4–15cm, shiny dark green, deeply lobed. Bracts tipped reddish-purple. Corolla *c.* 16mm long, reddish-purple. Seeds with a plume of *feathery hairs.*

GROWTH Perennial.
HEIGHT To 5cm (10cm).
FLOWERS Late Jun–Sep.
STATUS Native.
ALTITUDE To 440m.

Creeping Thistle
Cirsium arvense

Abundant in a wide variety of rough and grassy places. ❁ Leaves spiny but *stems unwinged, without spines*. Flower heads 8–20mm across, bracts tipped purple, *florets lilac*, sometimes white, scented. **DETAIL** *Grows from long, creeping rhizomes*; stems usually well branched. Leaves 8–20cm long, variably lobed, margins spiny, at least basal leaves cobwebby below. Corolla *c.* 15mm long. Seeds with a plume of *feathery hairs*.

GROWTH Perennial.
HEIGHT 20–120cm (200cm).
FLOWERS Jul–Sep.
STATUS Native.
ALTITUDE 0–700m.

Spear Thistle
Cirsium vulgare

Abundant. A wide variety of rough grasslands, arable fields and waste ground. ❁ *Stem and leaves viciously spiny*, dark green with a pale midrib, the stem with *discontinuous spiny wings. Flower heads globular, large*, solitary or in loose clusters, slightly woolly, *outer bracts tipped by yellowish spines*. **DETAIL** Stems branched, cottony. Leaves 15–30cm, lobed, *blistered above with weak spines*, hairy below; spines on leaf margins with a swollen yellowish base. Flower heads 20–50mm across, corolla *c.* 30mm long, reddish-purple. Seeds with a plume of *feathery hairs*.

GROWTH Biennial.
HEIGHT 30–150cm.
FLOWERS Jul–Oct.
STATUS Native.
ALTITUDE 0–685m.

Marsh Thistle
Cirsium palustre

Very common. A wide variety of damp or wet places, including heavily-grazed moorland, and often common in wet woodland. ❁ Often *very tall and slender*, little-branched below and very spiny; often flushed purple. Stems with very spiny wings running down their whole length. Flower heads relatively small (7–12mm across), in dense clusters. Florets usually purple but often white and sometimes pale pink. **DETAIL** Leaves 10–50cm, deeply lobed, with spiny edges. Bracts purplish. Corolla *c.* 13mm long. Seeds with a plume of *feathery hairs*.

GROWTH Biennial.
HEIGHT 30–200cm.
FLOWERS Late Jun–Sep.
STATUS Native.
ALTITUDE 0–760m.

Slender Thistle
Carduus tenuiflorus

Locally fairly common. Rough, dry grassland and sandy waste places near the sea. Casual away from coastal areas. ❀ Little branched. Stem with *spiny wings, 5–10mm wide, all the way to the flowers*. Flower heads *small and narrow* (5–10mm across excluding florets, longer than wide), more or less *stalkless*, in *dense clusters of 3–10*. Florets pale pinkish-purple, often barely longer than the spine-tipped bracts. **DETAIL** Stems cottony. Leaves to 15cm, lobed, spiny, *white-woolly below*. Corolla *c.* 10mm long, with 5 equal lobes. Seeds with a plume of *simple hairs*.

GROWTH Biennial (annual).
HEIGHT 15–60cm (80cm).
FLOWERS Jun–Aug.
STATUS Native.
ALTITUDE Lowland.

Welted Thistle
Carduus crispus

Locally fairly common. Grassy places on moist, fertile soils, often on chalk or chalky boulder-clay: verges, field borders, woodland rides, streamsides, waste ground. ❀ Well branched. Stem with *narrow, wavy, spiny wings continuous from the base to (or nearly to) the flowers*. Flower heads *globular, 15–20mm across, stalked*, more or less erect, in *clusters of 2–4* (1–5). **DETAIL** Stems cottony. Leaves 10–50cm long, lobed, spiny, *sparsely hairy below*. Bracts variably washed purplish. Corolla *c.* 14mm, 5-lobed, reddish-purple. Seeds with a plume of *simple hairs*.

GROWTH Biennial.
HEIGHT 30–120cm (150cm).
FLOWERS Jun–Aug (Sep).
STATUS Native.
ALTITUDE To 365m.

Musk Thistle
Carduus nutans

Locally common. Rough grassland, field margins and waste ground, especially on chalk and limestone or sandy soils. ❀ *Flower heads large, mostly single, nodding, very fragrant.* Stem with *discontinuous spiny wings*. **DETAIL** Stems cottony. Leaves 5–30cm, lobed, spiny, hairy below. Flower heads flattened-globular, 20–60mm across; bracts variably washed purplish, outer bracts bent outwards. Corolla 15–25mm long with 5 lobes, reddish-purple. Seeds with a plume of *simple hairs*.

GROWTH Biennial.
HEIGHT 20–100cm.
FLOWERS May–Sep.
STATUS Native.
ALTITUDE To 530m.

Woolly Thistle
Cirsium eriophorum

Local. Rough, dry grassland, open scrub, woodland clearings and old quarries on limestone or chalk. ❀ Distinctive, with *globular, cobwebby flower heads*, 40–70mm across, and dark reddish-purple florets. DETAIL *Stems unwinged*, branched, cottony. Leaves to 60cm, dark green with prominent midribs, *deeply cut into narrow, strap-shaped lobes tipped by vicious spines, bristly-hairy and spiny above*, white-woolly below; stem leaves clasping. Bracts tinged purple, tipped with straw-yellow spines. Corolla *c.* 40mm long. Seeds with a plume of *feathery hairs*.

GROWTH Perennial.
HEIGHT 60–150cm.
FLOWERS Jul–Aug.
STATUS Native.
ALTITUDE 0–310m.

Cotton Thistle
Onopordum acanthium

Increasingly common, often as a garden escape. Verges, field borders and waste ground, especially on dry, sandy or chalky soils. ❀ Striking, often *very tall. Whole plant silvery-grey* (due to cottony hairs), stems with *continuous, broad, spiny wings.* Flower heads solitary, 20–60mm wide, very spiny. DETAIL Stems branched. Leaves 10–50cm, toothed to shallowly-lobed, very spiny. Bracts spine-tipped. Corolla 15–25mm long, reddish-purple. Seeds with a plume of simple hairs.

GROWTH Biennial.
HEIGHT 50–300cm.
FLOWERS Jul–Aug.
STATUS Introduced in the Iron Age.
ALTITUDE 0–330m.

Milk Thistle
Silybum marianum

Scattered. A casual, usually appearing on tipped soil or rubble, also rough grassland and arable fields. ❀ Leaves shiny green, *conspicuously veined or marbled with white.* Flower heads solitary, 50–140mm across, *lower bracts tipped by very long, robust, yellow spines.* DETAIL Stems grooved, unwinged, branched. Leaves 15–60cm, often hairy below, variably lobed; upper leaves clasping. Florets reddish-purple. Seeds with a plume of simple hairs.

GROWTH Annual or biennial.
HEIGHT 40–100cm (250cm).
FLOWERS Jun–Aug.
STATUS Ancient introduction, formerly widely cultivated.
ALTITUDE Lowland.

Saw-wort
Serratula tinctoria

Locally common. A variety of grassy places on unimproved, damp, clay soils, including flushes and fens, also dry chalk and limestone grassland, damp grassy heathland, coastal heaths and rocky lake shores. ❀ Much like a slender Common Knapweed, but *leaves very variably lobed, with jagged, saw-toothed margins, each tooth tipped with a bristle.* Flower heads in loose clusters, relatively long and narrow (8–10mm across), with many purple-washed bracts *without lobes or teeth*, florets reddish-purple. May be very short in exposed places. DETAIL More or less hairless. Stem grooved, erect, usually well branched. Leaves 5–20cm, lower leaves toothed to deeply cut with strap-shaped side lobes and long-oval terminal lobe; upper leaves stalkless. Corolla *c.* 11–14mm long. Seeds with a plume of simple, straw-coloured hairs.

GROWTH Perennial.
HEIGHT 20–70cm (100cm).
FLOWERS Jul–Sep.
STATUS Native.
ALTITUDE 0–560m.

bracts not lobed

yellowish pappus

Carline Thistle
Carlina vulgaris

Locally common. Usually found in short, unimproved grassland on poor, dry soils on chalk, chalky-sand or limestone: downland, dunes, scree, cliffs, old quarries and other rocky places. ❀ Distinctive, with *'dead', brownish flowers.* Flower heads 15–30mm across, with a *halo of long, pale buffy-grey bracts* that fold inwards in wet weather. Florets very dark purple, but quickly withering – the *abundant orange-yellow scales in the flower head are more conspicuous and much longer-lasting.* DETAIL Stems not spiny, purplish. Leaves to 10cm, lobed, dull green with cottony hairs but soon more or less hairless; spiny; stem leaves clasping. Outer bracts leaf-like, green tinged purple, innermost strap-shaped, buffy-grey, chaffy, held at 90° in dry weather to form a ring around the florets. Flower heads in clusters of 1–6. Corolla *c.* 10mm long, intermixed with abundant strap-shaped scales. Seeds with a plume of feathery hairs.

GROWTH Biennial.
HEIGHT 10–60cm.
FLOWERS Jul–Oct.
STATUS Native.
ALTITUDE 0–455m.

Greater Knapweed
Centaurea scabiosa

Locally common on dry soils, usually over chalk or limestone, in rough, often scrubby, grassland on verges, railway embankments, cliff-tops, old quarries and waste ground.
❁ Tall and robust, with deeply divided leaves and large reddish-purple flowers, the outer florets much longer than the inner, with deeply forked tips. DETAIL Stems tufted, well branched. Leaves 10–25cm long,

GROWTH Perennial.
HEIGHT 30–90cm (120cm).
FLOWERS Jul–Sep.
STATUS Native.
ALTITUDE 0–320m.

hairy, all except uppermost deeply lobed (but on coasts of Wales and Scotland var. *succisifolia* may have undivided basal leaves). Bracts *pale, with a narrow, crescent-shaped, blackish-brown tip* and a fine, comb-like fringe. Flower heads 30–60mm across, solitary at the tip of the stems.

Common Knapweed *Centaurea nigra*

Very common in all types of rough grassland, often on damp or heavy soils, including verges, field borders and old meadows.
❁ Erect, with *reddish-purple, rather thistle-like flowers* growing *singly* at the tip of the stems. Leaves usually strap-shaped, more or less undivided. An uncommon variant has long, deeply-cut outer florets and could be confused with Greater Knapweed. DETAIL Stems tufted, well branched, hairy. Leaves hairy, basal leaves entire (rarely lobed), 5–20cm long, stalked, upper leaves sometimes lobed, stalkless. The bracts form a globular head below the florets,

GROWTH Perennial.
HEIGHT 20–80cm (100cm).
FLOWERS Jun–Sep.
STATUS Native.
ALTITUDE 0–580m.

mostly *15–20mm across*, with the *flower stalk immediately below distinctly swollen*. All bracts with *broad, oval blackish tips*, more or less concealing the pale bases, and a fringe of long, fine, hair-like teeth. Florets usually all tubular, but sometimes outer florets long and deeply forked; seeds with a plume of very short hairs. SIMILAR SPECIES Chalk Knapweed *C. debeauxii* is very similar but has *smaller flower heads* (mostly 9–14mm across), with the *stem only slightly swollen* immediately below, and the *blackish tips to the bracts at the base of the flower heads narrowly lanceolate*, often revealing the pale bases of those above. In addition, the leaves are more often deeply lobed and the plume is sometimes absent. This recently recognised species is common on light soils in the S but rare or absent in N England and Scotland.

Chalk Knapweed

Cornflower
Centaurea cyanus

GROWTH Annual.
HEIGHT 20–50cm (90cm).
FLOWERS Jun–Aug.
STATUS Introduced in the
Iron Age.
ALTITUDE 0–350m.

Once a common arable weed, improved seed-cleaning led to its near extinction in the wild. Now found in wild-flower seed mixes (including seeded arable margins), as a garden escape and possibly still when long-buried seed germinates. ❀ Flower heads *c.* 20mm wide with *bright blue florets, the outermost elongated into ray-like 6-lobed trumpets.* DETAIL Stems and leaves with whitish, cottony hairs, stems grooved, branched. Lower stem leaves 5–20cm, stalked, with narrow, well-spaced side lobes, upper leaves strap-shaped, stalkless. Bracts finely toothed at tip. Inner corollas *c.* 7mm long, outer corollas split at the mouth into 6 long lobes. Seeds with a plume of simple, bristly hairs.

Chicory Cichorium intybus

Occasional garden escape on rough grassland, including verges and field margins, often near houses. ❀ *Tall, with blue, dandelion-like flowers,* 25–40mm across. DETAIL More or less hairy. Stems erect, tough, grooved, well branched. Lower leaves stalked, deeply lobed, upper strap-shaped, sometimes with jagged teeth, clasping the stem. Flower heads short-stalked, in clusters in the leaf axils, with 2 rows of bracts. Corolla *c.* 16mm, strap-shaped, bright blue. *Seeds with a ring of minute, ragged scales on top.*

GROWTH Perennial.
HEIGHT 30–120cm.
FLOWERS Jun–Oct.
STATUS Ancient introduction,
formerly grown for fodder.
ALTITUDE Lowland.

Common Blue Sowthistle
Cicerbita macrophylla

A garden escape on verges and by ponds and rivers. ❀ Strongly patch-forming, with *pale mauve, dandelion-like flowers c.* 30mm across and *large, oval leaves, often with 2 side lobes at the base.* DETAIL Bleeds whitish sap. Stems branched, upper stem sometimes glandular-hairy. Basal leaves to 20cm, hairy on veins below, toothed, with a long, winged stalk. Flower stalks glandular-hairy. Bracts in several rows. Corollas all strap-shaped. Seeds with a plume of simple hairs.

GROWTH Perennial.
HEIGHT 60–150cm (200cm).
FLOWERS Jul–Sep.
STATUS Introduced to
gardens from Russia in 1823.
First recorded in the wild in
1915 and still spreading.
ALTITUDE 0–460m.

Catsear
Hypochaeris radicata

Abundant in grassy places.
❁ Flower heads *solitary*, on
leafless stems that may be
branched or unbranched, with
the leaves replaced by a few *tiny,
triangular dark-tipped scales.*
Upper stem and bracts *hairless*, the
*longer bracts with a 'cockscomb'
of short bristles near the tip.*
DETAIL Bleeds sparse white sap
that rapidly turns brown. Rosette
leaves 4–25cm long, usually
roughly hairy, wavy-edged to
deeply lobed.

GROWTH Perennial.
HEIGHT 20–40cm (60cm).
FLOWERS Jun–Sep.
STATUS Native.
ALTITUDE 0–710m.

'cockscomb'
at tip of bract

expands
abruptly

scales

basal leaves showing variation

Stem broadens *slightly* towards
base of flower, but junction with
flower head still *abrupt*. Flower head
20–40mm across, with long, slender,
translucent yellowish scales hidden
amongst the florets. Seeds usually *all
beaked*, with a plume of hairs in 2 rows:
outer simple, inner feathery.

Smooth Catsear
Hypochaeris glabra

Listed as Vulnerable. Very local.
Bare ground and sparse grassland
on poor, acidic, sandy soils,
including dunes. ❁ As Catsear
but whole plant smaller, slenderer
and often more sprawling. Flower
heads *much smaller* (sometimes
tiny), only opening in full sun in
the morning, with the *ray-florets
hardly longer than the few, purple-
tipped bracts.* DETAIL Leaves
c. 1–7cm long, often tinged
reddish, hairless apart from a sparse fringe along the
margins. Flower heads 4–15mm across, ray-florets about
twice as long as wide (4 times longer than wide in Catsear).
Outer seeds *lack a beak* (inner sometimes also not beaked).

GROWTH Annual.
HEIGHT 10–20cm (40cm).
FLOWERS Late May–Sep.
STATUS Native.
ALTITUDE Lowland.

bracts

expands
gradually

scales

Autumn Hawkbit
Scorzoneroides autumnalis

Abundant in wide variety of grassy
places. ✺ As Catsear (flower
heads *solitary*, on *leafless*, usually
branched stems, with *a few dark-
tipped scales*), but the stem *tapers
into the flower head, usually
without an abrupt junction*, and
the bracts *lack* a 'cockscomb'
and are often *much hairier*.
DETAIL Sap bluish, *not turning
brown*. Stem usually hairless, at
least above. Leaves 5–15(20)cm
long, variably hairy, with reddish
midrib, typically deeply lobed
with narrow central section
and long, fine side lobes, often
sticking out at right angles but
may be merely wavy-edged. Bracts
sometimes hairless. Flower heads 12–35mm across, underside of outer ray-
florets usually reddish. *No scales amongst florets.* Seeds *not* beaked, with a
dirty-white plume of a *single* row of feathery hairs.

GROWTH
Perennial.
HEIGHT To
60cm.
FLOWERS
Late Jul–Oct.
STATUS
Native.
ALTITUDE
0–1090m.

Goatsbeard
Tragopogon pratensis

Fairly common. Rough, often rank
grassland: verges, field margins,
waste ground. ✺ *Leaves narrow
and grass-like*, held upright.
Flower heads solitary, large
(18–30mm across), only opening
in the morning sunshine (hence
the old name 'Jack-go-to-bed-
at-noon'), with the *single row
of bracts usually conspicuously
longer than the ray-florets*. DETAIL
Leaves basal and on stems,

GROWTH Annual to perennial.
HEIGHT 30–80cm.
FLOWERS Jun–Jul (May-Oct).
STATUS Native.
ALTITUDE 0–365m.

20–40cm long × 0.6–1cm, hairless, grey-green, with a white
strip along the midrib. Stems hairless (may be woolly when
young), expanding slightly below the flower; when broken,
exudes milky sap. Seeds beaked, plumes feathery, forming
conspicuous clocks.

IDENTIFYING YELLOW 'DANDELIONS'
Dandelions, hawkbits, hawksbeards, sowthistles etc. have the reputation of
being 'difficult'. Their identification is, however, mostly straightforward. Start
by checking the following features:
1. Flowers solitary, with just 1 per stem, or several flowers per stem?
2. Flowering stems branched or unbranched?
3. Flowering stems leafless or leafy?
4. Bracts below flower heads in 1 row or several rows; outer bracts
obviously shorter, or bracts all the same?

Rough Hawkbit *Leontodon hispidus*

GROWTH Perennial.
HEIGHT 10–35cm (60cm).
FLOWERS Late May–Oct.
STATUS Native.
ALTITUDE 0–575m.

Fairly common but local in grassy places, mostly on calcareous soils, including ruins, quarries and rocky areas; sometimes in alkaline fens. ❀ Flower heads large, shaggy moptops, 25–40mm across, *solitary*, on *unbranched, leafless stems* (may have 1–2 tiny scales). Whole plant, including stems and bracts, *usually conspicuously hairy*, some hairs on stem and leaves *forked into a Y-shape at the extreme tip* (10× lens, best seen with leaf bent to show some hairs in silhouette), a feature *shared only with Lesser Hawkbit.* Leaves held more or less erect. DETAIL Leaves to 20cm long, wavy-edged to deeply-lobed. Junction of stem and flower head abrupt. Bracts 11–13mm (5–15mm) long. Seeds not beaked. Plume dirty white, comprising an inner row of feathery hairs and an outer row of simple hairs; seed heads form a 'clock'. Bleeds bluish-white sap.

Lesser Hawkbit *Leontodon saxatilis*

GROWTH Perennial.
HEIGHT 8–20cm (40cm).
FLOWERS Jun–Oct.
STATUS Native.
ALTITUDE 0–500m.

outer seeds
tipped with
a collar of
chaffy scales

inner
seeds
already
gone

Locally abundant in short grass on a variety of soils, from mown verges to dunes and dune slacks, moorland and rocky places; tolerates heavy grazing, frequent mowing and trampling. ❀ Similar to Rough Hawkbit in having the flower heads *solitary*, on *leafless, unbranched stems*, with *forked hairs on the leaves*, but *usually far less hairy* and rather smaller and more delicate, with slender flower stems, fewer bracts and fewer florets in each flower head; leaves held more or less flat to the ground. Most reliably separated by the 'clock': inner seeds with plumes of long hairs but *outer row of seeds tipped by a collar of pale chaffy scales*, remaining on the seed head long after the inner seeds have blown away (10× lens). DETAIL Leaves with hairs on the midrib often red-based (rarely so in Rough Hawkbit). Flower stalks typically hairless towards tip, bracts typically sparsely hairy to hairless. Flower head 12–25mm across, bracts (5) 7–11mm long.

Dandelion
Taraxacum agg.

Abundant in a wide variety of grassy places, especially in the years following soil disturbance. ✵ A complex of 232 microspecies that can only be identified by specialists (see p. 366), but easily recognisable as a group by the *large, solitary flower heads on hollow, leafless, hairless stems that bleed white sap when broken.* DETAIL Leaves in a basal rosette, sparsely hairy to hairless, often shiny, more or less sharply lobed. Flowers 20–60mm across. Seeds beaked, with a plume of simple white hairs forming a 'clock'.

GROWTH Perennial.
HEIGHT To 20cm.
FLOWERS Mar–Oct (peaks Apr-May).
STATUS Less than half the microspecies are native, but of these a substantial number are endemic to the British Isles.
ALTITUDE 0–1220m.

Mouse-ear Hawkweed *Pilosella officinarum*

Locally abundant. Short, sparse, grassland on dry or thin soils, including dunes, rocky places and grassy heaths. ✵ *Patch-forming*, with slender runners that produce small leaves at intervals. Easily identified by the solitary, *pale lemon-yellow flowers* on unbranched, leafless stems, and rosettes of *dark green leaves* with scattered *long hairs above* and *contrastingly white-felted undersides.* DETAIL Leaves 1–6cm

GROWTH Perennial.
HEIGHT 5–30cm (50cm).
FLOWERS May–Aug (Oct).
STATUS Native.
ALTITUDE 0–915m.

long, vaguely toothed. Flower heads 15–25mm across. Bracts with black-based glandular hairs and white non-glandular hairs. Florets striped red below. Seeds with 1 row of unbranched, dirty white hairs.

Hawkweeds *Hieracium* agg.

A group of 411 microspecies that can only be separated by specialists (like dandelions, they are apomictic, see p. 366). Locally abundant in grassy places, from woodland rides to verges, grassy heaths and dunes, but especially characteristic of rocky areas in the uplands. ✵ Very variable. Mostly hairy. Leaves variably toothed but *never* lobed; some have a basal rosette at flowering time, and most have leafy stems, but in some stems leafless. Most have several flower heads per stem, but others, mostly found in Scotland, have 1. DETAIL *Bracts in several rows* (unlike hawksbeards), seeds 10-ribbed, not beaked, with a *plume of 1 row of dirty white to pale brown hairs.* SEE ALSO Marsh Hawksbeard (p. 267).

GROWTH Perennial.
HEIGHT To 100cm.
FLOWERS Jun–Sep.
STATUS Native.
ALTITUDE 0–1220m.

outer
bracts
spreading

Beaked Hawksbeard *Crepis vesicaria*

Abundant. Grassy verges, unkept lawns, pastures and rough ground. ● The *earliest leafy-stemmed, branched,* dandelion-type to come into flower, very conspicuous on May roadsides. Usually hairy. **DETAIL** Bleeds bluish-white sap. Stems ridged, usually roughly hairy towards base. Leaves 10–35cm, densely hairy to almost hairless, with red-based hairs (at least on the

midrib above), sharply lobed, the stem leaves with triangular basal lobes *attached to the stem.* Flower heads 15–30mm across, bracts usually hairy (sometimes with glandular hairs), in 2 rows, the short outer row *spreading.* Differences from a robust, hairy Smooth Hawksbeard are the much longer, *beaked* seeds (5–9mm, including a beak as long as the seed) and the flattened silky white hairs on the inner faces of the bracts (20× lens).

GROWTH Usually biennial.
HEIGHT 15–80cm (120cm).
FLOWERS May–Jul.
STATUS Introduced. First recorded 1713 and rapidly spreading.
ALTITUDE Lowland.

Smooth Hawksbeard *Crepis capillaris*

Abundant. Verges and other unimproved grasslands, lawns (where may be very petite), rough and waste ground. ● Flower heads on *leafy, branched stems.* Very variable in size but usually *rather spindly,* hairless or only sparsely hairy, with *thin, shiny leaves* and *small flower heads that contract abruptly into a slender stem;* bracts in 2 rows, the short outer row *not spreading.* **DETAIL** Stems weakly ribbed. Leaves 5–15cm, toothed to deeply lobed, with pointed basal lobes that are *free* (i.e. not attached to the stem). Bracts with inner face hairless. Flower heads 10–15mm across. Seeds 1.4–2.5mm long, ribbed, *not* beaked, plume white, simple. Leaves and bracts sometimes glandular-hairy, especially in N, and robust, hairy plants must be carefully separated from Beaked Hawksbeard.

GROWTH Annual or biennial.
HEIGHT 10–120cm.
FLOWERS Jun–Oct.
STATUS Native.
ALTITUDE 0–490m.

stem leaf showing pointed lobes at base

basal leaves

Rough Hawksbeard *Crepis biennis*

Rather local. Road verges and other rough grassy places, often on chalky soils. ☸ Usually tall, stout, well branched and *roughly hairy* (but hairs *not* hooked; some plants may be almost hairless), with *large flowers*. DETAIL *Bleeds yellowish-white sap.* Stem ridged, obviously hollow. Leaves 15–30cm, with irregular, often sharply-triangular lobes, stalkless, lower more or less clasping. Flower heads 25–35mm

GROWTH Biennial.
HEIGHT 30–120cm.
FLOWERS Jun–Jul.
STATUS Probably native in S England, scattered introductions elsewhere, often short-lived.
ALTITUDE Lowland.

across. Bracts with dark bristly hairs, some often gland-tipped, *inner faces with short, silky, white hairs*; in 2 rows, the outer row spreading. Seeds 4–7.5mm, finely ribbed, *not* beaked, with a plume of simple white hairs.

Marsh Hawksbeard *Crepis paludosa*

Fairly common. Damp, rocky and often shady places in the uplands, often on water-splashed rocks, also damp grassland at lower altitudes. ☸ Usually tall and slender. *Stem and leaves hairless* (unlike some otherwise similar hawkweeds), *flower bracts woolly-hairy, often also with blackish glandular hairs.* DETAIL Patch-forming. Stem reddish towards base. Leaves *thin and shiny*, toothed to shallowly-lobed, basal leaves with a short winged stalk, *stem leaves clasping, with large, rounded basal lobes.* Bleeds white sap. Flower heads 15–25mm across; bracts in 2 rows, the outer row not spreading. Seeds 4–5.5mm long, *not* beaked, with a plume of *brittle, pale brown, unbranched hairs.*

basal leaf

GROWTH Perennial.
HEIGHT 30–90cm.
FLOWERS Jul–Sep.
STATUS Native.
ALTITUDE To 1000mm.

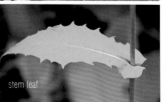

stem leaf

Bristly Oxtongue
Helminthotheca echioides

Locally abundant on disturbed and rough ground; can be dominant on late summer road verges, especially on calcareous clay soils. ❀ Robust, well branched and very bristly, with many hairs with tips like miniature grappling hooks (as harsh as an ox's tongue) and scattered, larger, thorn-like bristles with *swollen, blister-like whitish bases.* Flowers in clusters at the tip of leafy stems, with *outer 3–5 bracts broadly oval,* forming a loose pouch around the base of the flower head. **DETAIL** Basal leaves to 20cm, strap-shaped, variably toothed,

GROWTH Annual or biannual.
HEIGHT To 80cm.
FLOWERS Jun–Nov.
STATUS Ancient introduction.
ALTITUDE Lowland.

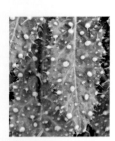

usually dead by flowering time; stem leaves clasping, with heart-shaped bases. Latex sparse, white. Flower heads 20–25mm across. Seeds with long beak and white plume, all hairs feathery. **SEE ALSO** The bristly leaves recall teasels (p. 295) and Green Alkanet (p. 201).

Hawkweed Oxtongue   *Picris hieracioides*

GROWTH Biennial to perennial.
HEIGHT 15–80cm (100cm).
FLOWERS Jul–Oct.
STATUS Native (introduced Ireland).
ALTITUDE Lowland.

Fairly local in rough, often scrubby, grassland, mainly on calcareous soils, but intolerant of heavy grazing: verges, railway banks, grassy pits. ❀ Robust, well branched and *roughly hairy* (many hairs with minute 2- or sometimes 3-pronged hooked tips). Flowers in *clusters* at the tip of *leafy* stems, all bracts *strap-shaped,* the outer row often spreading in a loose ruff. Seeds with an off-white parachute, *inner row of hairs feathery, outer row simple.* **DETAIL** Basal leaves to 20cm, strap-shaped, midrib reddish above, but dead by flowering time. Stem leaves clasping, with wavy, toothed edges. Latex sparse, white. Bracts dull green, with *whitish bristles on midrib.* Flower heads 20–30mm across. Seeds curved, wrinkled, either beakless or with a short beak.

Wall Lettuce
Mycelis muralis

Locally common
in woodland and
on walls, rocks
and hedge-
banks, often
in shady places
and especially
on calcareous
soils. ❁ Slender,
often tall, with
leafy stems; the whole
plant may be tinged
maroon. Lower leaves with
a distinctive *large, sharply-
angled terminal lobe. Flowers
small*, only a few open at any
one time, in open, well-branched
clusters. DETAIL Hairless; the whole
plant bleeds milky sap. Lower
leaves lobed, stalked, upper leaves
simpler, clasping, with arrow-
shaped bases. Flowers 7–10mm
across, with *c.* 5 florets; bracts in 2
rows, outer row very short. Seeds
black, flattened, ribbed, with a very short beak and a plume
of 2 unequal rows of simple white hairs.

GROWTH Perennial.
HEIGHT 25–100cm (150cm).
FLOWERS Jul–early Oct.
STATUS Native.
ALTITUDE 0–570m.

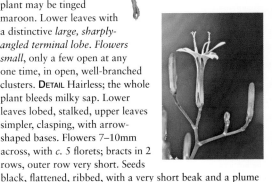

Nipplewort *Lapsana communis*

Common.
Open
woodland,
hedgerows,
verges, gardens,
arable fields,
waste ground; often in
half-shade. ❁ Usually
*tall, well branched and
leafy*, with loosely-branched
clusters of *small, pale yellow
flowers* on slender stalks. The
name refers to the *nipple-like
flower buds*. DETAIL Stem hairy, at least towards base. Leaves
hairy, lower to 15cm, stalked, with a large terminal lobe and
1(3) pairs of small side lobes, upper leaves oval to diamond-
shaped, coarsely toothed, short-stalked to stalkless. Bleeds
sparse white sap. Flower heads 10–15mm across, with a
single row of long, narrow bracts, an incomplete row of much smaller outer bracts, and 8–15 ray-
florets. Seeds ribbed, with neither beak nor hairy plume.

GROWTH Annual.
HEIGHT To 100cm.
FLOWERS Apr–Oct.
STATUS Native.
ALTITUDE 0–465m.

GROWTH Annual.
HEIGHT 20–80cm (150cm).
FLOWERS Apr–Oct (Dec).
STATUS Native.
ALTITUDE 0–435m.

Smooth Sowthistle
Sonchus oleraceus

Abundant. Disturbed ground, arable fields, verges, walls; a common garden weed.
❁ Foliage *waxy grey-green*, usually *hairless*. Flowers pale yellow to whitish, in *clusters* on *branched, leafy stems*. DETAIL Stems stout, hollow. Leaves usually deeply-lobed, *margins not or weakly spiny, basal lobes spreading, more or less pointed*. Bleeds white sap that eventually turns orange. May be a few glandular hairs on bracts. Flower heads 20–25mm across. Seeds 2.5–3.75mm, flattened, ribbed, with *cross-wise wrinkles*, not beaked, pale brown, with a plume of unbranched white hairs.

Prickly Sowthistle
Sonchus asper

GROWTH Annual.
HEIGHT 20–80cm (150cm).
FLOWERS Jun–Oct.
STATUS Native.
ALTITUDE 0–570m.

Abundant. Disturbed ground, arable fields, roadsides. ❁ Similar to Smooth Sowthistle, but leaves usually shiny dark green, rigid, with *spiny, thistle-like margins*, flowers deeper yellow. DETAIL Leaves with *rounded (but often very spiny) basal lobes, pressed to the stem*; bleeds white sap that quickly turns dirty orange. Seeds 2–3mm, orange-brown, ribbed.

basal lobe

Perennial Sowthistle
Sonchus arvensis

GROWTH Perennial.
HEIGHT 60–150cm.
FLOWERS Mid Jul–Oct.
STATUS Native.
ALTITUDE 0–490m.

Common on rich soils, often where disturbed: riverbanks, fens, sea walls, dunes, shingle, the strand-line of saltmarshes and beaches, also an arable weed, and conspicuous on verges in late summer. ❁ Usually tall, with *leafy, branched stems* and *large, rich yellow flowers*. Upper stem and bracts densely hairy, the hairs tipped with sticky yellow glands. DETAIL Leaves shiny green, edged with soft spines. Basal leaves variably lobed, narrowing to a winged stalk; stem leaves clasping, with rounded basal lobes. Bleeds white sap. Flower heads 40–50mm across. Seeds 2.5–3.5mm, flattened, not beaked, pale brown, with a plume of unbranched white hairs. SIMILAR SPECIES **Marsh Sowthistle** *S. palustris* is often *very tall* (to 3m), with *dark* glandular-hairs, stem leaves with *long, triangular side lobes*, and straw-coloured seeds. Locally common in fens in Norfolk, Suffolk and Kent.

Prickly Lettuce
Lactuca serriola

Fairly common. Disturbed soil on verges, waste ground, sea walls etc., occasionally shingle banks and dunes. ✿ Tall, with waxy grey-green, net-veined leaves with a *conspicuous white midrib, with irregular soft, spiny teeth along the margins and prickles on the midrib below.* Flowers *small*, only opening in the morning, a few at a time, in open, well-branched clusters, with the lower branches leafy. Leaves in 2 rows either side of the *pale stem*, often *all tilted towards the sun.* DETAIL Stem usually single, occasionally reddish at base. Basal leaves 7–15cm, stalked, strap-shaped. Stem leaves stalkless, usually oval-oblong,

Prickly   Great

but sometimes deeply-lobed, with clasping, arrow-shaped basal lobes; bleeds white sap. Bracts purplish. Flower heads 7.5–11mm across. *Seeds dull, dingy grey,* 3–4mm long (2.8–4.2mm), faintly winged near tip, with a white beak of equal length. Plume 2 rows of simple white hairs.

GROWTH Annual or biennial.
HEIGHT 30–150cm (200cm).
FLOWERS Jul–Sep.
STATUS Introduction. First recorded 1632 and still spreading, often following road developments.
ALTITUDE Lowland.

Great

Great Lettuce
Lactuca virosa

Uncommon as a native in coastal grassland, also an increasing introduction on rough and disturbed ground inland, often spreading with road developments. ✿ Much like Prickly Lettuce but *stems and especially leaf midribs tinged maroon*, seeds *deep purplish-black.* DETAIL Leaves usually tinged maroon, basal leaves 15–35cm, oval. Flowers in a more diamond-shaped, mostly leafless, cluster. Flower heads 14–20mm across. Seeds 4.2–4.8mm (4–5.2mm), *narrowly winged from top to bottom.*

GROWTH Annual or biennial.
HEIGHT To 200cm (250cm).
FLOWERS Jun–Sep.
STATUS Native and introduction.
ALTITUDE Lowland.

Common Cudweed
Filago vulgaris

GROWTH Annual.
HEIGHT 5–30cm (45cm).
FLOWERS Jul–Aug.
STATUS Native.
ALTITUDE Lowland.

Listed as Near Threatened. Locally common in E Anglia, scarce elsewhere, on light, often sandy soils: arable fields, heathland, forestry rides, dunes. ✺ Often well branched, forking at 45°. Greyish due to a covering of whitish, woolly hairs. Leaves small and strap-shaped, held erect. Flower heads small, in tight clusters in the forks and tip of the stems; florets tiny, almost hidden behind yellowish-tipped bracts. **DETAIL** Usually branching below each cluster of flower heads. Leaves 10–20mm, usually wavy-edged. *Flower heads (15)20–40 together* in spherical clusters *c.* 10–12mm across. *Bracts largely woolly, outer bracts more or less pointed, held erect in fruit.* Outer florets female, inner bisexual, corolla *c.* 3mm long, tubular, white with a reddish tip. Seeds with a plume of unbranched hairs.

Mountain Everlasting *Antennaria dioica*

GROWTH Perennial.
HEIGHT 5–20cm.
FLOWERS Late May–Jul.
STATUS Native.
ALTITUDE 0–910m.

Common in the far N and W, becoming ever scarcer to S and E and almost extinct in lowland England. Short, sparse grassland and rocky places on poor, alkaline to mildly acidic soils, both in the uplands and near the coast, including dunes, machair and heaths. ✺ Patch-forming, with rooting runners and tight rosettes, *leaves dark green above and contrastingly white-woolly below.* Flower heads *c.* 10mm across, in clusters of 2–8 at the tip of unbranched, leafy stems. *Male and female flowers on separate plants.* **DETAIL** Stems and runners white-woolly. Leaves to 3cm, narrowly spoon-shaped, hairless to sparsely hairy above; stem leaves clasping. In male flower heads outer bracts usually white, blunt and often spreading and petal-like; in female heads outer bracts narrower, erect, with chaffy pink tips. In both sexes florets vary from very pale pink to reddish, corolla *c.* 4.5mm long, tubular, wider in male flowers. Seeds with a plume of simple white hairs.

Heath Cudweed
Gnaphalium sylvaticum

Listed as Endangered. Very local on dry, acid soils, mostly sandy or gravelly, especially in forestry rides (on former heathland), also heaths and dunes. ✹ *Very upright*, with greyish-woolly stems and numerous narrow, strap-shaped, *dark green leaves*. Flower heads numerous, *c.* 6mm long, in an *elongated spike*. DETAIL Leaves pointed, tapering gradually to a narrow base, 2–8cm long, decreasing in size up the stem, hairless on upperside, woolly below. Bracts greenish with a broad, brownish or pinkish papery margin. Corolla *c.* 4mm long, pinkish. Plume reddish.

GROWTH Perennial.
HEIGHT 20–60cm (80cm).
FLOWERS Jul–Sep.
STATUS Native.
ALTITUDE 0–850m.

Marsh Cudweed
Gnaphalium uliginosum

Fairly common. Arable fields, tracks and woodland rides, especially where compacted and wet in winter, on more or less acid soils; avoids heavy clays. Also the drying margins of ponds etc. ✹ Low growing and *often well branched* from the base. Dull greyish due to a covering of white, woolly hairs (sparse on leaves, which are greener). *Leaves narrow and strap-shaped, spreading.* Flower heads 3–4mm long, in leafy clusters of 3–10 at the tip of the stems, *florets brownish-yellow*. DETAIL Leaves 1–5cm, stalkless. Bracts papery, pale brown, tipped darker. Outer florets female, corolla tubular, *c.* 1.5mm long, inner bisexual, corolla wider. Seeds with a plume of unbranched white hairs.

GROWTH Annual.
HEIGHT 4–25cm.
FLOWERS Jul–Sep.
STATUS Native.
ALTITUDE 0–590m.

Small Cudweed *Filago minima*

Rather local in barren places on light, dry, usually sandy soils: heathland tracks, forestry rides, dunes, sometimes rocky places. ✹ As Common Cudweed but slender and very inconspicuous, despite growing on open ground. Many plants are *matchstick-thin*, with the clusters of flower heads just 2–5mm across. DETAIL Stems branching irregularly. Leaves 4–10mm long, often pressed to stem, *Flower heads in clusters of 2–8. Outer bracts more or less blunt, held at 90° to the stem in fruit* (fruiting heads *star-shaped* when viewed from above). Corolla *c.* 2mm long.

GROWTH Annual.
HEIGHT 2–15cm (25cm).
FLOWERS Jun–Sep.
STATUS Native.
ALTITUDE 0–375m.

Golden Samphire
Inula crithmoides

Nationally Scarce. Sea cliffs and cliff-top grassland, especially on chalk and limestone. Also saltmarshes in SE England from the Solent to Suffolk. ☸ A large-flowered daisy with *succulent foliage*. DETAIL Hairless. Stems tufted, erect to sprawling. Leaves all on stem, stalkless, to 5cm long, *usually with 3 teeth at the tip*. Flower heads 15–25mm wide. Bracts green, hairless. Disc-florets orange-yellow, ray-florets *c.* 9mm long. Seeds downy, with a plume of white hairs.

GROWTH Perennial.
HEIGHT 15–80cm (100cm).
FLOWERS Jul–Aug (Oct).
STATUS Native.
ALTITUDE Lowland.

Ploughman's Spikenard *Inula conyzae*

Locally common. Sparse, rough, often disturbed grassland on dry soils over chalk, limestone or sands. ☸ Upright, with purplish stems and abundant flower heads in an untidy, well-branched cluster. Flowers yellow, *without ray-florets, bracts variably washed dull purple, conspicuous in bud*. DETAIL Densely softly-hairy. Basal leaves to 12cm, long oval, net-veined, very shallowly toothed, each tooth tipped purplish, tapering to a winged stalk; upper leaves narrower, near stalkless. Flower heads 7–12mm across. Ray-florets minute or absent, disc-florets *c.* 6mm long. Seeds downy, with a plume of rosy hairs. SEE ALSO Foxglove (p. 213); the leaf rosettes are rather similar.

GROWTH Perennial.
HEIGHT 20–80cm (125cm).
FLOWERS Jul–Sep.
STATUS Native.
ALTITUDE 0–305m.

Common Fleabane
Pulicaria dysenterica

Locally abundant. A wide variety of damp, rough grassland. ❀ Patch-forming, with *cottony-hairy, slightly greyish foliage* and large, golden-yellow flowers in *loose, flat-topped clusters*. DETAIL Leaves all on stem, wrinkled, 3–8cm long, strap-shaped, stalkless and clasping the stem at the base, with the edges rolled downwards. Bracts leafy. Flower heads 15–30mm across, ray-florets *c*. 10mm long. Seeds downy, with a plume of simple hairs.

GROWTH Perennial.
HEIGHT 20–60cm (100cm).
FLOWERS Late Jul–Sep.
STATUS Native.
ALTITUDE 0–325m.

Goldenrod
Solidago virgaurea

Locally common. Rough, dry grassland, mostly on acid soils: grassy heaths, woodland rides, hedgebanks, sea cliffs, and rocky places in the uplands. ❀ Varies from tall in the lowlands to very short in the mountains. A yellow-flowered daisy with *irregular clusters of rather small, untidy flowers*. DETAIL Stems variably hairy, usually unbranched, leafy, sometimes purplish. Basal leaves to 15cm, often crinkled, sparsely hairy, dull yellowish-green, often with purple margins, and with long, winged stalks; stem leaves narrower, stalkless. Bracts leaf-like. Flower heads 6–10mm across, ray-florets few, 4–9mm long. Seeds downy, with a plume of white hairs. SIMILAR SPECIES Canadian Goldenrod *S. canadensis* is a fairly common garden throw-out with sprays of very small yellow flowers. Several other alien goldenrods may escape.

GROWTH Perennial.
HEIGHT 5–70cm (100cm).
FLOWERS (Late May) Jul–Sep.
STATUS Native.
ALTITUDE 0–1040m.

GROWTH Perennial.
HEIGHT 3–12cm (20cm).
FLOWERS Mar–Oct (all year).
STATUS Native.
ALTITUDE 0–915m.

Daisy *Bellis perennis*

Abundant in almost any short grass, including, of course, garden lawns; avoids very acid and very dry soils. ❂ *Rosettes of leaves*, flattened to the ground, produce the familiar archetypical daisy flowers, each *solitary, on a leafless stalk*. Ray-florets usually white, but may be tipped or washed reddish. DETAIL Sparsely hairy. Leaves 2–5cm, spoon-shaped, with a few irregular, blunt teeth, tapering to a short, winged stalk. Bracts green, blunt. Flower heads 12–25mm across (larger and sometimes double in cultivars). Seeds downy, without a hairy plume.

GROWTH Perennial.
HEIGHT Stems to 50cm.
FLOWERS Apr–Oct.
STATUS Introduced to gardens in 1836 (Mexico).
ALTITUDE Lowland.

Mexican Fleabane *Erigeron karvinskianus*

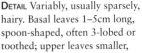

Fairly common garden escape, especially in the SW. Sheltered places on walls and other dry, often stony ground, particularly around houses. ❂ Sprawling, with *leafy stems* (often trailing over walls), each with *several Daisy-like flowers. Ray-florets white to mauve above and contrastingly pink to purple below.*

DETAIL Variably, usually sparsely, hairy. Basal leaves 1–5cm long, spoon-shaped, often 3-lobed or toothed; upper leaves smaller, strap-shaped, pointed, usually unlobed. Flower heads 15–30mm across.

GROWTH Annual or biennial.
HEIGHT 8–40cm (60cm).
FLOWERS Jul–Aug.
STATUS Native.
ALTITUDE 0–430m.

Blue Fleabane
Erigeron acris

Locally common, especially in the S and E. Rough, grassy places, often with some bare ground, on poor, dry soils, usually over chalk or limestone but also on dunes. ❂ Upright, with a single branched, purplish stem and clusters of small flowers with yellow centres and *short, erect, lilac ray-florets.*
DETAIL *Rather hairy.* Basal leaves to 5cm, spoon-shaped, with a purplish midrib; stem leaves numerous, alternate, strap-shaped, clasping. Bracts green, tipped purple. Flower heads 8–12mm across; ray-florets only 2–4mm longer than disc-florets. Seeds yellow, hairy, with a plume of reddish-white hairs.

Canadian Fleabane
Conyza canadensis

Abundant weed of disturbed ground and arable fields, especially on well-drained soils. ❀ Very variable in height but always *slender*, with many narrow leaves and a long, roughly cylindrical cluster of *tiny flower heads*. DETAIL Stems variably hairy. Leaves 1–5cm long, hairy, with a conspicuous fringe of long hairs. Flower heads 3–5mm across; central, tubular disc-florets yellowish-green, spreading at the tip into 4 triangular lobes (20× lens – a few may be 5-lobed); outer florets more numerous, with *tiny white rays, clearly visible* above the bracts, which are *hairless to sparsely hairy,* green, with pale chaffy margins. Seeds downy, pale yellow, with a plume of unbranched, dingy white hairs.

GROWTH Annual.
HEIGHT 8–100cm.
FLOWERS Jun–Nov.
STATUS Introduced. Recorded in the London area since 1690. Has spread greatly in recent decades (N America).
ALTITUDE Lowland.

Bilbao's Fleabane *Conyza floribunda*

GROWTH Annual.
HEIGHT To 150cm.

FLOWERS Jul–Oct?
ALTITUDE Lowland.

STATUS First recorded 1992, spreading rapidly (S America).

A recent introduction, spreading rapidly on disturbed ground, mostly around London. ❀ Very similar to Canadian Fleabane but hairs on stems more stiffly erect and *harsher* (especially towards the base). DETAIL Flower size as Canadian Fleabane, bracts similarly *more or less hairless*, but central disk florets *5-lobed* at tip, and outer florets with rays *tiny or absent* (as Guernsey Fleabane).

Guernsey Fleabane
Conyza sumatrensis

A recent introduction, spreading rapidly on disturbed ground, especially in urban areas. ❀ As Canadian Fleabane but subtly greyer, with upper stem, flower stalks and bracts *densely hairy*. Flower heads slightly larger, 5–8mm across. DETAIL Disk florets *5 lobed* at tip; outer florets with rays *tiny or absent* (merely dingy yellow-white plumes and thread-like white styles). SIMILAR SPECIES Argentine Fleabane *C. bonariensis*, another recent introduction, but as yet scattered records only. Seeds with a greyish-white plume; bracts often *conspicuously red-tipped*.

GROWTH Annual.
HEIGHT To 100cm (200cm).
FLOWERS Jun–Nov.
STATUS First recorded 1974, spreading rapidly (S America).
ALTITUDE Lowland.

Sea Aster *Aster tripolium*

Locally abundant. Salt-marshes, tidal rivers and other brackish places near the sea; in the W also coastal rocks and cliffs.

GROWTH Biennial (annual).
HEIGHT 15–100cm.
FLOWERS Jul–Oct.
STATUS Native.
ALTITUDE Lowland.

❋ *Leaves fleshy*, flower heads in dense clusters. Two varieties occur, sometimes side-by-side. Var. *flosculosus* has flower heads *c.* 10mm across with *yellow disk-florets only* (this is the dominant form on saltmarshes in E Anglia); var. *tripolium* has bluish-mauve ray-florets and looks like a Michaelmas Daisy. DETAIL *Hairless*. Basal leaves spoon-shaped, stalked; stem leaves 7–12cm, strap-shaped, stalkless. *Bracts few, blunt, unequal.* Ray-florets *c.* 13mm long. Seeds downy, with a plume of simple brown hairs. SIMILAR SPECIES **Common Michaelmas Daisy** *A.* × *salignus* is one of several garden species and hybrids that may be found as garden escapes or throw-outs.

Tansy *Tanacetum vulgare*

GROWTH Perennial.
HEIGHT 30–100cm (120cm).
FLOWERS Jul–Oct.
STATUS Ancient introduction.
ALTITUDE 0–380m.

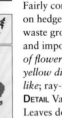

Fairly common. Rough grassland on hedge- and riverbanks and waste ground. ❋ Patch-forming, tall and imposing, with *flat-topped clusters of flower heads*, each 6–10mm across, the *yellow disk-florets tightly packed and button-like*; ray-florets absent. *Strongly scented.* DETAIL Variably sparsely hairy. Stems purplish. Leaves dotted with minute scented glands, 15–25cm, finely cut, lower stalked, upper clasping the stem. Seeds ribbed, without a hairy plume.

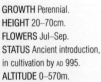

Feverfew *Tanacetum parthenium*

GROWTH Perennial.
HEIGHT 20–70cm.
FLOWERS Jul–Sep.
STATUS Ancient introduction, in cultivation by AD 995.
ALTITUDE 0–570m.

Fairly common garden escape: walls, pavement cracks, waste ground, usually close to houses or on tipped soil. ❋ Foliage yellowish-green, *strongly aromatic*. Flower heads 15–23mm across, in open, leafy clusters, with *short, broad, white ray-florets* (sometimes double). DETAIL Finely hairy. Stems well branched. Lower leaves 3–8cm, long-stalked. Seeds ribbed, without a hairy plume.

Mugwort
Artemisia vulgaris

Common. Rough grassland and bare ground on verges, hedgebanks and waste ground. ❁ Tall, with sprays of tiny flowers and *deeply-cut leaves, dark green and hairless above and contrastingly silvery below* (due to silky white hairs). **DETAIL** Weakly aromatic. Stems purplish, cottony-hairy. Leaves 5–8cm long, lower leaves stalked, stem leaves stalkless, clasping. Flower heads 1.5–3mm across, the cottony-hairy bracts largely concealing the tubular florets, which are yellow with purplish lobes. Seeds hairless. **SIMILAR SPECIES**

GROWTH Perennial.
HEIGHT 60–120cm (200cm).
FLOWERS Jul–Sep.
STATUS Ancient introduction.
ALTITUDE To 465m.

Chinese Mugwort *A. verlotiorum* is an increasing introduction, especially around London and in NE England. Patch-forming and aromatic. Leaves cut into *long*, narrow lobes, with branching sprays of flowers produced *Oct–Dec.*

Wormwood *Artemisia absinthium*

Rather local. Waste ground, gravel pits, quarries etc. ❁ Tall and *strongly aromatic* (it is one of the flavourings of absinth), with *deeply-lobed, silvery-grey foliage* and sprays of small *drooping* flowers. **DETAIL** Whole plant silky-hairy. Leaf segments *blunt*. Flower heads 3–5mm across, bracts pale green with chaffy brown margins, corolla greenish-yellow, styles purplish-brown, receptacle *hairy* (hairless in Sea Wormwood).

GROWTH Perennial.
HEIGHT 30–100cm.
FLOWERS Jul–Aug.
STATUS Ancient introduction. In cultivation by 1200 as a medicinal herb and flavouring.
ALTITUDE 0–375m.

Sea Wormwood
Artemisia maritima

Locally abundant. Saltmarshes, also other brackish places by the sea and nearby shingle, cliffs, walls and waste ground. ❁ Medium-tall and *strongly scented*, with *finely cut, silvery-grey foliage* and clusters of small, inconspicuous flowers. **DETAIL** *Whole plant white-woolly.* Stems woody towards base. Leaves 2–5cm long, lower stalked, upper stalkless. Flower heads 1.5–3.5mm across, corolla yellow to reddish.

GROWTH Perennial.
HEIGHT 20–50cm.
FLOWERS Aug–Sep.
STATUS Native.
ALTITUDE Lowland.

Yarrow *Achillea millefolium*

GROWTH Perennial.
HEIGHT 8–50cm.
FLOWERS Jun–Aug.
STATUS Native.
ALTITUDE 0–1210m.

Abundant in all kinds of grassy places, avoiding only very poor, wet or acid soils. ✿ *Finely cut, feathery foliage* distinctive. Flowers in *dense, flat-topped clusters* that recall an umbellifer, varying from pure white to deep dusty pink. DETAIL Weakly aromatic. Hairy. Leaves 5–15cm long, the thread-like lobes spreading in 3 dimensions; lower leaves stalked, upper stalkless. Flower heads 4–7mm across, with *c.* 5 ray-florets and dirty-white disc-florets. Seeds hairless.

Sneezewort *Achillea ptarmica*

GROWTH Perennial.
HEIGHT 20–60cm.
FLOWERS Late Jun–Sep.
STATUS Native.
ALTITUDE 0–770m.

Fairly common. A wide variety of damp, grassy places, including verges, wet heathland and hillside flushes. ✿ The loose clusters of daisy-like flowers with *white ray-florets and a dingy-cream centre* are unique. DETAIL Sparsely hairy. Leaves 1.5–8cm, strap-shaped, pointed, *finely saw-toothed*, stalkless. Flower heads 12–20mm across. Seeds hairless. 'Doubles' sometimes occur as garden escapes.

Hemp Agrimony
Eupatorium cannabinum

Common. A wide variety of damp or wet places on neutral to alkaline soils; occasional in dry habitats. ✿ Tall, with *3-lobed, cannabis-like leaves* (hence the scientific name) and *irregular, rounded clusters, up to 15cm across, of many dusty-pink flower heads*. DETAIL Hairy. Stems often purplish. Leaflets 5–10cm long; lower leaves stalked, uppermost stalkless. Flower heads 2–5mm across, each with 5–6 florets surrounded by pinkish-purple bracts. Corolla tubular, pinkish, with protruding forked white styles. Seeds black, with a plume of simple white hairs.

GROWTH Perennial.
HEIGHT To 150cm.
FLOWERS Jun–Sep.
STATUS Native.
ALTITUDE 0–350m.

Corn Marigold
Glebionis segetum

Listed as Vulnerable. A declining arable weed, still locally abundant in some areas, mostly on light sandy soils; avoids chalk.
❀ Daisy-like flowers, with an *orange-yellow centre and golden-yellow rays*, coupled with *lobed, grey-green, often slightly fleshy leaves*, are unique. DETAIL Sparsely hairy. Stems well branched, sometimes sprawling. Leaves 2–10cm, variably toothed, lower leaves cut into finger-like

GROWTH Annual.
HEIGHT 20–60cm.
FLOWERS Jun–Oct.
STATUS Introduced in the Iron Age.
ALTITUDE 0–410m.

lobes towards tip, narrowing to a vague winged stalk, upper leaves toothed only, stalkless, clasping. Flower heads solitary, 3–7cm across. Seeds ribbed, hairless.

Oxeye Daisy Leucanthemum vulgare

Common. A variety of grassy places, especially where periodically cut or lightly grazed, also waste ground; sometimes sown on the verges of new roads.
❀ The *large* daisy-like flowers grow singly on erect, little-branched, leafy stalks from a basal rosette. DETAIL Sparsely hairy. Stems purplish. Leaves dark green, lower leaves to 5cm, strap-shaped, toothed, *contracting abruptly* to a long stalk; stem leaves stalkless, clasping, *variably toothed or shallowly lobed all the way to the base*. Flower heads 2.5–6cm across. Seeds ribbed, hairless. SIMILAR SPECIES Shasta Daisy *L.* ×

GROWTH Perennial.
HEIGHT 20–75cm.
FLOWERS May–Sep.
STATUS Native.
ALTITUDE 0–600m.

basal leaf

superbum taller, with the basal leaves tapering into the stalk, flower heads 6–10cm across, often double. A fairly frequent garden escape.

Fox-and-cubs
Pilosella aurantiaca

Widely naturalised. Churchyards, verges and other grassy places.
❀ *Orange, dandelion-like flowers* distinctive. DETAIL Patch-forming, with leafy runners. Leaves 5–20cm, roughly hairy, strap-shaped, toothed. Flower stems leafy, with abundant long hairs. Flower heads *c.* 15mm across.

GROWTH Perennial.
HEIGHT To 40cm (65cm).
FLOWERS Jun–Oct.
STATUS Introduced to gardens by 1629 and recorded in the wild from 1793 (Europe).
ALTITUDE 0–490m.

Scentless Mayweed *Tripleurospermum inodorum*

An abundant weed of arable fields and other disturbed ground. ❀ Delicate, feathery foliage and large daisy-like flowers. *Not scented.* DETAIL Leaves more or less hairless, cut into very fine segments that are pointed or have a minute bristle at the tip. Bracts green with white borders and *fine, neat, dark brown edges.* Flower heads 20–45mm across, receptacle *solid,* without chaffy scales among the florets. Seeds 1.3–2.2mm long, covered with *minute pimples*, with *3 bold ribs* on 1 face and *2 dark, rounded oil glands* on opposite face (10× lens, but hard to see).

GROWTH Annual.
HEIGHT 10–60cm.
FLOWERS Jun–Nov.
STATUS Ancient introduction.
ALTITUDE To 590m.

Scented Mayweed *Matricaria chamomilla*

A local weed of arable fields and other disturbed ground. ❀ Very like Scentless Mayweed but usually has a *mild, pleasant scent* when crushed, flower heads often smaller, 10–25mm across, and receptacle *obviously hollow and 'squashy'.* DETAIL Bracts green with chaffy brownish-white borders. Seeds *very small*, with 4–5 obscure ribs on 1 face and *no* oil glands.

GROWTH Annual.
HEIGHT 10–50cm.
FLOWERS May–early Sep.
STATUS Ancient introduction.
ALTITUDE 0–470m.

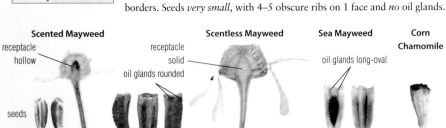

Scented Mayweed — receptacle hollow — seeds

Scentless Mayweed — receptacle solid — oil glands rounded

Sea Mayweed — oil glands long-oval

Corn Chamomile

Sea Mayweed *Tripleurospermum maritimum*

Common on sand, shingle, saltmarsh margins, cliffs etc. near the sea. ❀ Very like Scented and Scentless Mayweeds but perennial, *leaf segments fleshier,* more rounded in cross-section and often *blunt*; check seeds for certain identification. DETAIL Seeds 1.8–3.5mm long with *3 bold, closely-spaced ribs* on 1 face and *2 dark, long-oval oil glands* on opposite face (10× lens, but hard to see). Stems purplish in S England, where close to Scentless Mayweed in appearance.

GROWTH Perennial (biennial).
HEIGHT To 60cm.
FLOWERS May–Sep.
STATUS Native.
ALTITUDE Lowland.

Chamomile *Chamaemelum nobile*

Listed as Vulnerable. Very locally abundant in damp turf (especially old commons), on sandy, mildly acid soils, kept short by grazing, mowing, trampling or, on cliff-tops and other coastal grassland, exposure to the wind. ❀ *Strongly chamomile-scented*, with *prostrate stems that root at the nodes.* **DETAIL** Leaves grey-green, sometimes slightly hairy. Receptacle conical, with *yellowish, strap-shaped scales between the disk florets.* The base of corolla tube swells to form a 'hood' over the developing seed, which is weakly ridged on 1 face.

GROWTH Perennial.
HEIGHT Stems 10–30cm.
FLOWERS Jun–Sep.
STATUS Native.
ALTITUDE 0–465m.

Corn Chamomile *Anthemis arvensis*

Listed as Endangered. An uncommon and declining weed of arable fields and other disturbed ground, especially on sandy or chalky soils. ❀ *Pleasantly aromatic.* **DETAIL** Stems hairy, even woolly. Leaves finely hairy, segments bristle-tipped. Flower head with *numerous lanceolate scales between the florets,* which persist as the flower goes to seed, giving a ragged or tatty look to the seed head. Seeds conical, obscurely 10-ribbed.

GROWTH Annual or biennial.
HEIGHT To 60cm.
FLOWERS Jun–Sep.
STATUS Ancient introduction.
ALTITUDE Lowland.

Stinking Chamomile *Anthemis cotula*

Listed as Vulnerable. Uncommon and declining weed of arable fields and other disturbed ground, especially on heavy clay soils. ❀ *Obviously foul-smelling.* **DETAIL** Stem and leaves grey-green, hairless to sparsely hairy. Flower head with yellowish lanceolate scales among the *central* disk florets only. Seeds ribbed, with minute warts on the ribs.

GROWTH Annual.
HEIGHT To 50cm.
FLOWERS Jul–Sep.
STATUS Introduced in the Iron Age (perhaps earlier).
ALTITUDE Lowland.

Pineappleweed
Matricaria discoidea

An abundant weed of arable fields and trampled ground. ❀ *Strongly pineapple-scented* when crushed, flower head with *dull yellow disk florets only,* lacking white ray florets. **DETAIL** Leaves more or less hairless. Receptacle conical, hollow. Disk florets divided into 4 lobes at tip (5-lobed in other mayweeds).

GROWTH Annual.
HEIGHT To 35cm.
FLOWERS May–Nov.
ALTITUDE Mostly lowland.
STATUS Introduced to cultivation by 1781. First recorded in the wild in 1871, spreading rapidly (E Asia).

Common Ragwort *Senecio jacobaea*

Very common. A wide variety of grassy places, also waste ground and walls. ✿ Stem leaves *deeply and intricately lobed.* Flowers in *dense clusters*; bracts tipped black. Flowering peaks in July.

DETAIL Stems hairless. Leaves hairless to sparsely cobwebby-hairy on underside. Basal leaves in a rosette that usually dies off by flowering time, to 15cm long, stalked, with a large oval terminal lobe and several pairs of side lobes. Outermost bracts *c.* 0.25–0.4 times main bracts. Flower heads 15–25mm across, ray-florets 5–9mm long. *Seeds of ray-florets hairless, of disc-florets hairy*, all with a plume of simple white hairs.

GROWTH Perennial (biennial).
HEIGHT 30–150cm.
FLOWERS Mid Jun–Oct.
STATUS Native.
ALTITUDE 0–710m.

Oxford Ragwort

Senecio squalidus

Locally common. Pavement cracks, waste ground, railways etc., *mostly in urban areas.* As Common Ragwort but typically well branched and more spreading, with *looser clusters of slightly larger flowers*; comes into bloom *early.* **DETAIL** More or less hairless. Stem sometimes woody at base. Leaves bright, shiny green, usually deeply lobed, with *narrow, pointed, lobes,* but sometimes merely deeply toothed. No basal rosette, lower leaves held upright, with winged stalks, upper leaves stalkless, clasping. Bracts conspicuously tipped black. Ray-florets (6)8–10mm long. *All seeds hairy*, with a plume of white hairs.

GROWTH Annual to short-lived perennial.
HEIGHT To 50cm.
FLOWERS Apr–Oct (Dec).
STATUS Escaped from Oxford Botanic Garden in 1794, spreading rapidly via the railway network.
ALTITUDE Lowland.

Hoary Ragwort *Senecio erucifolius*

Locally common. Grassland, shingle banks, dunes, waste ground, railways, mostly on alkaline soils, especially clays. ✿ A late-flowing ragwort (peaking August), picked out by *pale yellow flowers,* rather open flower clusters, and leaves with *narrow, parallel-sided lobes.* DETAIL Patch-forming, stems sparsely cottony towards base. Leaves *densely hairy when fresh,* becoming dull green above, but still more or less *grey-hairy below;* lower leaves 5–12cm, stalked, upper leaves stalkless, clasping. Outermost bracts *about half the length of the main bracts.* Ray-florets 5–9mm long. *All seeds downy.*

GROWTH Perennial.
HEIGHT To 100cm (120cm).
FLOWERS Late Jul–Sep.
STATUS Native.
ALTITUDE Lowland.

Marsh Ragwort
Senecio aquaticus

Local. Damp grassland and the margins of ponds, streams and ditches. ✿ As Common Ragwort but flowers larger, in a *relatively loose, spreading cluster,* and leaves with *a large, oval terminal lobe and a few much smaller side lobes.* DETAIL Very sparsely hairy. Flower heads 20–25mm across (to 45mm in N Scotland). *Seeds of ray-florets hairless, of disc-florets hairless to sparsely hairy,* all with a plume of simple white hairs.

GROWTH Biennial (perennial).
HEIGHT To 80cm.
FLOWERS Jul–Sep.
STATUS Native.
ALTITUDE To 460m.

Niger *Guizotia abyssinica*

An alien, mostly growing from wild bird seed. ✿ Flower heads 2-4cm across, with *c.* 8 yellow rays. DETAIL Stems hairy, purplish towards base. Leaves opposite below, alternate above, lanceolate, toothed, stalkless, clasping the stem. Outer bracts green, inner row chaffy, shorter and narrower. Seeds without a plume.

GROWTH Annual.
HEIGHT To 80cm (200cm).
FLOWERS Sep–Nov.
ALTITUDE Lowland.

STATUS Introduced to gardens by 1806 and recorded from the wild by 1876. Has increased in recent years.

Heath Groundsel *Senecio sylvaticus*

GROWTH Annual.
HEIGHT 30–70cm (100cm).
FLOWERS Jun–Sep.
STATUS Native.
ALTITUDE 0–430m.

Locally common. Heathland (especially after disturbance), clear-fell, sea cliffs, dunes, on sandy, acid soils. ❁ Resembles Common Groundsel, but flower heads 7–9mm across, with very short ray-florets that rapidly curl under. *Whole plant glandular-hairy, although only slightly sticky.* **Detail** Stem straight, branched. Rosette leaves grey-green, to 10cm long, short-stalked; upper leaves stalkless, sometimes clasping the stem. *Outermost bracts very short* (less than 0.3 times main bracts), main bracts tipped purplish. Ray florets less than 6mm. Seeds minutely but densely hairy.

Sticky Groundsel
Senecio viscosus

Locally common. Disturbed ground on sand and gravel, often in man-made habitats – railways, waste ground etc. – also dunes and coastal shingle. ❁ Resembles a small-flowered ragwort, with flower heads 10–12mm across, but ray-florets quickly *curling downwards*, and *foliage very sticky*. **Detail** Densely covered in sticky glandular hairs. Stem wavy, well branched. Rosette absent. Lower leaves to 5cm, short-stalked, upper leaves stalkless, weakly clasping. Outermost bracts rather longer than Heath Groundsel (0.3–0.5 times main bracts), main bracts tipped dark purple. Ray-florets less than 8mm long. Seeds hairless (or hairy in grooves), with a plume of simple white hairs.

GROWTH Annual.
HEIGHT To 60cm.
FLOWERS Jul–Sep.
STATUS Introduced *c.* 1660, but may be native on coastal shingle.
ALTITUDE 0–570m.

Groundsel *Senecio vulgaris*

GROWTH Annual.
HEIGHT To 30cm (45cm).
FLOWERS All year.
STATUS Native.
ALTITUDE 0–570m.

An abundant weed of gardens, arable fields, pavement cracks, walls, etc., also dunes and coastal cliffs. ❀ Flower heads just *c.* 4mm across, *usually with no ray florets*, grouped into clusters that are *often drooping*. DETAIL Variably cottony-hairy. Stems purplish. Leaves variably lobed, lower 3–20cm, with blunt, toothed, strap-shaped lobes, narrowed to a winged stalk; mid and upper stem leaves clasping. Bracts often black-tipped. Ray-florets, when present, less than 5mm long. Seeds hairy, with a plume of simple white hairs.

GROWTH Perennial (biennial).
HEIGHT 7–30cm (60cm on Anglesey).
FLOWERS May–Jun.
STATUS Native.
ALTITUDE Lowland.

Field Fleawort
Tephroseris integrifolia

Nationally Scarce. Very local. Species-rich chalk grassland, especially on S-facing slopes, also grassy cliff-tops on Anglesey (where Sch 8). ❀ Whole plant woolly-hairy (losing hairs with age), with a basal rosette and a solitary, leafy stem with a loose cluster of a few, daisy-like yellow flowers. DETAIL Basal leaves hairy, white-woolly below, especially when young, 2–5cm long, oval to circular, occasionally sparsely toothed, with a short winged stalk, stem leaves narrower, stalkless, clasping. Flowers 15–25mm across, ray-florets 5–10mm long. Seeds hairy, with a plume of simple white hairs.

Gallant Soldier
Galinsoga parviflora

Locally common. Pavement cracks, waste ground, gardens and arable fields, especially on light soils. ❁ Flowers *small and daisy-like,* with a dull yellow centre and 5 (4–6) *well-spaced ray-florets, 3-lobed at the tip.* DETAIL *Hairless to sparsely hairy,* flower stalks may have glandular hairs. Leaves opposite, 2–7cm long, sparsely toothed, variably stalked. Flower heads 4–7mm across, with *tiny scales between the florets, mostly distinctly 3-lobed with the central lobe the largest.* Seeds black, bristly, with a crown of silvery scales, all fringed with hairs.

GROWTH Annual.
HEIGHT 10–25cm (80cm).
FLOWERS May–Oct.
STATUS Introduced to Kew Gardens by 1796, escaping by 1860 and still spreading (S America).
ALTITUDE Lowland.

Shaggy Soldier
Galinsoga quadriradiata

Locally common. Mostly in urban areas, on pavement cracks, waste ground, etc., but also arable fields. ❁ Rather like Gallant Soldier but usually *obviously hairy* and flowers slightly larger. Intermediate plants can only be identified by close examination of the flowers and seeds. DETAIL As Gallant Soldier but flower stalks with shaggy glandular and non-glandular hairs. Scales amongst florets mostly *unlobed or with 1–2 weak side lobes.* Seeds with crowning scales both fringed with hairs and with a *long, fine point at the tip.*

GROWTH Annual.
HEIGHT 10–25cm (80cm).
FLOWERS May–Oct.
STATUS First recorded 1909, and still spreading (Mexico).
ALTITUDE Lowland.

Leopard's Bane *Doronicum pardalianches*

A garden escape, very locally but widely naturalised in woods and shady verges. ❁ Patch-forming. Basal leaves long-stalked, with a heart-shaped base, stems leafy, with 3–8 flowers. DETAIL Hairy. Basal leaves to 8cm long; stem leaves clasping. Bracts in 2 rows. Flower heads 4–5cm across. Seeds black, in disc-florets hairy, with a plume of simple white hairs. SIMILAR SPECIES Other garden *Doronicums* may escape, but have fewer flowers per stem or differently shaped leaves.

GROWTH Perennial.
HEIGHT 30–80cm.
FLOWERS May–Jul.
STATUS Cultivated since the 16th century. Recorded in the wild from 1633 (W Europe).
ALTITUDE Lowland.

Nodding Bur Marigold

Bidens cernua

Local. The drying margins of slow-flowing rivers, streams, lakes and ponds, often where flooded in winter, also ditches and marshes. ❂ Flower heads button-like, rather *drab yellow, nodding* on long stalks, with disk-florets only (*rarely* has yellow ray-florets – var. *radiata*). The outer bracts are spreading and leaf-like, the inner row, together with numerous long scales amongst the florets, straw-coloured with blackish stripes. Leaves coarsely toothed but *not* lobed. DETAIL Sparsely hairy. Leaves opposite, stalkless. Flower heads 15–25mm across. Corolla golden-yellow. Seed broadening towards the tip, 4-angled, with downward-pointing barbs on the angles and on the pappus of 3–4 stout bristles (thus *bur* marigold).

GROWTH Annual.
HEIGHT 8–60cm (80cm).
FLOWERS Jul–Oct.
STATUS Native.
ALTITUDE 0–310m.

var. *radiata*

Trifid Bur Marigold *Bidens tripartita*

Local. The margins of lakes, gravel pits and slow-flowing rivers and streams, also ditches and peat workings. ❂ As Nodding Bur Marigold but flower heads *more or less erect*, even in bud, and *at least some leaves 3-lobed*. DETAIL Leaves 5–15cm, with *winged stalks* joining across the stem, many leaves with 2 side lobes (rarely, all leaves unlobed), end lobe stalkless or with a winged stalk. Seeds more strongly flattened, more rounded at tip, with 2–4 bristles. SIMILAR SPECIES **Beggarticks** B. *frondosa* from America is widely naturalised by canals in the Midlands. Leaves similarly lobed (often 2 pairs of side lobes), but leaf stalks *unwinged*, and terminal leaflet also with a *wingless* stalk; barbs on seed (*not* bristles) pointing *upwards*.

GROWTH Annual.
HEIGHT 8–60cm (80cm).
FLOWERS Jul–Oct.
STATUS Native.
ALTITUDE 0–365m.

Butterbur
Petasites hybridus

female flowers

Locally abundant. Damp, seasonally flooded ground by rivers and streams, wet, open woodland, and damp verges. ❀ Forms extensive patches. Flower heads in *robust leafless spikes* that come into bloom *just before the leaves*. Male and female flowers are mostly on separate plants and the rather taller female spikes may reach 100cm in fruit; female plants are found in central and N England, seldom elsewhere. The leaves grow through the summer to an *impressive size*. **DETAIL** Leaves 10–100cm across, toothed, whitish-cobwebby below, with stalks eventually up to 150cm. Flowering stems variably washed purplish or pinkish, with numerous triangular bracts. All florets pinkish, tubular, many together in tight flower heads; male florets 7–12mm long, female florets 3–6mm, much narrower. Seeds with unbranched hairs forming a conspicuous 'clock'.

GROWTH Perennial.
HEIGHT Flowering stems 10–40cm.
FLOWERS Mar–May.
STATUS Native.
ALTITUDE 0–380m.

male flowers

male florets

Winter Heliotrope *Petasites fragrans*

Locally abundant, naturalised on verges, rough ground and streamsides. ❀ Forms extensive patches of circular leaves, with the flowers in leafless spikes. The *flower heads are strongly almond-scented, and the outer florets have short, erect rays*. **DETAIL** Leaves 10–20cm across, finely toothed, sparsely hairy below, with a long, reddish, glandular-hairy stalk. Only male plants are found in Britain.

GROWTH Perennial.
HEIGHT Flowering stems 10–30cm.
FLOWERS Nov–Mar.
STATUS Introduced to gardens in 1806. Recorded from the wild by 1835 and still spreading (Italy.)
ALTITUDE Lowland.

Coltsfoot
Tussilago farfara

Locally common, colonising bare ground on dunes, shingle, soft cliffs, landslides, the margins of rivers and ponds, verges and waste ground. ❀ Patch-forming, the yellow, daisy-like flowers appear *well before the leaves, on short, scaly stalks.* DETAIL Leaves 4–30cm across, *polygonal*, heart-shaped at base, initially with felted white hairs both sides, but these quickly go from upperside. Flower stems with felted white hairs and numerous purplish, scale-like bracts. Flower heads 15–35mm across. Seeds with a plume of simple hairs forming a conspicuous 'clock'.

GROWTH Perennial.
HEIGHT 5–15cm.
FLOWERS Feb–Apr.
STATUS Native.
ALTITUDE 0–1065m.

DEWPLANTS: FAMILY AIZOACEAE

Hottentot Fig
Carpobrotus edulis

An invasive alien, locally dominant on cliffs, dunes and rocky places, but frost sensitive and restricted to mild areas by the sea. ❀ *Patch-forming,* with *fleshy leaves, triangular in cross-section,* and *large, yellow or purple, daisy-like flowers* (but not a member of the daisy family). DETAIL Stems angled, woody at base. Leaves to 8cm long, stalkless. Flowers solitary, to 10cm across, calyx with 5 lobes, petals and stamens numerous. Fruit fleshy and fig-like. SIMILAR SPECIES Several related species are found very locally on the S coast, especially the far SW.

GROWTH Perennial.
HEIGHT To about 15cm.
FLOWERS Mar–Aug (Dec).
STATUS In cultivation by the late 17th century, but not recorded from the wild until 1886 (S Africa).
ALTITUDE Lowland.

Common Valerian
Valeriana officinalis

GROWTH Perennial.
HEIGHT 30–150cm (200cm).
FLOWERS Jun–Aug.
STATUS Native.
ALTITUDE To 805m.

Fairly frequent. Rank, wetland vegetation: marshes, streamsides and wet woodland, also much less frequently on rough grassland on drier alkaline soils. ❃ *Tall*, with dense, irregular clusters, 5–12cm across, of *tiny pale pink flowers; buds bright pink. All leaves cut into smaller leaflets.* Resembles an umbellifer, but flowers funnel-shaped, with 3 stamens (not 5) and seeds with a feathery plume. DETAIL Sparsely hairy. Stem erect, usually unbranched. Leaves opposite, to 20cm, lower leaves long-stalked, upper leaves stalkless; leaves on central stem with 2–6 pairs of toothed leaflets in the wet-ground subspecies *sambucifolia*, 7–13 pairs of untoothed leaflets in the dry-ground subspecies *collina*. Calyx tubular, with feathery hairs in fruit, corolla tubular, slightly pinched at base, 4–5mm across at tip, where cut into 5 petals. Style 1, with 3 stigmas. SIMILAR SPECIES Pyrenean Valerian *V. pyrenaica* is sometimes naturalised in shady places, mostly in Scotland. Similarly tall, but flowers pinker and lower leaves heart-shaped.

Marsh Valerian
Valeriana dioica

GROWTH Perennial.
HEIGHT 15–30cm (40cm).
FLOWERS Late Apr–Jun.
STATUS Native.
ALTITUDE 0–780m.

Very locally abundant. Marshes, fens and wet meadows with relatively low, sparse vegetation on poor, alkaline to mildly acid soils, also wet woodland. ❃ Smaller and more delicate than Common Valerian, stem leaves similarly divided into leaflets, but basal leaves long-stalked, oval, undivided. Dioecious, with male and female flowers on separate plants; male flower heads *c.* 4cm diameter, female 1–2cm. DETAIL As Common Valerian but hairless, stems growing from creeping stolons. Basal leaves 2–4cm long, upper leaves with leaflets more or less untoothed. Corolla tube 3–5mm long, with male flowers 5mm across, female flowers 2.5mm.

female plant

Common Cornsalad *Valerianella locusta*

Locally common on thin, poor soils: dunes, shingle and rocky places, also walls, paths, pavement cracks, gardens etc. Sometimes cultivated as 'Lamb's Lettuce', a salad crop. ✿ Size variable, may be very small. Stems *repeatedly forked*, with a pair of bright green, short,

GROWTH Annual.
HEIGHT 3–15cm (40cm).
FLOWERS Apr–Jun.
STATUS Native.
ALTITUDE 0–365m.

spoon- or strap-shaped leaves at each fork. *Flowers tiny, 1–2mm across, 5-petalled, pale lilac*, in clusters 1–2cm wide. **DETAIL** Variably but sparsely hairy. Stems finely ridged. Leaves 2–9cm long, sometimes with 1–2 teeth, basal leaves often forming a rosette, stem leaves opposite, their bases connected across the stem. *Calyx minute or absent.* Corolla funnel-shaped, cut towards the tip into 5 lobes; stamens 3, style with 3 stigmas. Can only be reliably separated from other corn-salads by the fruit, which is *hairless, 1.8–2.5mm long, round in profile and slightly flattened, with a shallow groove on outer flat face.* **SIMILAR SPECIES Keel-fruited Cornsalad** *V. carinata* is widespread in the S and SW. Fruits oblong, 2–2.7mm long, square in cross-section, about twice as long as wide, with a deep groove on outer face. **Narrow-fruited Cornsalad** *V. dentata* is an arable weed on chalky soils in the S and E. Calyx present, toothed, remaining on the fruit which is 2mm long and flask-shaped. Two other species are much rarer.

Red Valerian *Centranthus ruber*

Commonly naturalised on walls, pavement cracks, cliffs and other rocky places. ✿ Very conspicuous, with large heads of small, red, deep pink or white, 5-petalled flowers. Foliage blue-green. **DETAIL** Hairless. Leaves opposite, 5–10cm, variably oval, unlobed, lower stalked, stem leaves stalkless. Calyx with a long, feathery plume in fruit. Corolla 5mm wide, tube 8–10mm long with a *fine, pointed, backward-directed spur 5–12mm long.* Stamen 1, stigma 1.

GROWTH Perennial.
HEIGHT 30–80cm (100).
FLOWERS May–Aug (Mar–Oct).
STATUS Introduced to cultivation from S Europe by 1597 and recorded from the wild from 1763.
ALTITUDE Lowland.

Field Scabious
Knautia arvensis

Common. Verges and other rough, grassy places on well-drained soils. ❀ Tall, with *flat* flower heads, 15–40mm across, made up of *bluish-lilac florets*, each with pink anthers and *4 unequal petals* (most obvious on the outer florets, which are larger). Below each flower head is a *double row of oval, leaf-like bracts*. **DETAIL** Roughly hairy. Leaves opposite.

stem leaf

GROWTH Perennial.
HEIGHT To 100cm.
FLOWERS Jul–Sep.
STATUS Native.
ALTITUDE 0–365m.

Basal leaves to 15(30)cm, long-oval, variably bluntly-toothed; stem leaves usually deeply lobed. Each floret has a 4-ridged, tubular epicalyx *c.* 5mm long at its base, tipped by a collar of dense hairs, from which project the *8 bristle-like calyx teeth*. Flower heads may be entirely female, or bisexual, with both male and female florets.

Small Scabious
Scabiosa columbaria

stem leaf

Locally very common. Unimproved permanent grassland on poor, dry, chalky soils. ❀ Rather like Field Scabious but shorter, slenderer and not as roughly hairy. *Upper leaves more finely cut*, flower heads 15–35mm across, with a *single row of narrow, bristle-tipped bracts just below*. Flowers with *5 unequal petals*. **DETAIL** Sparsely hairy. Basal leaves 5–15cm, spoon-shaped, toothed, long-stalked; upper rosette leaves deeply-lobed, with a large, oblong terminal lobe; leaves progressively smaller and more finely

GROWTH Perennial.
HEIGHT 15–70cm.
FLOWERS Late Jun–Sep.
STATUS Native.
ALTITUDE To 640m.

cut higher on stem. *Florets with several tiny, scale-like bracts at the base.* Epicalyx 2–3mm long, 8-ridged, with a chaffy funnel-shaped collar at the tip. *Calyx with 5 bristle-like teeth.*

Devil's-bit Scabious
Succisa pratensis

Locally abundant. Rough grassy places, often damp: woodland rides, grassy heaths and fens. ❀ Erect, with long, slender, often arching stems and *dense, rounded heads, 15–30mm across*, of tiny bluish-violet florets, *all equal in size, with 4 petals*. Leaves *unlobed*. **DETAIL** Roughly hairy. Basal leaves to 15(30)cm, sometimes toothed; stem leaves

few, narrower. A ring of narrow, leafy bracts below the flower head; florets with tiny leafy bracts, longer than calyx, at base. Epicalyx *c.* 3mm long, 4-angled, with 4 teeth; the 4–5 bristle-like calyx teeth are just visible. **SEE ALSO** Sheep's-bit (p. 252), which has the corolla *cut to the base into 5 petals*.

GROWTH Perennial.
HEIGHT To 100cm.
FLOWERS Jul–Nov.
STATUS Native.
ALTITUDE 0–970m.

Wild Teasel
Dipsacus fullonum

Fairly common. Rough, grassy places, often colonising recently disturbed places and waste ground. ❁ Tall and statuesque. Leaves bristly, *each bristle with a swollen, blister-like base, with a row of spines on the midrib below. Flower heads egg-shaped,* 4–8cm tall, loosely enclosed by long, narrow, spiny bracts. Individual flowers tiny, pinkish-purple to lilac, *opening in a horizontal band*, but the flower head is dominated by *numerous long, spine-tipped bracts*, and persists through the winter. DETAIL Stems prickly. Leaves oval-oblong, variably toothed, basal leaves 5–40cm, stem leaves narrower, fused at the base forming a water-retentive cup around the stem. Epicalyx 4-angled, calyx a very short cup, corolla *c.* 10mm, a long, slender tube split at the tip into 4 lobes. SEE ALSO Bristly Oxtongue (p. 268).

GROWTH Biennial.
HEIGHT 50–200cm (300cm).
FLOWERS Jul–Aug.
STATUS Native.
ALTITUDE 0–365m.

Small Teasel
Dipsacus pilosus

Rather local. Damp, lightly shaded places on calcareous soils: woodland rides, hedge-bottoms, stream and river banks, old quarries, waste ground. ❁ Tall and bristly-hairy, with tiny, *whitish flowers mixed with longer, spine-tipped bracts in globular heads,* 20–25mm across, with a collar of similar bracts immediately below. DETAIL Stem angled and grooved. Lower leaves 4–9cm, oval, narrowing to a long stalk, variably toothed, sometimes with basal lobes; underside of leaves with swollen hair bases and spines on midrib. Stem leaves to 17cm, often with 2 large leaflets at base; their winged stalks join across stem at base. Epicalyx *c.* 2mm long, 4-angled, calyx a very short cup, corolla 6–9mm, conical, split at the tip into 4 lobes. Anthers purple.

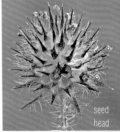

seed head

GROWTH Perennial.
HEIGHT 30–120cm (150cm).
FLOWERS Jul–Sep.
STATUS Native.
ALTITUDE Lowland.

Family Apiaceae: UMBELLIFERS

A large and often confusing family that includes some of
the commonest and most conspicuous wild flowers: Cow
Parsley, followed by Rough Chervil, Upright Hedge Parsley
and Hogweed line many roadsides from May through to
August. Umbellifers are so-called because of their *umbels*
of flowers. All the flower stalks, known as *rays*, originate at
the same point at the tip of a stem, and radiate outwards,
like the spokes of an umbrella. In turn, each ray ends in
a *secondary umbel* that has its own set of rays radiating
outwards, and these terminate at the flowers themselves. At
the tip of the stem there may be several *bracts*, while at the
base of the secondary umbels *bracteoles* may be found; the presence or
absence of bracts and bracteoles can be an important character.

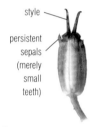

Most umbellifers have white flowers, in some species washed pink,
especially the buds. A few have yellow flowers. Apart from colour, the
size and structure of the individual tiny, 5-petalled flowers is of little help
in identification. Much more important are the ripe fruits, the leaves,
and the presence or absence of hairs on the stems, leaves and fruits.

The fruits are hairy or hairless, or may have hooked bristles that attach to fur or
feathers and help disperse the seed. They are variably flattened, and may have a
persistent style or persistent sepals.

Hemlock Water-dropwort

Sanicle has leaves that are palmately-lobed, but most umbellifers have leaves variably divided into
leaflets. The degree of division is an important identification feature, but is variable: in many species
the lower leaves are more finely cut than the upper leaves; in some species, such as Burnet Saxifrage,
the contrast between the size and shape of the lower and upper leaves is important The shape of the
leaflets is also noteworthy: whether they are toothed, or are cut into long, narrow lobes.

As its name suggests the carrot family includes some important crop plants, including carrots,
parsnips, celery, fennel, lovage and parsley. The family also includes, however, some of the most
toxic plants in Britain, including Hemlock and Hemlock Water-dropwort. *On no account should
you consume any part of an umbellifer unless you are 100 per cent confident that you are able to
identify the plant and its potential confusion species correctly.* Mistakes can be fatal.

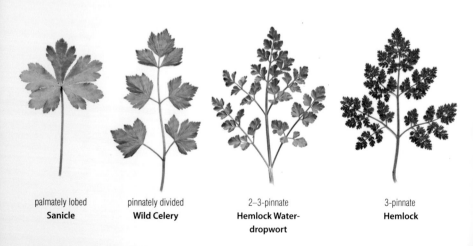

palmately lobed
Sanicle

pinnately divided
Wild Celery

2–3-pinnate
Hemlock Water-dropwort

3-pinnate
Hemlock

Cow Parsley
Anthriscus sylvestris

Abundant – the commonest umbellifer, dominating roadside verges in May, also woodland rides and clearings, gardens, waste ground, etc. ❀ Stems hollow, unspotted, downy towards base, either green or purple, leaves finely cut and fern-like. **DETAIL** More or less hairy (leaves typically hairy in the S but hairless, darker and shinier in the N). Stems furrowed. Bracts absent, bracteoles present. Umbels 2–6cm wide. Fruits 6–10mm long, flattened, smooth.

GROWTH Perennial.
HEIGHT 60–100cm (150cm).
FLOWERS Late Apr–Jun.
STATUS Native.
ALTITUDE 0–760m.

Rough Chervil *Chaerophyllum temulum*

Common. Rough grassland on verges, in clearings and woodland rides, often lightly shaded. ❀ As Cow Parsley but stems *swollen at the nodes* and *blotched purple* (lower stems may be uniformly dull purple, as in Cow Parsley), leaves fresh *matt* green, lacking any sheen, leaflets with *broad, blunt lobes.* Flower heads smaller and flowers purer white. **DETAIL** Whole plant obviously hairy. Stems solid. Bracts absent (rarely 1–2), bracteoles present. Fruits 5–6mm long, slender, ridged, hairless. **SEE ALSO** Hemlock (p. 298).

GROWTH Biennial.
HEIGHT To 100cm.
FLOWERS Late May–early Jul.
STATUS Native.
ALTITUDE 0–465m.

Upright Hedge Parsley *Torilis japonica*

Common on verges, hedgebanks etc; *the common roadside umbellifer in July.*
❀ As Cow Parsley but stems and leaf stalks *solid*, roughly hairy, and umbels typically smaller (2–4cm across). **DETAIL** Whole plant with flattened, rather bristly hairs. Leaves finely cut, leaflets deeply divided into toothed lobes. Bracts and bracteoles present. Flowers pinkish to purplish-white. Fruits 2–2.5mm long, slightly flattened, with *many curved spines*; styles turned downwards.

GROWTH Annual.
HEIGHT To 120cm.
FLOWERS Jul–Aug.
STATUS Native.
ALTITUDE 0–570m.

Giant Hogweed
Heracleum mantegazzianum

Rather local. Verges and rough ground, especially near rivers. ❀ A *huge*, unmistakable umbellifer with *stems up to 10cm thick*, *flower heads the size of dinner plates* and large, coarsely lobed and *sharply toothed* leaves. DETAIL Hairy. Stems hollow, ridged, usually spotted purple. Leaves up to 2.5m long, lobed or divided into 1–2 pairs of leaflets up to 50cm across. Bracts and bracteoles present. Fruits 9–14mm long, oval, flattened, ridged and broadly winged.

GROWTH Biennial or short-lived perennial.
HEIGHT To 4m (6m).
FLOWERS Jun–Jul.
STATUS Introduced to gardens by 1820 and first recorded from the wild in 1828 (SW Asia).
ALTITUDE Lowland.

Hogweed
Heracleum sphondylium

Abundant in rough grassy places. ❀ A robust, summer-flowering umbellifer, roughly hairy, with leaves cut into 2–3 pairs of large irregular leaflets, and large flat (or slightly convex) flower heads. DETAIL Stems hollow, ridged. Leaves 15–60cm long, stalk greatly broadened and inflated at the base and often purplish. Umbels 5–15cm across, bracts few or absent, bracteoles usually present. Flowers white or pale pinkish. Fruits 7–8mm long, nearly circular, strongly flattened with fine, broad wings.

GROWTH Biennial to perennial.
HEIGHT To 180cm (300cm).
FLOWERS Jun–Sep.
STATUS Native.
ALTITUDE 0–1005m.

Hemlock
Conium maculatum

Common. Verges, riverbanks and other rough grassland on damp soil, also waste ground. ❀ Much like Cow Parsley and Rough Chervil, but *hairless* and usually with *many purple spots and blotches on the stem*. Has a mousy smell. DETAIL Stems hollow. Leaves repeatedly divided, with *finely cut, fern-like leaflets*. Umbels 2–5cm across, bracts and bracteoles present. Fruits 2–3.5mm long, oval, with prominent thin wavy ridges.

GROWTH Biennial.
HEIGHT To 200cm (250m).
FLOWERS Jun–Aug.
STATUS Ancient introduction.
ALTITUDE 0–370m.

Bur Chervil
Anthriscus caucalis

Fairly common but local on dry, sandy or gravelly soils: verges, sea walls, field borders, waste ground.
❀ Resembles a small Cow Parsley, but leaves paler, more *greyish-green* and *more finely cut. Umbels very small.* Fruits covered in *hooked spines.* **DETAIL** Hairy. Stems hollow. Bracts absent (rarely 1), bracteoles present. Fruits *c.* 3mm long, flattened, with a short beak.

GROWTH Annual.
HEIGHT To 40cm (80cm).
FLOWERS May–Jun.
STATUS Native.
ALTITUDE Lowland.

Knotted Hedge Parsley *Torilis nodosa*

Rather local. Dry, sparsely-vegetated places, often on disturbed ground, especially near the sea; increasingly on lawns. ❀ Low growing (often more or less sprawling), with small, finely cut leaves, *very small, near-stalkess* umbels, and *tight knots of spiny fruits.* **DETAIL** Bristly-hairy. Stems solid. Umbel stalks usually less than 10mm long, rays 2–3, very short (less than 5mm, mostly hidden). Bracts absent, bracteoles strap-shaped. Flowers pinkish-white. Fruits 2.5–3.5mm long, oval, with long, hooked spines on 1 side.

GROWTH Annual.
HEIGHT Stems to 50cm.
FLOWERS May–Jul.
STATUS Native.
ALTITUDE Lowland.

Sweet Cicely
Myrrhis odorata

Common in the N on verges and other grassy places, rare in S. ❀ Similar to Cow Parsley, but easily recognised by fern-like, pale green, white-blotched leaves and *strong aniseed scent of the crushed foliage and seeds.* **DETAIL** Softly hairy. Stems hollow. Leaves to 30cm long. Bracts absent, bracteoles present. *Fruits 15–25mm long, slender,* narrowing to a short beak, with short, forward-pointing bristles on the angles, dark brown and shiny when ripe.

GROWTH Perennial.
HEIGHT 60–180cm.
FLOWERS May–Jun.
STATUS Long-cultivated, and recorded in the wild from 1777 (Europe).
ALTITUDE To 500m.

Shepherd's Needle
Scandix pecten-veneris

Listed as Critically Endangered. A scarce and very local arable weed, especially on chalky clay soils. ❀ Flowers inconspicuous, in small umbels, but *long, needle-like fruits unique.*

DETAIL Sparsely hairy. Leaves finely cut. Umbels with 1–3 rays, each 0.5–4cm long, bracts usually absent, bracteoles large. Fruits 3–7cm long.

GROWTH Annual.
HEIGHT 15–50cm.
FLOWERS May–Jul.
STATUS Ancient introduction.
ALTITUDE To 320m.

Fool's Parsley *Aethusa cynapium*

Locally common weed of arable fields and waste ground. ❀ Easily identified by the *long, narrow bracteoles that hang down from the umbels.* **DETAIL** Hairless. Stems hollow. Leaves with wedge-shaped leaflets cut into narrow lobes. Bracts usually absent. Fruits 3–4mm long, oval, somewhat flattened, ridged, with very short styles.

GROWTH Annual.
HEIGHT 5–50cm (150cm).
FLOWERS Jun–Sep.
STATUS Native (introduced Ireland).
ALTITUDE Lowland.

Wild Carrot
Daucus carota

Common. Rough grassland, verges and scrub, especially on dry, chalky soils. ❀ Note *large, finely divided bracts* below the flower heads, which in fruit *ball into a distinctive 'fist'.*

DETAIL Whole plant roughly hairy. Stems solid, ridged. Leaves repeatedly divided, the leaflets deeply lobed, toothed or cut into irregular segments. Bracteoles present. Umbels 3–7cm across, the central flower *often reddish-purple.* Fruits 2–3mm long, egg-shaped, flattened, *with long, hooked spines.* **Sea Carrot** subspecies *gummifer,* local on W coasts, has fleshier foliage, is less hairy, and the umbels do not 'ball' in fruit.

GROWTH Biennial.
HEIGHT To 100cm.
FLOWERS Jun–Aug.
STATUS Native.
ALTITUDE 0–400m.

Pepper Saxifrage
Silaum silaus

Rather local. Unimproved meadows, pastures, commons and verges, especially on damp clay soils. ❁ Flowers *pale yellowish-cream*. DETAIL Hairless. Stems solid. Basal

GROWTH Perennial.
HEIGHT To 100cm.
FLOWERS Jun–Aug.
STATUS Native.
ALTITUDE Lowland.

leaves long-stalked, repeatedly divided, leaflets 10–15mm long with *narrow, strap-shaped lobes*; on upper stem leaves small, sometimes undivided. Bracts 0–3, bracteoles numerous. Fruits 4–5mm long, long-oval, ridged, the short styles turned downwards.

Stone Parsley
Sison amomum

Locally common. Verges and other grassland, mostly on clay (especially chalky boulder-clay). ❁ Stems finely branched, with many *small* umbels of 3–6 rays, often including a *single much shorter ray*; secondary umbels *tiny*, 4–10mm wide, with *few flowers*. Uniquely,

GROWTH Biennial.
HEIGHT 50–100cm.
FLOWERS Late Jul–Sep.
STATUS Native.
ALTITUDE Lowland.

the crushed foliage *smells of petrol* (obvious when fresh but rapidly lost). DETAIL Hairless. Stems solid. Lower leaves with *2–5 pairs of leaflets*. Upper leaves very finely cut. Bracts and bracteoles present, very fine. Fruits 1.5–3mm long, globular, flattened, ribbed.

Corn Parsley *Petroselinum segetum*

Very local. Rough, dry grassland near the sea (riverbanks, sea walls), also inland on dry chalky soils; formerly an arable weed, but much-declined in this habitat. ❁ Much like Stone Parsley,

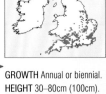

but flower heads even sparser, with *all* rays *unequal*, and rays of secondary umbels longer – the tiny white flowers are even less obvious. Stems and foliage greyer, with a *weak celery scent*. DETAIL Hairless. Lower leaves cut into 4–12 pairs of small leaflets, all

GROWTH Annual or biennial.
HEIGHT 30–80cm (100cm).
FLOWERS Aug–Sep.
STATUS Native.
ALTITUDE Lowland.

held in a horizontal plane like the steps of a ladder. Umbels with 3–10 rays. Bracts and bracteoles present, bristle-like. Fruits 2.3–3mm long, egg-shaped, flattened, strongly ribbed.

Burnet Saxifrage
Pimpinella saxifraga

Locally common. Well-drained old grassland, especially on chalk or limestone, both grazed and ungrazed, including churchyards, old lawns and sometimes woodland rides. ✿ Very variable in size, from tiny plants in grazed, coastal grassland to robust, medium-tall plants in lush, riverside grassland. Usually downy, with the basal leaves *rather different* from the few, finely cut stem leaves, which have *long, winged, sheathing stalks*. The umbels *lack* bracts and bracteoles. **DETAIL** Stem more or less solid, slightly ridged, hairless or covered with very short fine hairs, often purplish.

GROWTH Perennial.
HEIGHT 30–70cm
(5–100cm).
FLOWERS Jul–Sep (May).
STATUS Native.
ALTITUDE To 810m.

stem leaves   basal leaf

Lower leaves stalked, with 4–5 (2–7) pairs of oval, coarsely-toothed, stalkless leaflets up to 2.5cm long, variably hairy, stem leaves cut into very narrow leaflets; intermediate between the 2 leaf types occur. Umbels 20–50mm across. Flowers 2mm across with very short styles. Fruits rounded, flattened, ridged, shiny, 2–3mm long.

Greater Burnet Saxifrage
Pimpinella major

Locally common. Rough grassland on verges, railway banks and woodland edges, mostly on chalk and limestone but also on clays. ✿ Tall, with the leaves divided into 3–4 pairs of *shiny*, dark green, *rather coarsely toothed leaflets*, the lower pairs short-stalked. Flowers white or, especially in the N, pink; *bracts and bracteoles absent*. **DETAIL** Stems hollow, grooved, *hairless*, often purplish towards base. Leaflets 2–5(10)cm long, often hairy on veins below, terminal leaflet may be 3-lobed; lower leaves with a sheathing, hollow stalk. Umbels 30–60mm across, flowers 3mm across, with long styles. Fruits 3–4mm long, egg-shaped, flattened, ridged.

GROWTH Perennial.
HEIGHT 50–100cm
(200cm).
FLOWERS Mid Jul–mid Aug (late Jun–Sep).
STATUS Native.
ALTITUDE To 320m.

Sanicle *Sanicula europaea*

Common in deciduous woodland and old, shady hedgebanks. ❁ The long-stalked, *palmately-lobed* leaves and *irregular clusters of tiny white or very pale pink umbels* are distinctive. **DETAIL** Hairless. Leaves dull above, shiny below, mostly at base of the stem, long-stalked, 2–8cm across, with 3–5 toothed lobes; stem leaves few, short-stalked or stalkless. Umbels small, at the tip of branches of various lengths, each with unequal rays, secondary umbels ball-shaped, *c.* 5mm across; bracts and bracteoles present, petals rolled inwards. Fruits *c.* 3mm long, slightly flattened, covered in forward-pointing hooked bristles, with long (*c.* 3mm) curved styles.

GROWTH Perennial.
HEIGHT 20–40cm (60cm).
FLOWERS May–Aug.
STATUS Native.
ALTITUDE To 500m.

Pignut

Conopodium majus

Common in old grassland (notably some churchyards and traditional hay meadows) and deciduous woodland. ❁ Rather slender, with *stem leaves cut into long, narrow (1–2mm wide) strap-shaped lobes*. **DETAIL** *Hairless*. Basal leaves repeatedly divided, soon withering, stem leaves few. Umbels 3–7cm across, bracts absent (rarely 1–2), bracteoles present. Fruits 3–4.5mm long, oval, flattened, finely ridged, with short erect styles. The stems grow from a single dark brown tuber – hence the name.

basal leaf

GROWTH Perennial.
HEIGHT 8–50cm (90cm).
FLOWERS May–Jun.
STATUS Native.
ALTITUDE 0–710m.

Rock Samphire *Crithmum maritimum*

Locally common on rocky coasts, sometimes on man-made structures, and occasionally on the upper part of sand or shingle beaches. ❁ Distinctive, with yellowish-green flowers and *stiff, obviously fleshy leaves*, strong-smelling when crushed. **DETAIL** Hairless. Stems woody at base, ridged. Leaves cut into narrow lobes 2–5cm long. Umbels 3–6cm across, bracts and bracteoles present. Fruits 3.5–5mm long, egg-shaped, ridged.

GROWTH Perennial.
HEIGHT To 45cm.
FLOWERS Jun–Aug.
STATUS Native.
ALTITUDE Lowland.

Alexanders *Smyrnium olusatrum*

Abundant, especially near the sea. Verges and other grassy places.
⚫ Robust and *early-flowering*, with *dull yellow-white flowers* and *glossy dark green* leaves, divided into *broad, 3-lobed leaflets*. DETAIL Hairless, celery-scented. Leaves to 30cm long, appearing in late autumn. Bracts and bracteoles few or absent. Petals dull cream; ovary large, yellow-green. Seeds 7–8mm long, ridged, black.

GROWTH Biennial.
HEIGHT 50–150cm.
FLOWERS Mar–May.
STATUS Introduced in the Roman period and long naturalised.
ALTITUDE Lowland.

Fennel
Foeniculum vulgare

Locally common garden escape near houses, also well-naturalised on disturbed ground, especially near the sea.
⚫ Distinctive tall, late-flowering umbellifer with large heads of *yellow* flowers and *very finely divided leaves* that smell strongly of aniseed when crushed. DETAIL Hairless. Stems solid. Leaves cut into fine, thread-like lobes. Flower heads 4–8cm across, bracts and bracteoles absent. Fruits 4–6mm long, long-oval, strongly ribbed.

GROWTH Perennial.
HEIGHT To 250cm.
FLOWERS Jul–Oct.
STATUS Introduced by the Roman period.
ALTITUDE Lowland.

Wild Parsnip
Pastinaca sativa

Locally common. Rough grassland, verges and open scrub, especially on chalk and limestone.
⚫ The only *yellow-flowered* umbellifer with the leaves divided into *broad, more or less oval* leaflets. Smells strongly of parsnip. DETAIL Rather hairy. Stems solid or hollow, ridged. 5–11 pairs of leaflets, sometimes lobed at the base, with coarsely toothed margins. Flower heads 5–10cm across, bracts and bracteoles few or none. Fruits 5–9mm long, oval, flattened, finely ribbed and narrowly winged on the edges.

GROWTH Biennial.
HEIGHT To 180cm.
FLOWERS Jul–Aug.
STATUS Native.
ALTITUDE 0–380m.

Wild Angelica
Angelica sylvestris

Very common. Damp, open or lightly shaded places: fens, streamsides, woodland rides, also sea-cliffs and upland grassland. ❁ Tall and robust, with distinctive *bulbous, inflated sheaths* where the leaf and flower stalks join the stem. Flowers white

GROWTH Perennial.
HEIGHT To 200cm (250cm).
FLOWERS Jun–Sep.
STATUS Native.
ALTITUDE 0–855m.

or pinkish, *whole plant often washed purple*. **DETAIL** Stems stout, hollow, purplish, downy near base. *Leaf stalks wide and gutter-shaped, broadly sheathing stem at base*. Leaves usually *hairless*, 30–60cm long, leaflets toothed, pointed. Umbels 3–15cm across, strongly domed, with no bracts and few bracteoles, rays densely but finely hairy. Fruits 4–5mm long, flattened, ridged, with 2 broad wings.

Ground Elder *Aegopodium podagraria*

Fairly common. Verges, churchyards and woodland margins, especially near houses, originating as a garden throw-out – it is a notoriously persistent

GROWTH Perennial.
HEIGHT 40–100cm.
FLOWERS May–Jul.
STATUS Probably introduced in the Roman period.
ALTITUDE 0–470m.

garden weed. ❁ *Strongly patch-forming*, with the leaves divided into large, oval leaflets and pure white flowers. **DETAIL** Hairless. Stems hollow, grooved, growing from slender rhizomes. Leaves long-stalked, 10–20cm long, leaflets toothed. Umbels 2–6cm across, bracts and bracteoles usually absent. Fruits 3–4mm long, long-oval, ridged, flattened.

Wild Celery
Apium graveolens

Local on damp, brackish ground near the sea – the edge of coastal reedbeds etc. Occasional inland in marshes and other

GROWTH Biennial.
HEIGHT 30–60cm (100cm).
FLOWERS Jun–Aug.
STATUS Native.
ALTITUDE Lowland.

damp places. ❁ Robust, with *shiny* green leaves, the lower divided into *1–3 pairs of large, toothed leaflets*, each often cut into 3 lobes; upper leaves 3-lobed. Umbels with *unequal* rays, those at tip of stems short-stalked, but in leaf axils more or less *stalkless*. Has a *strong smell of celery*. **DETAIL** Hairless. Stems solid, strongly ridged. Umbels 3.5–4.5cm across, *bracts and bracteoles absent*. Fruits 1–1.5mm long, circular, flattened, finely ridged. **SEE ALSO** Hemlock Water-dropwort (p. 308).

Greater Water Parsnip
Sium latifolium

Nationally Scarce, listed as Endangered. Very local in or beside shallow, still or slow-flowing, alkaline water, typically species-rich grazing marsh ditches, also reed swamp and tall herb fen. ❀ *Robust.* Leaves cut into 3–8 pairs of *finely-toothed* leaflets, each *up to 15cm long*. **DETAIL** Hairless. Stems to 2cm across, hollow, 7-ridged. Leaf stalks hollow, sheathing stem. Submerged leaves (only present in spring) finely divided. Umbels 6–10cm across, with large, leafy bracts and bracteoles. Fruits oval, 2.5–4mm long (*longer than wide*), *flattened, well ridged*, with *tiny persistent sepal teeth*.

GROWTH Perennial.
HEIGHT To 200cm.
FLOWERS Jun–Aug (Sep).
STATUS Native.
ALTITUDE Lowland.

Cowbane Cicuta virosa

Nationally Scarce. Very local in shallow water or on floating mats of vegetation on the margins of ditches, ponds, lakes and slow-flowing streams and rivers, sometimes in fens or wet pastures. ❀ Tall, with leaves cut into distinctive *long, narrow, saw-toothed lobes* and flower heads divided into *obvious 'pom-poms'*. **DETAIL** Hairless. Stems and leaf stalks hollow, leaves to 30cm long. Umbels 7–13cm across, with many bracteoles; bracts absent. Fruits 1.2–2 mm long, oval, *wider than long*, slightly flattened, ridged, with long curved styles and *prominent triangular sepal teeth*.

GROWTH Perennial.
HEIGHT 50–150cm.
FLOWERS Jun–Aug.
STATUS Native.
ALTITUDE Lowland.

Milk Parsley
Thyselium palustre

Nationally Scarce, listed as Vulnerable. Fairly common in the Norfolk Broads; very rare elsewhere. Tall herb fen and adjacent scrub and wet woodland. ❀ Tall, with *delicate, finely cut foliage*. The younger parts bleed a watery white sap when broken. **DETAIL** Near-hairless. Stems often purplish, strongly ridged, hollow. Leaf stalks hollow, leaflets cut into several pointed lobes *c.* 2mm wide. Upper leaves smaller, short-stalked. Umbels 5–10cm across, rays downy, bracts and bracteoles present, angled downwards. Fruits 3–5mm long, oval, flattened, ridged, with 2 broad wings. **SEE ALSO** Parsley Water-dropwort (p. 308).

GROWTH Perennial.
HEIGHT To 150cm.
FLOWERS Jul–Sep.
STATUS Native.
ALTITUDE Lowland.

Lesser Water Parsnip *Berula erecta*

Common in ditches, ponds, rivers and streams, both as a submerged aquatic and as an emergent, also marshes and other wet ground. ❀ Rather like Fool's Watercress but umbels with both bracteoles and 4–7 *obvious bracts*; stalks of lower leaves usually with a *distinct whitish ring* near the base. DETAIL Hairless. Patch-forming, stems often sprawling, hollow. 5–10 pairs of *dull* yellow-green leaflets, to 6cm long, *sharply toothed*. Umbels 3–6cm across, on *relatively long stalks*. Fruits 1.3–2mm long, globular, slightly flattened, ridged.

GROWTH Perennial.
HEIGHT Stems 30–100cm.
FLOWERS Jul–Sep.
STATUS Native.
ALTITUDE Lowland.

Fool's Watercress *Apium nodiflorum*

Common along streams and ditches and in other wet places, often forming dense masses. ❀ Separated from Lesser Water Parsnip by the *absence of bracts*. DETAIL Hairless. Stems often sprawling, rooting at the lower nodes, hollow. 2–6 pairs of leaflets, *shiny* green, to 6cm long, with *finely toothed* margins. Umbels *short-stalked or almost stalkless*, bracteoles 4–7, prominent, rarely also 1–2 bracts. Flowers greenish-white. Fruits 1.5–2.5mm long (longer than wide), oval, flattened, ribbed.

GROWTH Perennial.
HEIGHT Stems to 80cm.
FLOWERS Jul–Aug.
STATUS Native.
ALTITUDE 0–335m.

Lesser Marshwort *Apium inundatum*

Scarce and local in ponds, ditches, streams, dune slacks and other places with temporary or permanent shallow water. ❀ Small and inconspicuous, often with *most or all leaves submerged*, only the *tiny flower heads* projecting above the water. DETAIL Hairless. Stems hollow, rooting at nodes. Submerged leaves with fine, thread-like segments, grading into uppermost leaves, which have pairs of 3-lobed leaflets *c.* 5mm long. Umbels short-stalked with 2(4) short rays, bracts absent, bracteoles 3–6. Flowers white, *c.* 2mm across. Fruits 2.5–3mm long, long-oval, flattened, ribbed.

GROWTH Perennial.
HEIGHT Stems to 50cm.
FLOWERS Jun–Aug.
STATUS Native.
ALTITUDE 0–500m.

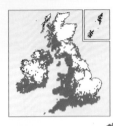

Hemlock Water-dropwort
Oenanthe crocata

GROWTH Perennial.
HEIGHT To 150cm.
FLOWERS Jun–early Aug.
STATUS Native.
ALTITUDE 0–320m.

Common to abundant, especially in the W, by standing or running water, also damp verges, marshes, wet woodland, wet cliffs and boulder beaches. ❀ Tall and robust. Overall blue-green, *all leaves* with roughly oval, notched or lobed leaflets 1–2cm long, flower heads *large*, with well-separated 'pom-pom'-like flower clusters. DETAIL Hairless. Stems hollow, grooved. Bracts 3–6, bracteoles present. Fruits 4–5.5mm long, ridged, styles about half as long again.

Tubular Water-dropwort *Oenanthe fistulosa*

GROWTH Perennial.
HEIGHT 30–80cm.
FLOWERS Jul–Sep.
STATUS Native.
ALTITUDE Lowland.

Rather local, listed as Vulnerable. Fens and unimproved meadows and pastures, typically where flooded in winter and remaining wet in summer, also in standing water in ponds and ditches. ❀ Stems hollow, *often appearing slightly swollen* but narrowing at the nodes. Leaves small, with *very long, hollow stalks*. Flower heads with just 2–4 relatively *small, domed* secondary umbels, which become *spiny, marble-sized balls* in fruit. DETAIL Hairless. Lower leaves (sometimes submerged) cut into tiny, wedge-shaped, lobed leaflets, upper leaves with a few very narrow lobes. Flowers white or pale pink; bracts absent, bracteoles very fine. Fruits 3–4mm long, tapering to the base, with long styles.

Parsley Water-dropwort *Oenanthe lachenalii*

Local in damp grassland by the sea or tidal rivers, increasingly scarce inland in calcareous fens. ❀ Slender, with *large, open flower heads*. Most leaflets narrowly lobed, those on mid stem *very narrow* (1–2.5mm wide). DETAIL Hairless. Stems solid to hollow. Bracts 1–5, bracteoles also present. Fruits 2.5–3mm long, barrel-shaped, ridged, with persistent sepal teeth, style *shorter* than fruit.

stem leaf

GROWTH Perennial.
HEIGHT 30–60cm (100cm).
FLOWERS Jun–Sep.
STATUS Native.
ALTITUDE Lowland.

Corky-fruited Water-dropwort
Oenanthe pimpinelloides

Rather local but increasing. Pastures and verges on both damp and dry soils. ❀ Rather like Parsley Water-dropwort but umbels crowded, becoming *dense and flat-topped* in fruit, with flower stalks and umbel rays *thickened*. Fruits with *long styles*. **DETAIL** Lower leaves with small, lobed, wedge-shaped leaflets (often withered by flowering), stem leaves with narrowly-lobed leaflets (lobes 20–55mm long × 1–2.5mm wide). Fruits 3–3.5mm long, cylindrical, ridged, with the styles *about as long again*.

stem leaf — lower leaf

GROWTH Perennial.
HEIGHT 30–80cm (100cm).
FLOWERS Jun–Jul.
STATUS Native.
ALTITUDE Lowland.

Fine-leaved Water-dropwort *Oenanthe aquatica*

lower aerial leaf

GROWTH Annual or biennial.
HEIGHT 30–100cm (150cm).
FLOWERS Jun–Sep.
STATUS Native.
ALTITUDE Lowland.

Rather local in ponds and ditches, especially those which dry out over the summer, sometimes in waterside vegetation or marshes. ❀ Usually grows *in water*, with a stout, *erect* stem. **DETAIL** Hairless. Stems hollow. Lower leaves repeatedly cut, with *very fine, thread-like lobes when submerged*. Aerial leaves finely cut, with small, lobed leaflets. Umbels short-stalked (stalks *shorter* than the umbel rays), bracts usually absent, bracteoles present. Fruits 3.5–4.5mm long, ribbed, with short styles.

River Water-dropwort *Oenanthe fluviatilis*

Rather local in clean, calcareous streams and rivers, also canals and ditches. ❀ Grows submerged and can form large beds, appearing above the water to flower, but may be non-flowering in faster-flowing water. Much like Fine-leaved Water-dropwort, but stems more or less floating rather than erect, submerged leaves not as finely cut, and fruits *slightly longer*. **DETAIL** Submerged leaves cut into *long, narrow, wedge-shaped lobes*. Fruits *5–6.5mm long*, cylindrical, ridged, with very short styles.

submerged leaf

GROWTH Perennial.
HEIGHT To 100cm.
FLOWERS Jul–Sep.
STATUS Native.
ALTITUDE Lowland.

Scots Lovage
Ligusticum scoticum

Almost confined to Scotland and N Ireland, where local on rocky coasts, shingle and dunes out of the reach of sheep. ❀ Attractive summer-flowering umbellifer with greenish-white flowers and *glossy-green leaves divided into 3 lobed leaflets; stems often reddish.* DETAIL Hairless. Stems hollow, ridged, often with fine reddish stripes or reddish towards the base. Leaves divided into 3 long-stalked, coarsely-toothed leaflets 2–5cm long, these in turn often 3-lobed. Bracts and bracteoles present. Fruits 4–7mm long, oblong-oval, with prominent winged ribs.

GROWTH Perennial.
HEIGHT 20–50cm (90cm).
FLOWERS Jun–Jul.
STATUS Native.
ALTITUDE Lowland.

Spignel *Meum athamanticum*

Rather local in dry, unimproved grassland: pastures, meadows, roadside banks. ❀ *The combination of white flowers with feathery, very finely cut, strongly aromatic leaves is distinctive.* DETAIL Hairless. Stems hollow, surrounded at the base by a mass of fibres. Basal leaves repeatedly divided into thread-like lobes *c.* 5mm long. Stem leaves few, much smaller. Umbel with unequal rays, bracts few or absent, bracteoles present. Flowers white, sometimes tinged pink. Fruits *c.* 7mm long, egg-shaped, ribbed, with short downturned styles.

GROWTH Perennial.
HEIGHT To 60cm.
FLOWERS Jun–Jul.
STATUS Native.
ALTITUDE To 300m (610m).

Whorled Caraway
Carum verticillatum

Very locally abundant in a variety of damp, unimproved grasslands, fens, wet heaths and upland flushes, avoiding chalky soils. ❀ A hairless, rather greyish-green umbellifer with large, open umbels of white flowers and unique leaves – long and narrow in outline with whorls of thread-like leaflets. DETAIL Hairless. Stems hollow, surrounded at the base by a mat of fibres. Leaves mostly basal, divided into 20 or more pairs of leaflets *c.* 10mm long, these in turn cut into thread-like segments. Bracts and bracteoles present. Fruits 2.5–3mm long, egg-shaped, flattened, finely ribbed.

GROWTH Perennial.
HEIGHT 30–60cm.
FLOWERS Jul–Aug.
STATUS Native.
ALTITUDE 0–465m.

Slender Hare's Ear
Bupleurum tenuissimum

Nationally Scarce, listed as Vulnerable. Uncommon and very local in grassland by the sea where there is some influence of salt water. ❁ An inconspicuous and very atypical umbellifer, grey-green, with wiry stems straggling through rough grass and a *few tiny orange-yellow flowers in umbels just 5mm wide*. DETAIL Hairless. Stems well branched. Leaves strap-shaped, pointed, 1–7cm long. Umbels very short-stalked or stalkless, with 2–3 unequal rays up to 10mm long ending in a secondary umbel of 1–4 flowers each 1.5mm wide. Bracts and bracteoles present, leafy. Petals tiny, curled. Fruit *c.* 2mm long, circular, flattened, ridged, covered with minute warts.

GROWTH Annual.
HEIGHT Stems to 50cm.
FLOWERS Jul–Sep.
STATUS Native.
ALTITUDE Lowland.

Sea Holly *Eryngium maritimum*

Locally common. Sand dunes, less often shingle. ❁ The stunning plant is an atypical member of the carrot family (the umbellifers, Apiaceae), and is very different from most of its relatives. DETAIL Well branched and bushy, whole plant spiny; the blue-green holly-like leaves have whitish veins and margins and are immediately distinctive. Flowers powder-blue, forming dense, egg-shaped heads which recall a teasel rather than an umbellifer. Fruits egg-shaped, with hooked bristles. The roots may extend 2m into the ground.

GROWTH Perennial.
HEIGHT 30–60cm.
FLOWERS Jun–Sep.
STATUS Native.
ALTITUDE Lowland.

PENNYWORTS: FAMILY HYDROCOTYLACEAE

Marsh Pennywort *Hydrocotyle vulgaris*

Common to abundant. A wide range of damp or wet places: marshes, dune slacks, unimproved meadows. ❁ Forms *mats of circular leaves*, each held up on a slender stem that is attached to the *underside of the leaf blade*.

Flowers in tiny heads, *c.* 3mm across. DETAIL Stems creeping, rooting at the nodes. Leaves 8–35mm in diameter, all veins radiating from the centre, subtly wavy-edged; leaf-stalks slightly hairy. Flowers in clusters of 3–6 on *short stalks* (thus *below* the leaves), greenish-white tinged pink; sepals minute. Fruits *c.* 2mm across, oval, ridged, strongly flattened, with many brown, resinous dots. SEE ALSO Navelwort (p. 42).

GROWTH Perennial.
HEIGHT Leaf stalks to 25cm.
FLOWERS Jun–Jul.
STATUS Native.
ALTITUDE 0–530m.

Family Orchidaceae: ORCHIDS

One of the largest families of flowering plants, with as many as 26,000 species worldwide. All orchids have the same flower structure, with 3 sepals and 3 petals. The lower petal, known as the *lip*, almost always differs in size and shape from the other petals and from the sepals, which are generally all similar to each other. The lip may be very elaborate, and in many orchids is extended back into a tubular *spur* (which may or may not contain nectar). The flower 'stalk' is, in fact, the ovary. Male and female reproductive structures are fused into a *column*, which has a single *stigmatic surface*, with the pollen usually aggregated into 2 relatively large, rugby-ball-shaped masses, the *pollinia*, which may sit on top of the column (as in the helleborines), or be stalked and held in protective pouches until removed by an insect. Most orchids have sticky pads, the *viscidia*, which glue the pollinia to visiting insects, with sophisticated mechanisms to ensure that the pollinia are in the correct position to ensure cross-pollination. Note that all British wild orchids are perennials, and all those included here are native species.

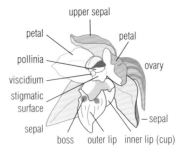

A typical helleborine
(see also p. 13)

HEIGHT 25–80cm
(10–120cm).
FLOWERS Mid Jul–mid Aug.
ALTITUDE 0–350m.

Broad-leaved Helleborine *Epipactis helleborine*

Local. Deciduous woodland, especially rides, clearings and edges. Also scrub, hedgebanks, plantations and, in the N and W, limestone pavements and other open, rocky places. A garden weed in some northern cities. ❀ The most widespread helleborine. Very variable in size, with 4–10 long-oval leaves, usually obviously 'pleated', spread along the stem. Flower spike with just 1–2 flowers in the smallest plants to 100 crowded together in the largest. Flowers opening widely, almost always with some purple tones and with the tip of the lip curled under. Detail Upper stem and ovaries hairy, leaves dull, mid green, *base of flower stalk washed purple.* Flowers a variable mixture of pale green, pink and purple. Cross-pollinated, with intact pollinia and a functional viscidium when fresh – both structures are *removed intact* by visiting insects.

Narrow-lipped Helleborine *Epipactis leptochila*

Nationally Scarce. Deciduous woodland, especially beechwoods, on chalk or limestone, in shady places with little ground cover. ❀ Overall green, with drooping flowers that have a *long, pointed lip.* Detail Upper stem and ovaries hairy, leaves typically in 2 opposite rows, *base of flower stalk greenish-yellow.* Outer portion of lip longer than wide. Self-pollinated, with the pollinia *crumbling messily* onto the stigmatic surface below, and *no viscidium.*

HEIGHT 30–60cm
(15–75cm).
FLOWERS Mid Jul–mid Aug.
ALTITUDE Lowland.

Dark Red Helleborine *Epipactis atrorubens*

Nationally Scarce. Very local. Rocky places on limestone and limestone pavements. ❀ *Reddish-purple* flowers distinctive. Insect pollinated. DETAIL Broad-leaved Helleborine is the only other helleborine in these rocky habitats, but note Dark Red's *dark* green leaves, strongly folded and keeled, in 2 *opposite* rows, *very hairy* ovary and large, rough bosses on the lip.

HEIGHT 12–60cm (100cm).
FLOWERS Jun–early Aug.
ALTITUDE 0–610m.

Violet Helleborine
Epipactis purpurata

Rather local. Shady deciduous woodland, often on chalky clay soils. ❀ Dusky green stems and leaves contrast with the *wide open, relatively large, pale green flowers* with pinkish bosses on the lip. Several stems often grow from the same rootstock. *Insect pollinated.* DETAIL Leaves washed purplish on underside, the lowest longer than wide. Sepals and petals pale greenish-white, lacking pink or purple tones. Bosses on lip smoothly pleated.

HEIGHT 20–70cm (90cm).
FLOWERS Mid Jul–early Sep.
ALTITUDE Lowland.

Green-flowered Helleborine *Epipactis phyllanthes*

Nationally Scarce. Scattered in deciduous woods, including wet riverside woodland and birches on mine spoil; on the NW coast also in dune slacks. ❀ Upper stem and ovaries *always hairless (or almost hairless)*. Flowers *often green, drooping and never really opening, with a prominent swollen ovary*. In a minority the flowers open widely, are held erect and may have a pink flush. DETAIL Lip variable: may be almost identical to the petals, or fully formed, with an inner cup and outer heart-shaped portion, the tip turned down and under; interior of cup usually *pale. Self-pollinated.*

HEIGHT 15–50cm (5–75cm).
FLOWERS Mid Jul–mid Aug
(late Jun–early Sep).
ALTITUDE Lowland.

Dune Helleborine *Epipactis dunensis*

Nationally Scarce. Locally common in dune slacks and nearby plantations on Anglesey and in NW England, also scattered inland from Lincs to central Scotland in woodland, especially birch, on mine spoil, clinker or areas contaminated by heavy metals. ❀ Leaves *yellow-green*, held erect in 2 rows. Flowers yellow-green, petals and lip washed pink, with the lip curled under at the tip; they do *not open widely*. Inland plants *often almost identical* to Narrow-lipped Helleborine, best separated by range and habitat. DETAIL Stem and ovaries hairy. Base of flower stalk tinged violet in dune plants, green inland. *Self-pollinated.*

HEIGHT 20–50cm.
FLOWERS Late Jun–mid Aug.
ALTITUDE Lowland.

Marsh Helleborine
Epipactis palustris

Very locally abundant in species-rich alkaline fens and dune slacks. ✿ Often patch-forming, with crowded spikes of very distinctive flowers with a *frilly white lip*. DETAIL Stems hairy, often purplish-brown. Leaves 5–15cm long, narrowly oval, pointed, 'pleated', with prominent veins. Sepals and petals green, variably washed purplish and fringed whitish. Lip with a dish-shaped inner portion, *white with delicate purple veins*; outer portion, attached by an *elastic hinge*, roughly circular, white, with crimped edges and a yellow boss at the base. Var. *ochroleuca* lacks purplish in the stem and flowers.

HEIGHT 20–45cm (80cm). FLOWERS Late Jun–early Aug. ALTITUDE Lowland.

Sword-leaved Helleborine *Cephalanthera longifolia*

Nationally Scarce, listed as Vulnerable. Very local in deciduous woodland (mostly on chalk in S England), occasionally scrub or even grassland. ✿ Upright, with *long, sword-shaped leaves* and a *more or less leafless spike of white, egg-shaped flowers*. DETAIL Leaves 7–20cm long. Flowers as White Helleborine but *bracts shorter than ovary on upper flowers* (always longer in White Helleborine). Pollinated by solitary bees.

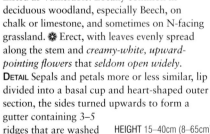

HEIGHT 15–65cm (5cm). FLOWERS Mid May–mid Jun. ALTITUDE Lowland.

White Helleborine
Cephalanthera damasonium

Listed as Vulnerable. Locally common in deciduous woodland, especially Beech, on chalk or limestone, and sometimes on N-facing grassland. ✿ Erect, with leaves evenly spread along the stem and *creamy-white, upward-pointing flowers* that *seldom open widely*. DETAIL Sepals and petals more or less similar, lip divided into a basal cup and heart-shaped outer section, the sides turned upwards to form a gutter containing 3–5 ridges that are washed golden-yellow. Self-pollinated.

HEIGHT 15–40cm (8–65cm). FLOWERS Mid May–late Jun. ALTITUDE Lowland.

Birdsnest Orchid
Neottia nidus-avis

Listed as Near Threatened. Local in deciduous woodland, generally on calcareous soils, showing a strong association with Beech trees. ❀ Parasitic on fungi-tree partnerships, it has no green parts and only appears above ground to flower; the honey-brown spikes are unique. DETAIL Stem with a few, brownish scale leaves. Flowers yellowish-brown, the sepals and petals forming a loose, fan-shaped hood; lip strap-shaped, dividing at the tip into 2 broad, rounded, spreading lobes. SEE ALSO Yellow Birdsnest (p. 188).

HEIGHT 20–40cm (15–50cm).
FLOWERS May–Jun.
ALTITUDE Lowland.

Common Twayblade
Neottia ovata

Locally common. Deciduous woodland, grassland, fens, dune slacks and limestone pavement, mostly on calcareous soils. ❀ The 2(3) large, oval leaves at the base of the stem (hence the English name) and long, slender flower spike, with up to 100 *tiny, green, man-like flowers*, are distinctive. DETAIL Stem with a few tiny bract-like leaves. Upper sepals and petals forming a loose hood, lateral sepals spreading horizontally, lip strap-shaped, tip deeply cut into 2 narrow, spreading lobes, and usually held angled inwards towards the stem.

HEIGHT 20–60cm (10–75cm).
FLOWERS Late Apr–Jul.
ALTITUDE 0–670m.

Lesser Twayblade
Neottia cordata

Locally fairly common but often hard to find. Damp moorland and bogs, on cushions of *Sphagnum* moss, often under the edges of Heather and other shrubs, also damp woodland, both deciduous and pine, on acid soils. ❀ *Stems with 2 heart-shaped leaves opposite each other a short distance above the ground.* Flowers tiny, *resembling elfin figures.* Both stem and flowers often washed reddish. DETAIL Leaves 1–2cm across. Sepals and petals spreading widely, lip with 2 short, narrow side lobes at the base ('arms') and 2 long, narrow, pointed lobes ('legs'). The ovary swells rapidly and is almost as large as the flower.

HEIGHT 5–10cm (3–25cm).
FLOWERS Mid May–mid Jul.
ALTITUDE 0–1065m.

Autumn Lady's-tresses *Spiranthes spiralis*

HEIGHT 3–20cm.
FLOWERS Aug–Sep.
ALTITUDE Lowland.

Near Threatened. Very locally common in short, dry, nutrient-poor turf (including uncut lawns), usually over chalk or limestone. ❀ Slender, with very small white flowers arranged in a *vertical line* that more or less obviously *spirals around the stem*. DETAIL Stem finely glandular-hairy. Rosette leaves oval, pointed, very hard to see and withered by flowering time (but already replaced by a fresh rosette immediately adjacent); stem leaves tiny and bract-like. Flowers glandular-hairy. Sepals and petals white, forming a trumpet, lip slightly larger, pale green in the centre, with the sides curved up to form a gutter and the tip rolled downwards and crimped.

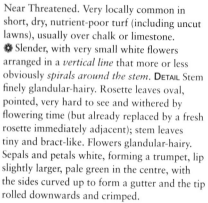

Creeping Lady's-tresses
Goodyera repens

Locally common in native pinewoods and sometimes also mature pine plantations. ❀ Slender, with very small, *densely hairy*, white flowers spread around the stem. DETAIL Creeping and patch-forming, with rhizomes that produce small rosettes of evergreen, privet-like leaves. Stem glandular-hairy with a few tiny, bract-like leaves. Flowers and bracts glandular-hairy, the upper sepal and petals forming a hood, the lateral sepals slightly spreading. Lip with a pouched base that tapers to a downward-curving, wedge-shaped tip.

HEIGHT 7–20cm (35cm).
FLOWERS Jul (late Jun–late Aug).
ALTITUDE 0–350m (740m).

Small White Orchid
Pseudorchis albida

Listed as Vulnerable. Very local in well-drained, unimproved grassland on mildly acidic to alkaline soils. ❀ Distinctive dense spikes of small, creamy- or greenish-white flowers, bell-shaped with a deeply 3-lobed lip. Beware albino spotted and fragrant orchids; check flower size and shape. DETAIL Basal leaves oval, shiny green, held erect; stem leaves tiny, bract-like. Flowers 2–4mm across, sepals and petals forming a loose hood, lip short, broader than long, deeply cut into 3 lobes and extended at the base into a short, blunt, sack-like spur.

HEIGHT 8–20cm (40cm).
FLOWERS Late May–mid Jul.
ALTITUDE 0–550m.

Greater Butterfly Orchid
Platanthera chlorantha

Listed as Near Threatened. Local, almost always on alkaline soils. Deciduous woods, especially ancient woodland, rough chalk grassland and, in the N and W, damp, unimproved pastures and meadows. ❀ Flower spike often rather open, with 10–40 relatively large flowers, each with a *narrow, strap-shaped lip* and an *extremely long, down-curved, translucent spur*, with nectar often clearly visible at its tip. DETAIL Basal leaves oval and rather shiny, stem leaves small and bract-like. Upper sepal and petals forming a hood, with lateral sepals widely spread. The lip extends back into the 19–35mm long spur, with the entrance, clearly visible at the base of the lip, framed by 2 stalked pollinia that are *angled inwards towards the tip*.

HEIGHT 20–40cm (65cm).
FLOWERS Jun–mid Jul (mid May–late Jul).
ALTITUDE 0–460m.

Lesser Butterfly Orchid *Platanthera bifolia*

Listed as Vulnerable. Rather local. Wet heathland, moorland, the less acid parts of bogs (where flushed by springs), damp hill pastures and, much less often, deciduous woods and chalk grassland. ❀ Very like Greater Butterfly Orchid but smaller in all parts, with the 2 pollinia on either side of the mouth of the spur closer together and *strictly parallel*. DETAIL Flower spike with 5–25 flowers, spur 13–23mm long.

HEIGHT 15–30cm (45cm).
FLOWERS Late May–Jul.
ALTITUDE 0–365m.

Frog Orchid *Coeloglossum viride*

Scarce and local in the SE, in short, species-rich grassland on chalk and limestone. Rather commoner to the NW in a wide range of short-grass habitats, often damp, on neutral to calcareous soils. ❀ Small and inconspicuous. The flowers face *downwards* and *range in colour from yellowish-green to reddish*. The sepals and petals form a cowl and the lip is *tongue-shaped with a small notch at the tip* (and a small tooth in the notch); the lip often points down and inwards. DETAIL Basal leaves 3–5, long-oval, held 35°–45° above the horizontal. Each flower has a leaf-like bract at the base, sometimes rather longer than the flowers. Spur extremely short.

HEIGHT 5–15cm (25cm).
FLOWERS Jun–Jul.
ALTITUDE 0–915m.

Pyramidal Orchid
Anacamptis pyramidalis

Locally common. Grassland on poor, often dry, neutral to alkaline soils, especially chalk, limestone and chalky boulder clay, also sand dunes. Often colonises man-made habitats. *Dense, conical spikes of cerise pink flowers distinctive.* **DETAIL** Leaves keeled, pointed, with 3–4 sheathing leaves at the base of the stem, grading into bract-like leaves towards the tip. Flower spike conical at first, eventually cylindrical, with 30–100 flowers. Upper sepal and petals form a hood, with the lateral sepals spreading horizontally and the *lip deeply cut into 3 equal lobes, with 2 vertical ridges at either side of the base.*

HEIGHT 20–60cm. (10–75cm).
FLOWERS Mid Jun–mid Jul.
ALTITUDE 0–350m.

Chalk Fragrant Orchid *Gymnadenia conopsea*

Locally common on species-rich chalk and limestone grassland. Fragrant orchids are characterised by pink flowers in tall, narrow spikes, each with the upper sepal and petals forming a hood, the lateral sepals held stiffly out to the sides, and the lip more or less divided into 3 lobes and extended back into a *very long, slender spur*. All are *strongly scented*, especially in the evening, recalling carnations. **DETAIL** Flowers with lip *distinctly 3-lobed, with the central lobe the longest*, and the lateral sepals narrow, parallel-sided, and usually held drooping at 30° below the horizontal. **SIMILAR SPECIES Marsh Fragrant Orchid** G. *densiflora* is typically larger (30–60cm) with a *long, dense flower spike*. Flowers larger, with the *lip broader than long, the side lobes the largest*, while the lateral sepals are not so strictly parallel-sided and typically held horizontally. Flowers Jul–mid Aug in species-rich alkaline fens scattered throughout Britain and Ireland. **Heath Fragrant Orchid** G. *borealis* is usually small (15–25cm), with a *short, often flat-topped spike*. Its flowers are small, with a *small lip that is longer than broad and only obscurely lobed (it is more of a diamond shape)*. The lateral sepals are more oval, and held drooping. Flowers Jun–Aug in mildly acid to base-rich grassland and base-rich flushes on heathland, mostly in the N and W, but also in the New Forest (where quite common) and Ashdown Forest.

map combines all 3 fragrant orchids

HEIGHT 20–40cm.
FLOWERS Early Jun–mid Jul.
ALTITUDE Lowland?

Heath Fragrant Orchid ▲

◀ **Marsh Fragrant Orchid**

Common Spotted Orchid
Dactylorhiza fuchsii

Locally abundant. A wide variety of grassy places, both wet and dry, including dune slacks, machair and woodland rides, on slightly acidic to alkaline soils. Does best in the sun, but sometimes flowers in shady spots. ❀ Leaves *usually boldly marked with dark spots and bars*. Flowers with the *lip always deeply cut into 3 more or less equal lobes*, but flower colour and markings very variable, from nearly unmarked white to lilac-pink with bold magenta lines and loops. **DETAIL** Stem with 3–6 sheathing leaves at the base and a few small, bract-like leaves towards the flowers. Flower spike

HEIGHT 15–50cm (7–70cm).
FLOWERS Jun–Jul (mid May–early Aug).
ALTITUDE 0–600m.

initially conical, elongating into a cylinder as all 20–70(150) flowers open. Upper sepal and petals form a hood, lateral sepals spread sidways. Lip 8–12mm across, extended backwards into a slender, more or less straight spur. **SEE ALSO** Southern and Northern Marsh Orchids (p. 320), which have more purplish flowers with a spur that is short, fat and bag-like.

Spotted and marsh orchids are notorious for their ability to hybridise. The commonest hybrids are between either Common or Heath Spotted Orchids, and Southern or Northern Marsh Orchids. First generation hybrids are sometimes obvious because they are very robust – unusually tall, leafy plants with very large flower spikes. They are fertile, however, and may back-cross with either parent to produce plants which do not have 'hybrid vigour' and are less obvious. After a few generations a 'hybrid swarm' may result, where almost all the plants are hybrids, with few if any of the pure species remaining. Identifying hybrids and determining their parentage is often a matter of guesswork, and can only be undertaken if you are very familiar with the range of variation in the parent species.

Heath Spotted Orchid *Dactylorhiza maculata*

Locally abundant. A variety of grassland, from bog and moorland to damp hill pastures, usually on mildly acidic soils; infrequent in wooded habitats. ❀ Much like Common Spotted Orchid and similarly variable in colour and leaf spotting, but *lip not as deeply cut*, with the *central lobe usually much smaller than the side lobes*, and typically marked with finer

HEIGHT 10–25cm (4–50cm).
FLOWERS Mid May–Jul.
ALTITUDE 0–915m.

dots. **DETAIL** As Common Spotted but leaves typically rather narrower and more obviously keeled, with the dark spots smaller and more rounded. Flower spike usually rounded or conical, even when all the flowers are open, with 5–20(60) flowers. Lip roughly circular or circular-diamond shaped, with the side lobes much bigger than the central lobe and often with wavy or toothed margins.

Southern Marsh Orchid
Dactylorhiza praetermissa

Locally common. Marshy meadows, fens and dune slacks, also wet 'brownfield' sites; very occasionally chalk grassland. ❋ Flowers relatively large, a *washed-out purplish-pink*. Leaves usually *unspotted*. DETAIL Flowers with a broad, rounded lip divided into 3 lobes (the side lobes may be turned downwards, making the lip appear narrower), marked in the centre with fine dots and dashes. 'Leopard Marsh Orchid', an uncommon variant, has black rings on the leaves and bolder dark markings on the lip.

HEIGHT 20–50cm (70cm).
FLOWERS Jun–Jul.
ALTITUDE Lowland.

Northern Marsh Orchid
Dactylorhiza purpurella

Locally common. Damp grassland, from road verges to marshy meadows, fens, dune slacks and machair; also wet 'brownfield' sites. ❋ Flowers a *deep, 'velvety' magenta* with a deep crimson tone when fresh. Lip usually *diamond-shaped* (margins variably folded to form straight edges, especially to the base of the diamond), marked with lustrous dark crimson loops and dashes. DETAIL Leaves unspotted or very finely freckled (may have dark spots in W Wales and NW Scotland, also locally in N England). SIMILAR SPECIES Irish Marsh Orchid *D. occidentalis* is scattered in Ireland. Lip heavily marked, with 3 *rounded* lobes. Leaves can be spotted.

HEIGHT 10–30cm (5–45cm).
FLOWERS Jun–Jul.
ALTITUDE 0–610m.

Early Marsh Orchid *Dactylorhiza incarnata*

HEIGHT 20–40cm (7–65cm).
FLOWERS Late May–Jun.
ALTITUDE 0–610m.

ssp. incarnata
alkaline fens & meadows

ssp. pulchella
bogs

ssp. coccinea
dune slacks, slumped cliffs

Uncommon in wet, marshy meadows, fens and dune slacks, sometimes also mildly acidic bogs. ❋ Very variable in size, with crowded spikes of *small, narrow flowers*. DETAIL Leaves *unspotted* (except in NW Scotland and W Ireland). Flowers typically with the *sides of the lip folded downwards* and the lateral sepals held upwards, *dark markings on lip enclosed within a solid looping line*. There are several subspecies, differing in flower colour and partly also habitat.

Green-winged Orchid *Anacamptis morio*

Listed as Near Threatened. Very locally common or even abundant. Unimproved grassland on neutral or calcareous soils, usually kept short and open by grazing or mowing; nowadays mostly in reserves or on greens and commons. ❀ Often rather petite, with a few, relatively large, sumptuously deep violet-purple flowers (rarely white or pink). The *lateral sepals always show fine, green, parallel lines*. Leaves *never* spotted. **DETAIL** Flowers occasionally rose-pink or white. The sepals and petals form a hood, the lip is folded downwards at the sides, the centre with dark spots on a white ground, and extended backwards into a long, slightly curved, spur.

HEIGHT 7.5–30cm.
FLOWERS Mid Apr–mid Jun.
ALTITUDE 0–305m.

Early Purple Orchid *Orchis mascula*

Fairly common. Deciduous woods, especially ancient woodland, hedgerows and a variety of old grassland, from chalk downs to damp hill pastures. Commonest on chalk, limestone and chalky clays. ❀ One of the earliest orchids to flower, with *well-spotted leaves* (occasionally unspotted) and flowers of various shades of purple (rarely white). **DETAIL** Lip with scattered dark spots and a white 'throat' leading to a gently up-curved spur; the lateral sepals are held upright.

HEIGHT 10–45cm (60m).
FLOWERS Apr–Jun.
ALTITUDE 0–880m.

Man Orchid
Orchis anthropophora

Nationally Scarce and listed as Endangered. Very local in old grassland on chalk or limestone, sometimes in scrub or on the edge of woodland. ❀ Flowers very distinctive: green, yellow or reddish with the *lip deeply divided to form 'arms' and 'legs'.* They have an uncanny resemblance to miniature marionettes. **DETAIL** Leaves in a basal rosette of 3–4, with 1–2 sheathing leaves on the stem. The sepals and petals form a hood, concealing the column, and the lip is extended into a very short spur.

HEIGHT 15–30cm (45cm).
FLOWERS May–Jun.
ALTITUDE Lowland.

Burnt Orchid *Neotinea ustulata*

HEIGHT 2.5–15cm (25cm).
FLOWERS Mid May–mid Jun;
late Jun–early Aug (*aestivalis*).
ALTITUDE Lowland.

Nationally Scarce. Listed as Endangered. Species-rich short grassland on chalk or limestone. ❁ *Small*, with '*burnt*', *reddish-purple buds* at the tip of the flower spike. *Lip white*, marked with fine dark spots. DETAIL Sepals and petals reddish-purple, forming a hood, lip divided into short 'arms' and 'legs' and extended back into a short spur. The very scarce, late flowering var. *aestivalis* has the lip rimmed or washed rose-purple.

Lady Orchid
Orchis purpurea

HEIGHT 20–50cm (100cm).
FLOWERS May.
ALTITUDE Lowland.

Nationally Scarce. Listed as Endangered, but very locally common in Kent. Deciduous woods on thin, chalky soils. ❁ Usually tall and statuesque. Unopened flower buds *dark reddish-purple*. The sepals and petals form a *dark 'bonnet'*, contrasting with the *broad, white, finely spotted lip*, which forms the 'arms' and 'skirt' of the lady. DETAIL Leaves shiny green, the lower broadly oval. Lip extended back into a short, curved spur.

Military Orchid
Orchis militaris

Nationally Rare, listed as Vulnerable (Sch 8). Confined to 2 sites on scrubby grassland in woodland in the Chilterns and an old chalk pit in a plantation in Suffolk. ❁ Flowers distinctive.
DETAIL Petals and sepals form a 'helmet', *purple-striped on the interior*. Lip with 4 lobes, 2 for the long, straight 'arms' and 2 short 'legs'. The rows of purple spots down the centre of the lip are *reminiscent of buttons on a soldier's tunic*. The lip extends back into a curved spur.

HEIGHT 20–45cm.
FLOWERS Mid May–mid Jun.
ALTITUDE Lowland.

Monkey Orchid *Orchis simia*

HEIGHT 10–30cm.
FLOWERS End May–early Jun.
ALTITUDE Lowland.

Nationally Rare and listed as Vulnerable (Sch 8). Confined to 2 chalk grassland sites in Kent and 1 in Oxfordshire. ❁ Flowers distinctive, resembling a Spider Monkey. DETAIL Flowers quite similar to Military Orchid, but 'arms' and especially 'legs' longer, thinner and *distinctly curved*. The flower spike is more disorganised, and the flowers open in sequence from the bottom up.

Lizard Orchid
Himantoglossum hircinum

Nationally Scarce and listed as
Near Threatened (Sch 8), with *c.* 20
sites in Britain. Grassland on chalk,
limestone, boulder clay or sand
– the largest British population is
on dunes. ✿ Flowers unique, and
smell of billy goat. DETAIL Leaves
grey-green, forming a basal rosette
that is often tattered and partially
withered by flowering time, with a
few narrower leaves on the stem.
Sepals and petals forming an open
hood, lip curled up like a spring in
the bud, the central lobe extending
to 25–60mm.

HEIGHT 25–60cm (90cm).
FLOWERS Jun–mid Jul.
ALTITUDE Lowland.

Musk Orchid
Herminium monorchis

Nationally Scarce. Listed as
Vulnerable. Short, species-rich
grassland on chalk or limestone. ✿ Short, slender and often rather hard to
spot, with a crowded spike of *small, bell-shaped, yellowish flowers without
an obvious lip.* DETAIL Basal leaves 2, oval-
oblong, keeled, with 1–3 tiny bract-like leaves
on the stem. Flowers *c.* 2.5mm across, lip with
a narrow central lobe and 2 short side lobes;
the central lobe of the lip is similar in shape to
the sepals and petals,
and all point forward,
giving the impression
of 6 narrow petals.

HEIGHT 3.5–15cm (20cm).
FLOWERS Early Jun–early Jul.
ALTITUDE Lowland.

Fly Orchid *Ophrys insectifera*

Listed as Vulnerable. Varied habitats, mostly
on chalk or limestone: open woodland (often
beechwoods in the S), old, species-rich grassland
on old pits, quarries and spoil heaps, limestone
pavements and, in W Ireland and Anglesey, fens.
✿ Surprisingly hard to spot, but flowers unique.
DETAIL Leaves mostly at base of stem, with 1–2
smaller leaves towards the flowers. Lip purplish-
brown, forming the
'body' of the fly, with
a lustrous slate-blue
patch forming the
'wings', while the 2
wire-like petals form the
'antennae'; all 3 sepals
are green. Pollinated by
solitary bees.

HEIGHT 15–60cm.
FLOWERS Late May–early
Jun (late Apr–early Jul).
ALTITUDE 0–390m.

Bee Orchid
Ophrys apifera

Locally common in
grassland on poor
soils, often over chalk,
limestone or sand,
including unkept lawns
and verges.
❀ *Bumblebee-like
flowers distinctive.*
DETAIL Leaves strap-
shaped, mostly basal.
Sepals pinkish, petals
green or pinkish-
brown, much smaller
and narrower. Lip
velvety, maroon-brown
with a horseshoe-
shaped pattern of
yellow lines. Stamens
and stigmas fused

into a green column in the centre of the flower, with the 2 yellow pollinia
released to dangle on slender stalks until blown onto the stigmatic surface.

HEIGHT 10–45cm (65cm).
FLOWERS Jun–early Jul (late
May–late Jul).
ALTITUDE 0–335m.

Bee, Early Spider and Fly Orchids belong to the genus *Ophrys*, a Mediterranean group of orchids
that have an elaborate pollination mechanism. Each species of orchid emits chemicals that mimic
the pheromones produced by female solitary bees to attract a mate. Male bees, freshly emerged
in spring, are thus attracted to the flowers and the deception is continued by both the appearance
and the feel of the flower (they really have evolved to look like a bee). The male bee attempts to
copulate with the flower, picking up pollinia in the process, before moving to another flower to
repeat the process. Bee Orchid is the only exception to this pattern, being self-pollinated.

Early Spider
Orchid
Ophrys sphegodes

Nationally Scarce (Sch 8), but very
locally abundant in species-rich
grassland on chalk or limestone.
❀ Flowers distinctive, with a *large,
furry, dark purplish-brown lip.*
DETAIL Stem and leaves yellowish-
green, with 3–4 basal leaves and
1–2 narrow, sheathing stem leaves.
Petals and sepals yellowish-green,
lip marked with a lustrous greyish
'H', 'X' or 'π'. The flowers rapidly
fade to yellowish-brown.

HEIGHT 5–20cm (45cm).
FLOWERS Apr–May.
ALTITUDE Lowland.

Lords-and-ladies *Arum maculatum*

GROWTH Perennial.
HEIGHT 20–55cm.
FLOWERS Apr–early Jun.
STATUS Native.
ALTITUDE To 425m.

The flowers lie on the spadix, hidden at the bottom of the cowl. The female flowers are clustered together at the base, the male flowers above; both male and female flowers lack petals. A ring of thread-like sterile male flowers trap midges attracted by the fetid smell produced by the flowers but then wither to release the insects, which fly off covered in pollen to pollinate another plant.

Common in hedgebanks, woodland and other shady places. ❀ Very distinctive, with a large pale green 'cowl' (the spathe) surrounding the upright, pencil-like spadix. The red berries are conspicuous in late summer. **DETAIL** Growing from swollen, tuber-like roots, the leaves appear from *late Dec*; long-stalked, 7–20cm long, arrow-shaped with pointed basal lobes, often with *blackish blotches*. Spathe 10–25cm tall, sometimes *dark-spotted*, spadix tip club-shaped, *dull purplish* (sometimes yellow).

Italian Lords-and-ladies *Arum italicum*

Nationally Scarce and listed as Near Threatened. Very local near the S coast from the Isle of Wight to the Isles of Scilly, in hedgebanks, woods, scrub and rough pastures, also the 'hanger' woodlands of W Sussex and E Hants. *Much more widespread as a garden escape.* ❀ As Lords-and-ladies but larger and more strongly patch-forming. Leaves 10–35cm

long, appearing *from late Sep*, seldom dark-spotted, basal lobes rounded. Spathe 15–40cm tall, *never dark-spotted, spadix always tipped creamy-yellow* (but many have the flowers bitten off). **DETAIL** Subspecies *neglectum* is native (also a rare garden escape). *Leaf veins and midrib slightly paler*, basal lobes converging or overlapping. Subspecies *italicum*, common in gardens, is a regular garden escape. Leaves more numerous (usually 4–8), with *conspicuous whitish veins*.

ssp. *italicum*

GROWTH Perennial.
HEIGHT 20–30cm.
FLOWERS Late Apr–Jun.
STATUS Native.
ALTITUDE Lowland.

Bog Asphodel *Narthecium ossifragum*

Locally abundant in the N and W in wet peaty places in bogs, and in gravelly hillside flushes; absent from most of S and E England. ❀ Very conspicuous in flower, with spikes of up to 20 *bright yellow, star-like flowers*. The whole plant often turns orange as the fruits develop, giving colour to the bog in late summer. **DETAIL** Grows from creeping rhizomes. Hairless. Leaves mostly basal, curved, 5–30cm × 0.2–0.5cm, with hairless

GROWTH Perennial.
HEIGHT 10–45cm.
FLOWERS Jul–Aug.
STATUS Native.
ALTITUDE 0–1005m.

margins, *held in the same plane, edge-upwards*, in a fan-like spray (like a tiny garden iris). Sepals 3, petals 3, identical, 6–9mm long, stamens 6, with dense yellow hairs on the filaments and orange anthers; style 1.

FAMILY **TOFIELDIACEAE**: SCOTTISH ASPHODEL

Scottish Asphodel
Tofieldia pusilla

Local in wet (but not waterlogged) areas by streams and in calcium-rich flushes, mainly in the uplands. ❀ Small and rather inconspicuous, with short, dense spikes of 5–12 tiny, creamy-white flowers. **DETAIL** Grows from creeping rhizomes. Hairless. Leaves mostly basal, 1–4cm × 0.1–0.2cm, with *fine teeth* along one edge, *all held in the same plane*, edge-upwards, in fan-like sprays (like miniature Bog Asphodel leaves). Flowers very short-stalked, sepals 3, petals 3, identical, blunt, 1.5–2.5mm long, stamens 6, ovary relatively large, green, with *3 short styles*.

GROWTH Perennial.
HEIGHT 5–20cm.
FLOWERS Jun–Aug.
STATUS Native.
ALTITUDE 40–975m, but mostly in uplands.

Marsh Arrowgrass
Triglochin palustris

GROWTH Perennial.
HEIGHT 15–40cm.
FLOWERS Jun–Aug.
STATUS Native.
ALTITUDE 0–970m.

Locally common in short, sparse vegetation in wet places, usually on calcareous soils: damp meadows, fens, flushes and, in the uplands, river shingle. ✿ Slender, grass-like and easy to overlook, even when present in good numbers. Flowers numerous, but very small, in a long spike. DETAIL Hairless. Leaves all basal, to 150mm × 0.8–1.5(2.5)mm, semi-cylindrical, grooved on the upperside *near the base*, and giving off a soapy odour when bruised. Flowering stem cylindrical, 1.2mm wide, elongating as the fruits ripen. Sepals 3, petals 3, identical, *c.* 1.5–2mm long, green with a pale or purplish fringe. Stamens 6, anthers more or less stalkless, yellow; ovary much larger, green, the stigmas a bottlebrush-like tuft. Fruits 7–10mm long, very slender, club-shaped, splitting *from the base* into 3 spear-shaped segments.

Sea Arrowgrass
Triglochin maritima

GROWTH Perennial.
HEIGHT To 60cm.
FLOWERS May–Sep.
STATUS Native.
ALTITUDE Lowland.

Common on saltmarshes, spray-washed cliff ledges and other saline, damp ground by the sea. ✿ Whole plant *fleshy*. Leaves in a dense tuft, flowering stems erect, with numerous small flowers varying in colour from green to purple. DETAIL Leaves all arising from the base of the stem, 15–60cm × 2–3mm, very narrow, semi-cylindrical with a groove above, but *flat* near the base, and giving off a soapy scent when bruised. Flowers stalked, sepals 3, petals 3, very similar, *c.* 2mm long, green with a fine pale fringe and sometimes a purple tip. Stamens 6, anthers more or less stalkless, yellow; ovary much larger, green, topped by up to 6 stigmas, each stalkless, tufted and rather brush-like. Ripe fruits conspicuous, erect, *oval*, 3–5mm long, *breaking up into 6 segments*. SEE ALSO Sea Plantain (p. 223), which has broader leaves and denser, more compact flower spikes, and, when fresh, obviously stalked anthers.

Herb Paris *Paris quadrifolia*

GROWTH Perennial.
HEIGHT 15–40cm.
FLOWERS Late Apr–Jun.
STATUS Native.
ALTITUDE To 360m.

Rather local in damp deciduous woodland on calcareous soils, mostly ancient woodland, also the grikes of limestone pavement in Cumbria. ❀ Very distinctive, with 4 pointed-oval leaves arranged in a cross near the tip of the otherwise leafless stem, and a single greenish flower. DETAIL Hairless. Leaves to 15cm, typically 4, but may be 3–8. Sepals and petals 20–35mm long, usually 4 of each, but may be up to 6; sepals green, lanceolate, petals greenish-yellow, very narrow; stamens 8(12), ovary purplish-black, with 4–5 stigmas. Fruit a black berry. SEE ALSO Chickweed Wintergreen (p. 181).

> The English and scientific names derive from the Latin pars (equal), and refer to the parity in the number of leaves, sepals and petals, rather than any association with the city of Paris.

FAMILY COLCHICACEAE: MEADOW SAFFRON

Meadow Saffron *Colchicum autumnale*

Near Threatened. Locally common. Woodland glades and rides and damp grassy places, especially unimproved meadows. An occasional garden throw-out. ❀ The crocus-like flowers appear in the *autumn*, *without leaves*, and have 6 *stamens*.

GROWTH Perennial.
HEIGHT 8–25cm.
FLOWERS Aug–Oct.
STATUS Native.
ALTITUDE Lowland.

DETAIL Grows from an underground corm. Leaves 2–5, hairless, strap-shaped, 15–35cm long, glossy green above, *present Feb-Jul*. Flower with 3 sepals and 3 identical petals, 30–45mm long, fused at the base to form the white 'stem' (actually still the flower and containing 3 long styles extending upwards from the underground ovary). Fruit a green, egg-shaped capsule, 3–5cm long, appearing with the leaves.

Spring Crocus *Crocus vernus*

Locally common in churchyards, road verges and meadows. ❀ Crocuses are familiar garden plants. All have fine grooved leaves with a central whitish stripe and flowers that only open fully on a sunny day. None are native to Britain. Spring Crocus is the most commonly naturalised. **DETAIL** Grows from an underground corm. Leaves *4–8mm wide*. Flowers white to deep purple, often with dark stripes on the outside of the petals, and a *mauve to purple throat* (*never paler* than the petals, and only white in white-flowered forms); there is a single papery spathe and a single white bract. **SIMILAR SPECIES Early Crocus** C. *tommasinianus* has *narrower* leaves (2–3mm wide) and more delicate flowers which *combine* mauve to pale purple petals with a white throat and tube. **Yellow Crocus** C. × *stellaris* has yellow flowers and leaves 1–4mm wide.

GROWTH Perennial.
HEIGHT 3–5.5cm.
FLOWERS Feb–Apr.
STATUS Introduced to cultivation from Europe before 1600 and first recorded in the wild in 1763.
ALTITUDE Lowland.

Early Crocus

Montbretia *Crocosmia × crocosmiiflora*

Common in gardens and an increasing garden throw-out on verges, waste ground, cliffs and in woods; especially abundant in the SW. Some viable seed is produced, but mostly spreads via rhizomes to form extensive patches. ❀ Produces dense tufts of *narrow*, strap-shaped leaves, with spikes of large, orange funnel-shaped flowers. **DETAIL** Grows from a corm. Leaves 15–17mm (5–20mm) wide, yellowish-green, with a prominent midrib. Sepals 3, petals 3, 25–40mm long, spreading at the tips but united into a *curving* tube for about *half their length*, stamens nearly as long as petals.

GROWTH Perennial.
HEIGHT To 60cm.
FLOWERS Jul–Oct.
STATUS A hybrid produced in France and introduced to gardens in 1880. First recorded from the wild in 1911 and still spreading.
ALTITUDE To 410m.

GROWTH Perennial.
HEIGHT 40–150cm.
FLOWERS May–Jul.
STATUS Native.
ALTITUDE 0–570m.

Yellow Iris *Iris pseudacorus*

Common in shallow water in ponds, ditches, streams and rivers, as well as in swamps, fens, wet meadows and carr woodland. In the N and W also in upper saltmarshes and other damp places on the coast. Tolerant of poor water quality, and in favourable sites may form dense stands. By far the commonest native iris. ✿ Flowers *yellow* with fine dark spots and veins on the sepals. **DETAIL** Grows from thick, tough rhizomes. Leaves long, *flat*, sword-shaped, 15–50mm wide. Flowers 7–10cm across, in groups of 2–3 but opening one at a time; the base of their stalks is enclosed in a sheathing bract. Sepals ('falls') 3, with a broad tip and narrow base; the sepals do not have a compact mass of hairs and therefore not a 'bearded' iris. Petals 3, narrower, upright, style split into 3 upright petal-like branches; stamens 3. Seed pods large and egg-shaped, seeds *brown*.

Stinking Iris *Iris foetidissima*

GROWTH Perennial.
HEIGHT 30–80cm.
FLOWERS May–Jul.
STATUS Native.
ALTITUDE Lowland.

var. *citrina*

Locally common in woods, hedgebanks and on scrubby sea cliffs, usually on calcareous soils, and also a fairly frequent garden escape. *Evergreen*, the leaves have a fetid, 'raw meat' smell when crushed. ✿ Flowers pale ochre-brown in the centre, pale mauve, with conspicuous purplish-blue veining towards the tips of the sepals and petals (rarely yellowish in var. *citrina*). Sepals without a compact mass of hairs (thus not 'bearded'). Seed pods egg-shaped, 5–7cm long, splitting open to reveal *bright orange-red seeds*. **DETAIL** Leaves grow in *dense tufts* from the rhizomes, dark green, long, *flat*, sword-shaped, 15–25mm wide. Flowering stems flattened. Flowers 5–7cm across, in groups of 2–3 at the tip of the stems, opening one at a time, the base of their stalks is enclosed in a sheathing bract.

Fritillary *Fritillaria meleagris*

GROWTH Perennial.
HEIGHT 20–50cm.
FLOWERS Apr–May.
STATUS In cultivation by
1578 but not recorded from
the wild until 1736 and now
considered an introduction.
ALTITUDE Lowland.

Nationally Scarce. Very local in damp, unimproved grassland, especially meadows that are cut for hay and prone to winter flooding; found in large numbers at a few long-established sites. Elsewhere deliberately planted and sometimes naturalised in small numbers. ❀ Flowers large and distinctive, held nodding at the tip of the slender stem, either with a *chequered* pattern of pale and dark purples, or off-white (sometimes with a ghost of the purple chequers). **Detail** Grows from a bulb, with 3–6 grey-green, grass-like leaves (6–20 × 0.4–1cm), alternately on the stem. Flowers typically solitary, 30–50mm long. Sepals 3, petals 3, identical, stigmas 3. Fruit an erect, subtly 3-cornered capsule.

Yellow Star of Bethlehem *Gagea lutea*

GROWTH Perennial.
HEIGHT 8–25cm.
FLOWERS Feb–May.
STATUS Native.
ALTITUDE To 340m.

Very local, in
woodland, shady
riversides, hedgerows,
rough grassland
and on limestone
pavements, on damp,
base-rich soils. Can be
hard to find as often
only a few, scattered
plants, and may not flower in heavy shade. ❀ The solitary leaf is much like a Bluebell's, but the star-like flowers are distinctive, yellow with a green stripe on the outside of the 6 petals. **Detail** Almost hairless. Grows from a bulb, with 1 basal leaf, 10–45 × 0.5–1.5cm, and 2–3 shorter, leaf-like bracts. Flowers 1–5 in an umbel-like cluster, with 3 sepals and 3 identical petals 10–18mm long, and 6 stamens.

GROWTH Perennial.
HEIGHT To 40cm.
FLOWERS Mar–Apr.
STATUS Native.
ALTITUDE 0–375m.

Wild Daffodil *Narcissus pseudonarcissus*
subspecies
pseudonarcissus

Daffodils are amongst the most familiar plants. ❀ The native Wild Daffodil is medium-sized, with a single flower to each stem, creamy-yellow petals and a *darker* yellow trumpet, and is locally abundant in deciduous woodland and old pastures. Garden varieties are, however, commonly planted in the 'wild', especially on verges, as well as growing from dumped garden rubbish or as relicts of commercial crops, and can be difficult to separate from Wild Daffodils. Furthermore, odd plants or small, out-of-range populations of wild-type daffodils, however convincing, are usually of garden origin. DETAIL Grows from a bulb. Leaves flat, 5–15mm wide. Sepals and petals identical, creamy-yellow, *more or less triangular, twisted,* 20–40mm long and *at least 10mm wide* at base. 'Trumpet' (corona) *about as long as petals* (18–40mm), *parallel-sided*; base of trumpet between petals and ovary 15–22mm long, distinctly *broadening* towards petals, *less than twice as long as greatest width*. Stamens 6, *all the same length*. SIMILAR SPECIES Tenby Daffodil *N. obvallaris* is confined to SW Wales and has a uniformly deep yellow flower, the petals 25–30mm, not twisted, and the base of the trumpet 12–15mm long. Of garden origin and now scarce.

FAMILY LILIACEAE: LILIES

GROWTH Perennial.
HEIGHT To 50cm.
FLOWERS Apr–May.
STATUS Introduced to cultivation by 1596 and recorded from the wild by 1790. Soon widely naturalised but has since vanished from many places.
ALTITUDE Lowland.

Wild Tulip *Tulipa sylvestris*

Very local. Woods, hedgerows and rough, grassy places, mostly where deliberately naturalised or as a relict of cultivation; occasionally a garden throw-out. ❀ The single yellow flowers *droop* in bud. DETAIL Grows from a bulb, with 1 basal leaf and 1–3 stem leaves, strap-shaped, to 30 × 1.2–2.5cm. Flowers *taper* into the stem, sepals 3, petals 3, identical, 20–60mm long, stamens 6, filaments *hairy* near base, stigmas 3, very short. SIMILAR SPECIES Many other tulips are grown in gardens and may escape but even if yellow-flowered they have broader leaves, erect buds, flowers that are rounded at the base, and hairless filaments.

Snowdrop
Galanthus nivalis

Locally common in damp woodland and other shady places. ❀ One of the most familiar flowers, but not native to Britain and mostly found as a relict of old gardens, a garden throw-out or where deliberately planted and naturalised. **DETAIL** Grows from a bulb, leaves *flat*, strap-shaped, 5–15 × 0.5–1cm, *uniformly bluish-green*. Flowers *solitary*, the flower stalk cowled by a green, leaf-like bract, sepals 12–25mm long, spreading when the flower is fully open, petals 5–12mm, tipped green and forming a cup. Fruit an oval capsule. **SIMILAR SPECIES** There are many cultivars of Snowdrop, including doubles, and several other species in cultivation, and these occasionally escape or are planted on verges, in churchyards etc.

GROWTH Perennial.
HEIGHT 15–20cm.
FLOWERS Jan–Mar.
STATUS Known from gardens by 1597 and recorded from the wild by 1778.
ALTITUDE 0–410m.

Summer Snowflake *Leucojum aestivum*

Nationally Scarce as a native in central-S England and S Ireland, in riverside woodland liable to flood in winter. Much more widespread as a garden throw-out, often near habitation. ❀ Tall, slender, flowering stems carry a few cup-shaped flowers drooping on long stalks. **DETAIL** Grows from a bulb. Leaves narrow and strap-shaped, 30–50 × 0.5–1.8cm, flower stalk with 2 sharp edges, petals and sepals *identical*, each with a green spot near the tip. Native populations (and some garden escapes) are subspecies *aestivum*, with scattered minute teeth along edges of the stem (especially towards the base), the bract at the base of the flower stems 7–11mm wide, 3–5 (2–7) flowers, and petals 13–22mm long. Subspecies *pulchellum* occurs only as a garden escape. Smaller and slenderer, with smooth-edged stems, (1)2–4 flowers, a much narrower bract (4–7mm) and shorter petals (10–15mm). **SIMILAR SPECIES Spring Snowflake** *L. vernum* is common in gardens and occasionally escapes (or is planted on verges and in churchyards). Shorter, with 1–2 flowers per stem, the petals 15–25mm long. Flowers Jan–Apr.

GROWTH Perennial.
HEIGHT 30–60cm.
FLOWERS Mar–Apr (with leaves).
STATUS Native.
ALTITUDE Lowland.

GROWTH Perennial.
HEIGHT 30–80cm (120cm).
FLOWERS Jun–Jul.
STATUS Native.
ALTITUDE 0–455m.

Wild Onion
Allium vineale

Fairly common but inconspicuous in sparse, dry grassland on verges and field margins, and on dunes and other grassy places near the sea. ❀ Stems *round*, finely ridged, with 2–4 *slender*, more or less *cylindrical* leaves that are *hollow* and channelled on one side. Flower spike often made up solely of *greenish-purple bulbils* (var. *compactum)*, but may be a mixture of bulbils and flowers (var. *vineale*), or rarely flowers only (var. *capsuliferum*), all enclosed by a *single chaffy bract* that narrows to a fine point. DETAIL Leaves 20–60cm long × 0.2–0.5cm wide, bluish-green. Flowers long-stalked, sepals and petals identical, 3–5mm long, pink or greenish-white. Stamens *longer* than petals, inner 3 stamens with the filaments *cut into 3 prongs*, the outer prongs much longer than the short central prong carrying the purple anther.

GROWTH Perennial.
HEIGHT To 80cm.
FLOWERS Jul–Aug.
STATUS Native.
ALTITUDE 0–365m.

Listed as Vulnerable. Rather local on dry, grassy slopes over chalk or limestone, often on verges. ❀ Leaves narrow, more or less *semi-circular* in cross-section. Stems *rounded*, slightly ridged, flower spike with a *mixture* of flowers and bulbils (rarely bulbils only), enclosed by 2 *persistent leaf-like bracts* that taper to *very long, narrow points* (*much* longer than flower spike). DETAIL Leaves 2–4mm wide, ribbed. Flowers 5–40, long-stalked, petals and sepals identical, 5–7mm long, white washed pink, green or brown, stamens *shorter* than petals, their filaments *not* forked.

Field Garlic *Allium oleraceum*

Three-cornered Garlic *Allium triquetrum*

Locally abundant in the S and especially the SW, on verges, hedgebanks, woods, field margins and waste ground. ❀ *Much like a white-flowered Bluebell.* Flower stems *sharply triangular* in cross-section, leaves *flat*, narrow and strap-shaped. Flowers bell-like, long-stalked, nodding. *No bulbils.* DETAIL Leaves 2–5, 4–14mm wide. Sepals and petals identical, 10–18mm long, white with a *fine green stripe* on the outside.

GROWTH Perennial.
HEIGHT 20–50cm.
FLOWERS Mar–May.
STATUS Introduced to gardens from the Mediterranean by 1759. Well naturalised and still increasing.
ALTITUDE Lowland.

Few-flowered Garlic *Allium paradoxum*

Rather local but sometimes invasive garden escape or throw-out in rough grassy places on verges, riverbanks etc., spreading by means of bulbils to form dense patches. ❀ Flower stems sharply *triangular* in cross-section, leaves *flat*, usually *only 1 per bulb*. The flower head is a *mixture* of green bulbils and flowers, *typically mostly bulbils and just 1 flower*, but sometimes *only bulbils*, all enclosed by 2 papery bracts. DETAIL Leaves 6–18mm wide, growing from the base, dull grey-green. Sepals and petals identical, 10–12mm long, white.

GROWTH Perennial.
HEIGHT To 40cm.
FLOWERS Mar–Jun.
STATUS Introduced to cultivation from SW Asia in 1823 and first recorded in the wild in 1863; still increasing.
ALTITUDE To 375m.

Sand Garlic
Allium scorodoprasum

Uncommon on rough grassland on verges, railway banks and waste ground, also sandy riverbanks, the upper fringes of saltmarsh and open woodland. Spreads mainly via bulbils. ❀ Stems *round*, leafy, leaves *flat*, narrow and strap-shaped. Flower spike with a *mixture* of flowers and bulbils (rarely bulbils only), enclosed by 2 papery bracts that taper to *short points*; flowers *pinkish-purple*. DETAIL Leaves 2–5, to 27cm long × 1–2.4cm wide, dull green. Petals and sepals 4–8mm long. Stamens *shorter* than petals, the inner with *3-pronged filaments* (as Wild Onion).

GROWTH Perennial.
HEIGHT To 80cm.
FLOWERS May–Aug.
STATUS Native.
ALTITUDE Lowland.

Ramsons
Allium ursinum

GROWTH Perennial.
HEIGHT 10–45cm.
FLOWERS Apr–Jun.
STATUS Native.
ALTITUDE To 450m.

Common in damp deciduous woodland, especially around streams and flushes, on shady verges and riverbanks, and sometimes the grikes of limestone pavement and other exposed rocky places. ✿ May dominate large areas of woodland, and *often smelt* before seen (all parts are *strongly* garlic-scented). Leaves 2–3, stalked, *long-oval*, flowers in loose clusters of 6–20. DETAIL Stems rounded to subtly 3-angled in cross-section. Leaves flat, 10–25cm × 1.5–7.5mm. Sepals and petals identical, 7–12mm long.

FAMILY ASPARAGACEAE: ASPARAGUS, BLUEBELLS ETC.

Grape Hyacinth *Muscari neglectum*

Nationally Rare, listed as Vulnerable. A scarce native in Breckland, mostly in dry grassland on verges and field margins. Scattered records elsewhere as a garden throw-out. ✿ Much like the familiar Garden Grape Hyacinth but flower spikes more delicate, the individual grape-like flowers *much darker*, more blackish-blue. DETAIL Grows from a small bulb. Leaves 3–6, all basal, 15–30 × 0.2–0.8cm, but curled into a slender cylinder, grooved on one side. Flower spike 1.5–5cm, each flower just 3.5–7.5mm long, blackish-blue with 6 small white teeth at the mouth of the corolla. Upper flowers smaller, paler blue and sterile. SIMILAR SPECIES Garden Grape Hyacinth *M. armeniacum* is commonly planted on verges near houses and also occurs as a garden throw-out.

GROWTH Perennial.
HEIGHT 10–30cm.
FLOWERS Apr–early May.
STATUS Native.
ALTITUDE Lowland.

Garden Grape Hyacinth

Solomon's Seal *Polygonatum multiflorum*

GROWTH Perennial.
HEIGHT Stems 30–80cm.
FLOWERS May–Jun.
STATUS Native.
ALTITUDE Lowland.

Rather local in ancient woodland, mostly over chalk and limestone. More widespread as a garden escape. ❀ Distinctive leafy, *arching* stems, with clusters of small, bell-like flowers hanging from the base of each leaf. DETAIL Patch-forming, with *rounded* stems growing from a rhizome. Leaves 5–12cm long, oval, alternate, stalkless, hairless, shiny green above. Flowers in *groups of 2–4 (1–6)*, petals and sepals fused into a tube 9–15(20)mm long and 2–4mm wide, *pinched* in the centre, with 6 short teeth at the tip, stamens with a *sparsely hairy* filament; *not* scented. Fruit a blue-black berry. SIMILAR SPECIES Garden Solomon's Seal *P. × hybridum*, the more or less sterile hybrid with Angular Solomon's-seal, is much commoner in gardens and thus a much commoner garden throw-out (or deliberately naturalised, e.g. in churchyards). Often taller and more robust, forming denser patches. Stems often round at base, becoming *slightly angled* towards tip, flowers *larger (15–22 × 3–6mm)*, *slightly pinched* in the centre; *fruit rarely produced*.

Angular Solomon's Seal
Polygonatum odoratum

GROWTH Perennial.
HEIGHT Stems 15–40cm.
FLOWERS Late May–Jul.
STATUS Native.
ALTITUDE To 485m.

Nationally Scarce. Very local in rocky ancient woodland, typically on limestone, and, in N England, the grikes of limestone pavement. ❀ Stems shorter and often more erect than Solomon's-seal, *distinctly 4-angled throughout*, with only *1 (2) flowers per cluster*, each relatively large and scented. DETAIL Leaves dull grey-green. Flowers 15–30 × 4–9mm, *not* pinched at the centre, stamens with *hairless* filaments. Fruit a blue-black berry.

Garden Solomon's Seal ▶

Lily of the Valley *Convallaria majalis*

Very locally common in Ash woodland on limestone in the N and W and deciduous woodland on free-draining, acid soils in E Anglia and the SE. Very persistent, may survive under Bracken or in the grikes of limestone pavements long after woodland has gone. Elsewhere an occasional garden escape in woods, scrub and hedgebanks. Grows from creeping rhizomes and in favoured places carpets the woodland floor.
✽ A pair of dull green, long-oval leaves grow from the base of the stem, with a distinctive 1-sided spike of small, drooping, bell-like flowers, very sweetly scented. DETAIL Hairless. Leaves long-stalked, up to 12 × 5cm (to 30 × 10cm in robust, cultivated varieties). Flower spikes leafless, with 6–12 flowers, each 5–10mm across, the corolla with 6 blunt, outwardly-curving teeth at the mouth. Fruit a red berry.

GROWTH Perennial.
HEIGHT 10–25cm.
FLOWERS May–Jun.
STATUS Native.
ALTITUDE 0–470m.

Spiked Star of Bethlehem

Ornithogalum pyrenaicum

Nationally Scarce. Locally common within its limited range on verges and hedgebanks, especially along green lanes, and in rough grassland and woodland.
✽ The tall, elongated flower spikes, with numerous star-like flowers, are distinctive. DETAIL Hairless. Leaves 5–8, all basal, arising from an underground bulb: 30–75 × 1–1.5cm, strap-shaped, grey-green, withering early. Bracts shorter than flower stalks, white. Petals 3, sepals 3, identical, 6–13mm, yellowish-cream, washed greenish along a broad central stripe. Fruit an oval capsule.

GROWTH Perennial.
HEIGHT 50–80cm (100cm).
FLOWERS Jun–Jul.
STATUS Native, also a rare garden escape.
ALTITUDE Lowland.

Confined to two discrete areas: N Somerset–W Gloucestershire–Wiltshire–SW Berkshire and N Bedfordshire–S Huntingdonshire. Formerly known as 'Bath Asparagus', the young shoots were harvested, sometimes on a large scale, and sold as a substitute for Asparagus.

Star of Bethlehem *Ornithogalum umbellatum*

GROWTH Perennial.
HEIGHT To 40cm.
FLOWERS Mid Apr–Jun.
STATUS Introduced to gardens by 1548 and first recorded from the wild in 1650.
ALTITUDE Lowland.

Local. Rough grassland, including verges and roadside banks, favouring light, dry, sandy or gravelly soils. ❁ The umbel-like clusters of star-like flowers, with the petals striped green on the outside, are distinctive. The flowers only open fully in the sun. DETAIL Grows from an underground bulb. Leaves all basal, to 30cm long and 0.6cm wide, with a *central white stripe* on the upperside. Bracts 20–30mm long, *shorter* than flower stalks; lower flowers much longer-stalked than upper. Sepals 3, petals 3, identical. Fruit a 6-angled capsule. Subspecies *umbellatum* is certainly introduced. Robust, with up to 20 flowers, the lowest with a stalk up to 11cm long; sepals 20–30mm. Subspecies *campestre* may be native in Breckland. Smaller overall, with fewer flowers (up to 12), the lowest with a stalk no more than 5cm long; sepals 15–20mm. The leaves may have died off by flowering time.

Ssp. *campestre*

Drooping Star of Bethlehem
Ornithogalum nutans

An occasional garden escape or throw-out (or planted) on woodland edges, hedgerows, verges and other rough, grassy places, and in churchyards. ❁ Flowers in a *1-sided spike*, relatively large and striking, *bell-shaped*, held spreading at first but eventually more or less *drooping*. Sepals and petals identical, greyish-white with a *broad pale grey-green stripe on the outside*. DETAIL Grows from a bulb. Leaves all basal, 30–60 cm long, with a central white stripe. Flowers 2–12,

GROWTH Perennial.
HEIGHT To 60cm.
FLOWERS Mar–mid May.
STATUS Introduced by 1648 and recorded from the wild by 1805 (SE Europe and Turkey).
ALTITUDE Lowland.

bracts *longer* than flower stalks, petals 3, sepals 3, 15–30mm long. Fruit a capsule, held drooping.

Spring Squill
Scilla verna

Locally abundant. Short turf and heathland on cliffs and rocky slopes near the sea. ✿ Flowers in the spring, *with the leaves*. Spikes of up to 12 pale bluish-lilac flowers, the petals and sepals *spreading widely*, with a conspicuous powder-blue ovary and dark blue anthers. DETAIL Grows from a bulb, the 2–6 leaves appear Oct–May. Leaves flat to more or less cylindrical, wavy, 2–5mm wide and up to 130mm long. Bracts *present*, bluish-purple, 5–15mm long. Sepals 3, petals 3, identical, 5–8mm long.

GROWTH Perennial.
HEIGHT 5–15cm.
FLOWERS Apr–early Jun.
STATUS Native.
ALTITUDE To 415m (on Shetland).

seed heads in August

Autumn Squill
Scilla autumnalis

Nationally Scarce. Very locally abundant on dry grassland and heathland in rocky places near the sea in Devon and Cornwall, with isolated populations at the Avon Gorge and at a few sites on short grassland on floodplain gravels near the Thames.
✿ Flowers mid to late summer, after the leaves have died off, the bare flower stems carrying 4–20 small pale lilac flowers. DETAIL Grows from a bulb, with 4–8 thread-like leaves present Oct–Jul, up to 120mm long and 1–2mm wide. Bracts *absent*. Sepals 3, petals 3, identical, 3–6mm long, spreading widely, anthers purple.

GROWTH Perennial.
HEIGHT 4–25cm.
FLOWERS Jul–Sep.
STATUS Native.
ALTITUDE Lowland.

Bluebell *Hyacinthoides non-scripta*

Locally abundant in deciduous woodland, along shady banks and hedgerows and, especially in the N and W, under Bracken and on unimproved grassland, particularly by the sea. ❀ An iconic and instantly recognisable flower, especially *en masse*, but beware Hybrid Bluebells. Flowers occasionally white. DETAIL Leaves strap-shaped, 7–15mm wide.

GROWTH Perennial.
HEIGHT 20–50cm.
FLOWERS Late Apr–Jun.
STATUS Native.
ALTITUDE Mostly lowland, but to 685m.

Flower spike *1-sided, drooping at the tip*. The flowers typically all *hang downwards* and each is made up of 6 tepals, 14–20mm long, that form a *parallel-sided tube*, with the tepals *strongly curved out and down at the tips*. Anthers *cream*. The flowers have a strong, sweet scent.

Spanish Bluebell *Hyacinthoides hispanica*

A scarce garden throw-out on verges, waste ground and woodland margins (usually where accessible by car), especially close to habitation. Sometimes planted in churchyards etc. ❀ Very like Bluebell but flower spike *erect, not* 1-sided, with the flowers held *erect* or *spreading*. Flowers *bell-* or even *saucer-shaped, broadening* towards the mouth, with the tips of the tepals *not* curved out and down; anthers *blue* in blue flowers, creamy in white or pink flowers. DETAIL Leaves 10–35mm wide, tepals 12–18mm long. SIMILAR SPECIES Hybrid Bluebell *H. × massartiana* is the cross between Bluebell and Spanish Bluebell. The commonest Bluebell in gardens, and *much more commonly naturalised than Spanish Bluebell*. Can appear spontaneously where its parents meet. Usually closer to Spanish Bluebell, with broad leaves, the flower spike erect and not 1-sided, the flowers held spreading, bell-shaped, with the tips of the tepals curving out but not down, and the anthers blue (often creamy in pink or white flowers).

GROWTH Perennial.
HEIGHT To 40cm.
FLOWERS Apr–Jun.
STATUS Introduced to gardens from Iberia by 1683; recorded from the wild by 1909.
ALTITUDE Lowland.

Hybrid Bluebell is fertile and able to back-cross with either parent to form a complete range of intermediates. Concern has been raised that Hybrid Bluebells are more vigorous than native Bluebells, and will swamp the native Bluebell. Small, isolated populations of native Bluebells may be at risk, but the effect on a population of millions of native plants in a 'Bluebell Wood' is not known.

Bluebell   Hybrid Bluebell   Spanish Bluebell

Butcher's Broom
Ruscus aculeatus

Local. Dry woods and hedgerows, also cliffs and other rocky places near the sea. Grown in gardens and sometimes naturalised in churchyards and other places near houses. ❀ A well-branched, evergreen shrub with *stiff, spiny, glossy, dark green 'leaves'*. The tiny yellowish-green flowers grow in the *centre* of the leaves, followed by *scarlet berries*. **DETAIL** Hairless. True leaves reduced to small papery scales on the stem; what appear to be leaves are highly modified flattened stems (cladodes) arising from their axils: flattened, oval, 10–40 × 4–10mm, *tapering to a sharp spine*. Flowers 1–2 together, sepals 3, petals 3, pale green. Dioecious, with male and female flowers on separate plants. Fruits 10–15mm across.

GROWTH Perennial.
HEIGHT 25–100cm.
FLOWERS Jan.–May.
STATUS Native.
ALTITUDE Lowland.

Garden Asparagus *Asparagus officinalis*

A familiar garden plant, cultivated since Roman times and well naturalised on dunes and heathy grassland, usually on dry, sandy soils. ❀ Tall, with delicate, *needle-like* foliage and tiny brownish-cream flowers. Fruit a red berry. **DETAIL** Hairless. True leaves reduced to chaffy scales on the stem; the 'leaves' are flexible, green, thread-like stems (cladodes), up to *c.* 30mm long, growing in clusters from their axils. Flowers 1–4 together on slender stalks, growing from the axils of the true leaves. Bell-shaped, *c.* 3–6mm long, sepals 3, petals 3. Male and female flowers are on separate plants. **SIMILAR SPECIES Wild Asparagus** *A. prostratus* is confined to a few spots on cliffs and dunes on the coast of SW England, S Wales and SE Ireland. Stems shorter, more or less prostrate, with shorter, more rigid, blue-green 'leaves'. SEE ALSO Fennel (p. 304).

GROWTH Perennial.
HEIGHT 1–2m.
FLOWERS Jun.–Sep.
STATUS Ancient introduction.
ALTITUDE Lowland.

Aquatic Plants

Aquatics are plants that almost always grow in water. Some are free-floating, with their roots dangling, but most are rooted into silt or stones at the bottom of the pond, lake, river or ditch in which they grow. Aquatics can be difficult to identify without experience, especially when the flowers are small or seldom produced and it is the leaves and stem that are the key features. Others grow in deep water and can only be reached by boat, or by throwing a small grapple out from shore to pull off pieces of plant, or identified from fragments washed up along the shore. Only aquatic plants with reasonably conspicuous floating leaves and aerial flowers have been covered here. Note that several umbellifers, including Cowbane (p. 306) and the water-dropworts (pp. 308–309), often behave as aquatics.

Flowering Rush *Butomus umbellatus*

Local. Shallow water along the margins of rivers, lakes, canals, ditches, usually alkaline and often nutrient-rich, also swamps. Often planted. ❀ The umbels of pretty pale pinkish flowers are unique amongst water plants.
Detail Leaves reddish at base, usually slightly shorter than flower stem, 4–15mm wide, sharply triangular, becoming flat towards tip. Flower stem leafless, round, with an umbel of flowers on stalks of varying lengths, the longest up to 10cm; several chaffy bracts. Sepals 3, purplish, tinged green, 10–15mm long; petals 3, pinkish, 10–15mm, long, stamens 9, carpels 6, dark red, each with a single style.

GROWTH Perennial.
HEIGHT To 150cm.
FLOWERS Jul–Sep.
STATUS Native.
ALTITUDE Lowland.

White Waterlily *Nymphaea alba*

WATERLILIES: FAMILY NYMPHAEACEAE

Widespread in lakes, ponds, river backwaters and larger ditches, from the shallows to water up to 1.5m deep, sometimes even swamps. Avoids water with very high levels of nutrients and quickly eradicated by heavy boat traffic. ❀ Distinctive, with plate-sized floating leaves and large white flowers with conspicuous yellow stamens. Cultivated forms, often with larger flowers washed pink or yellow, may occur in apparently 'wild' places. **Detail** Leaves shiny

GROWTH Perennial.
FLOWERS Jun–Sep.
STATUS Native.
ALTITUDE To 520m.

dark green above, often reddish below, circular, 9–35cm across, veins palmate, almost all radiating out from the point where the stalk joins the leaf; lacks submerged leaves. Flowers 9–20cm across with 4 greenish sepals, 12–23 petals and many stamens. They are slightly fragrant and only open fully in sunshine. Fruits rounded, warty.

Yellow Waterlily *Nuphar lutea*

Common in lakes, ponds and gravel pits, and large, slow-flowing ditches, canals and river backwaters, typically in water 0.5–2.5m deep. Better able to cope with damage from boat traffic than White Waterlily, and thus more widespread along the edges of rivers, albeit sometimes merely as submerged leaves. ❀ Flowers 3–6cm across, held a few cm above the water, larger than those of any other yellow-flowered floating aquatic. **DETAIL** Grows from a massive fleshy rootstock. Submerged leaves crumpled and translucent, floating leaves leathery, shiny green, oval, to 40 × 30cm, with the veins pinnate, radiating outwards at intervals along the midrib. Flowers with 5–6 yellow sepals, *c.* 20 shorter, narrower petals and many stamens. Fruits flask-shaped.

GROWTH Perennial.
FLOWERS Jun–Aug.
STATUS Native.
ALTITUDE To 505m.

FAMILY MENYANTHACEAE: BOGBEAN, FRINGED WATERLILY

Fringed Waterlily *Nymphoides peltata*

Nationally Scarce as a native, but locally common as an introduction. Still and slow-flowing water up to 2m deep. ❀ Leaves and flowers *much smaller* than Yellow Waterlily, the petals with a *fine, comb-like fringe*. **DETAIL** Grow from buried rhizomes. Leaves all floating, 3–11cm across, slightly angular, with a heart-shaped base, dark green above, with purplish glands below. Flowers long-stalked, in small clusters growing from a leaf axil. Corolla 3–4cm across, cut into 5 petals; stamens 5, style 1, stigma 2-lobed. Each flower lasts for 1 day, and individual plants may be 'pin' or 'thrum' (see p. 182). **SEE ALSO** Frogbit (p. 353), which has similar but smaller leaves, with the main veins circling around and converging towards the tip.

GROWTH Perennial.
HEIGHT Stems to 150cm.
FLOWERS Jun–Sep.
STATUS Native to the Fens and the Thames valley, introduced elsewhere.
ALTITUDE To 300m.

WATER CROWFOOTS

A group of 10 white-flowered buttercups. Three grow on bare mud or in very shallow water and have normal laminar leaves. The remaining 7 grow in still or running water and either have both floating laminar leaves and submerged, thread-like, capillary leaves, or submerged capillary leaves only. The aquatic species are very variable and can often only be identified after careful study. They may become conveniently stranded as water levels fall in summer, but inconveniently these terrestrial forms are not typical and mostly impossible to name accurately. Only three of the more obvious species are included here; full details of all can be found in the *Plant Crib*, available on the BSBI website (see p. 8).

Common Water Crowfoot *Ranunculus aquatilis*

Locally common. Shallow water on the edge of ponds, ditches and slow-flowing streams, mostly in alkaline, fairly nutrient-rich water. Leaves both floating and submerged. Flowers 12–18mm across, with a circular nectar pit at the base of the petals.
DETAIL Floating leaves up to 30mm wide, deeply cut into *c.* 5 lobes, submerged leaves spreading into a rough sphere, with the leaves shorter than the stem internodes.

GROWTH Annual or short-lived perennial.
FLOWERS May–Sep.
STATUS Native.
ALTITUDE To 530m.

Ivy-leaved Crowfoot *Ranunculus hederaceus*

Fairly common. Shallow water and wet mud on the margins of ponds, streams and ditches, also muddy tracks. Leaves shiny, with 3–5 angular lobes, broadest at the base. Flowers rather small, with obvious gaps between the petals, which are hardly longer than the sepals. **DETAIL** Leaves 3–35mm across. Petals 2.5–3.5mm long. Seeds hairless.

GROWTH Annual (short-lived perennial).
HEIGHT Stems 10–50cm.
FLOWERS Mar–Aug.
STATUS Native.
ALTITUDE 0–770m.

River Water Crowfoot *Ranunculus fluitans*

Local. Rapidly-flowing rivers with stony beds and clean, alkaline, relatively nutrient-rich water. Forms dense beds of long, hair-like leaves and stems peppered with white flowers 20–30mm across. **DETAIL** Leaves all submerged, thread-like, to 30cm long (longer than internodes), forking *c.* 4 times. Petals 7–13mm long, receptacle sparsely hairy to hairless. **SIMILAR SPECIES** Stream Water Crowfoot *R. penicillatus* subspecies *pseudofluitans* has the leaves forking 6 or more times and a densely hairy receptacle.

GROWTH Perennial.
HEIGHT Stems to 6m.
FLOWERS Jun–Aug.
STATUS Native.
ALTITUDE Lowland.

GROWTH Perennial.
HEIGHT Underwater stems to 250cm.
FLOWERS Jun–Sep.
STATUS Native.
ALTITUDE 0–380m, but mostly lowland.

Spiked Water-milfoil *Myriophyllum spicatum*

Locally common. A wide range of standing or flowing water, often in moderately nutrient-rich, chalky water; tolerates slightly brackish conditions. ✿ Produces small, *leafless, aerial flower-spikes, with tiny pinkish-red flowers.* Leaves feathery, submerged. DETAIL Stems rooted in the mud. Leaves 1.5–3cm long, *mostly in whorls of 4 (3–5)*, typically with 13–38 long, very narrow, side lobes. Flower stem erect, *even in bud. Flowers in whorls of 4*, lower female, central bisexual, upper male, each cupped by a *tiny toothed bract and minute bracteole*, with 4 minute sepals, 4 reddish petals (minute or absent in female flowers), 4 styles with tiny brush-like stigmas, and 8 stamens with large, yellowish anthers (absent in female flowers).

GROWTH Perennial.
HEIGHT Underwater stems to 300cm.
FLOWERS Jul–Aug.
STATUS Native.
ALTITUDE Lowland.

Whorled Water-milfoil
Myriophyllum verticillatum

Listed as Vulnerable. Rather local. Ponds, ditches, lakes, canals and slow-flowing streams with clear chalky water. ✿ Differs from Spiked Water Milfoil in *much larger, feathery, leaf-like bracts* at the base of the dull yellow flowers. DETAIL Leaves *mostly in whorls of 5 (4–6)*, typically with 24–35 side lobes. *Flowers mostly in whorls of 5*, sometimes leafy, lower female, central bisexual, upper male, each cupped by a deeply toothed bract and minute toothed bracteole. Petals greenish-yellow, rarely reddish. SIMILAR SPECIES Alternate Water-milfoil *M. alterniflorum* is widespread in the N and W, up to 780m, typically in acid water and sometimes in fast-flowing streams. Leaves *mostly in whorls of 4*, mostly with 6–18 side lobes. *Flower spikes drooping in bud.* Flowers yellowish, with *tiny toothed bracts*, lower flowers in whorls of 2–4, *upper (male) flowers opposite or alternate.*

GROWTH Perennial.
HEIGHT Underwater stems to 200cm.
FLOWERS May–Aug.
STATUS Introduced in 1878 as an oxygenator, first recorded in the wild 1960 (S America).
ALTITUDE Lowland.

Parrot's Feather *Myriophyllum aquaticum*

Locally common aquarium or garden throw-out in ponds, ditches, canals, reservoirs and gravel pits. ✿ Stems emerge from the water, often forming a mass of feathery, blue-green leaves. DETAIL. Emergent leaves with *many stalkless glands* (10× lens). Leaves mostly in whorls of 5 (4–6), 1.5–4.5cm long, typically with 8–30 side lobes. Flowers mostly in whorls of 4–6, whitish; only female plants are present in Britain.

Marestail
Hippuris vulgaris

GROWTH Perennial.
HEIGHT Underwater
stems to 100cm
(200cm).

FLOWERS Jun–Jul.
STATUS Native.
ALTITUDE 0–900m.

Locally common. Shallow water
on the margins of lakes, ponds,
ditches and slow-flowing streams
and rivers; also permanently wet
swamps. Prefers alkaline water.
❀ Trailing underwater stems
throw up erect, pale green aerial
stems with whorls of 6–12 shiny
dark green leaves, with tiny
flowers at the base of the upper
leaves. When growing on mud

the plant is smaller in all parts. DETAIL Grows from rhizomes rooted to the
bottom. Leaves narrow and strap-shaped, to 2(8)cm long. Flowers 1 per
axil, 1–2 mm long, no petals or sepals. On the same plant the flowers can
be bisexual, with the pale green ovary carrying both stigma and stamen;
female, with a single white style, or male, with a relatively large, single,
red-flushed stamen. Fruit a tiny green nut.

Water Violet *Hottonia palustris*

PRIMROSES: FAMILY PRIMULACEAE

GROWTH Perennial.
HEIGHT Flower spikes
20–40cm.
FLOWERS May–Jun.
STATUS Native.
ALTITUDE Lowland.

Local in sheltered ditches, ponds and gravel-pits in shallow, clear water, usually rich in calcium and
with moderate levels of nutrients. Tolerates some shade and sometimes found in woodland ponds.
❀ The masses of pale lilac flowers with yellow 'eyes' hovering over the water are unique.
DETAIL Stems to 100cm or more, rooted at the nodes into the bottom. Leaves 2–10cm long, cut
into many very narrow, flattened segments and group into whorls of 2–6 along the stem, with a
cartwheel-like rosette just below the water's surface that supports the aerial flower spike. Flowers
short-stalked, in whorls of 3–8 at intervals up the stem, 15–25mm across, calyx split into 5 narrow
sepals, corolla into 5 petals. Flowers of two types: 'pin' and 'thrum' (see p. 182). Fruit a globular
capsule. Can survive for short periods as a terrestrial form, with more rigid leaves.

Water Starworts *Callitriche* spp.

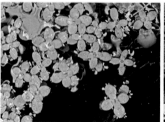

GROWTH Annual to perennial.
FLOWERS Apr–Oct.
STATUS Native.
ALTITUDE 0–915m.

A genus of 7 species of small plant that have both floating and submerged leaves. They grow in rivers and streams, ponds and the backwaters of larger water bodies, or in temporary pools such as flooded wheel ruts, wet woodland rides and large puddles. As these dry out, the plants become stranded on the mud and change their appearance. Although easy to identify as a group, the individual species are hard to separate; the leaf shape is very variable, the minute petal-less flowers do not help and the fruits, while more useful, are produced irregularly in some cases. Stems grow up to 100cm long, with many pairs of opposite leaves, variously rounded to strap-shaped. Characteristically, there is often a rosette of leaves at the tip of the stem.

Duckweeds *Lemna & Spirodela* spp. Family LEMNACEAE

◀ Greater and Least Duckweeds

Ivy-leaved Duckweed

A group of small, simple, perennial plants. Most float on the water's surface and are merely a rounded leaf-like frond with one or more thread-like roots. The minute, petal-less flowers are rarely seen; most reproduction is vegetative, with new plants budding from the old. Duckweeds are only found in still water or river backwaters, and in the right conditions can form a solid green carpet.
Common Duckweed *Lemna minor* Fronds egg-shaped, 1–8mm long with 1 root and *3 veins* (veins may branch on larger fronds); hold a frond against the light, but even then the veins are usually hard to see. *Bright green*, sometimes showing red above, but not below. Common throughout, to 500m.
Fat Duckweed *L. gibba* Fronds (2) 4–8mm long, subtly drop-shaped (widest towards the tip), with 1 root, whitish-green, or with dull red developing from the margins above and below. Sometimes has a *swollen, bulbous underside* in late summer. Larger fronds have *4–5 veins, all originating at the same point*, but most populations are medium and small plants with 1–3 veins, only distinguished from Common Duckweed by larger air spaces in fronds and vague network of veins on upper surface. Common in lowlands.
Least Duckweed *L. minuta* Introduced from America and first recorded in 1977, it is now widespread and very common. Fronds *thin and translucent*, pale greyish-green (*never* showing red), egg-shaped, 0.8–3mm long with 1 root and 1 vein.
Ivy-leaved Duckweed *L. trisulca* Grows just below the surface. Very distinctive. Fronds 3–15mm long, *narrowing to a stalk*, with plants budding one from another to *form a 'daisy-chain'*. Common in the lowlands.
Greater Duckweed *Spirodela polyrhiza* Distinctive. Fronds 1.5–12mm long, roughly circular, often with a purplish rim to the frond above and reddish-purple below. *Always has more than 1 root*, sometimes 20 or more; veins similarly numerous. Local, in lowlands.

Greater Duckweed

Greater Bladderwort *Utricularia vulgaris*

Rather local. Still or very slow-moving water, in lakes, ponds and ditches, also swampy pools hidden in the wet fens. Prefers alkaline water with moderate level of nutrients. ✿ A rootless, free-floating carnivorous plant. Leaves finely cut into thread-like segments, with tiny, animal-catching bladders, forming a dense mat that floats below the water's surface. The distinctive rich yellow flowers are produced on slender, often reddish, upright stems. DETAIL Stems all floating, with both leaves and bladders. Leaves with fine, bristle-like hairs. Lower lip *c.* 12–15mm long × 14mm wide, edges turned downwards, spur 7–8mm long. SIMILAR SPECIES Several other bladderworts are found in Britain, but their identification is complex. **Bladderwort** *U. australis* has stems as Greater Bladderwort but paler flowers. Three other species have stems of 2 sorts: either free-floating with fine green leaves and few or no bladders, or usually buried in the substrate with a few non-green leaves but abundant bladders. The members of this latter group rarely (if ever) produce flowers.

GROWTH Perennial.
HEIGHT Flowering stems 10–20cm, floating stems to 100cm.
FLOWERS Jun–Sep (only flowers regularly in the south of its range).
STATUS Native.
ALTITUDE Lowland.

Bladderworts produce tiny traps or 'bladders' and these catch prey via an ingenious mechanism. There is a negative pressure inside the trap and when a small creature touches the bristles on the outside of the trap a trapdoor opens and the creature is sucked inside. In a single season up to 15,000 traps may be produced by each Greater Bladderwort plant, catching and digesting 230,000 minute crustaceans, larvae and worms. The animal component of the plant's diet may account for 50% of its biomass.

Lesser Bladderwort *Utricularia minor*

Locally common in shallow, nutrient-poor, acidic water in bog pools, ditches and amongst emergent vegetation on the margins of lakes, in some areas (e.g. the Norfolk Broads, the Burren) in alkaline fens, amongst low vegetation or in shallow water. ✿ Flowers *smaller and paler* than Greater Bladderwort, with a *very short spur*. Flowering erratic, and when not in flower the network of fine leaves and stems is unlikely to be noticed. DETAIL Stems often of 2 types, floating, with finely cut leaves, and thread-like and submerged/buried in peat; they lack teeth or bristles. Flowering stems with 2–6 flowers. Lower lip of flower *c.* 7mm long × 5–7mm wide, with the edges turned downwards, spur 1–2mm long.

GROWTH Perennial.
HEIGHT Flowering stems 4–15cm, floating stems to 40cm.
FLOWERS Jun–Sep.
STATUS Native.
ALTITUDE 0–685m.

Sweet Flag *Acorus calamus*

Local. The shallow margins of still or slow-flowing, nutrient-rich, alkaline water. ✿ Sometimes forming dense stands but inconspicuous when growing amongst other emergent vegetation and shy to flower. Leaves iris-like, when crushed has a *pleasant, strong orange small. Flower spikes unique.*

DETAIL Leaves bright green, 7–25mm wide, pointed, usually patchily wrinkled, with a well-defined midrib and numerous fine veins. Flowers closely-packed on a cigar-shaped spadix, 5–9cm long, that emerges at an angle to the side of the stem, each tiny, with 6 yellow tepals and 6 stamens.

GROWTH Perennial.
HEIGHT 50–125cm.
FLOWERS May–Jul.
STATUS Introduced in the 16th century and found in the wild by 1668. Native to Asia and N America, all European plants are sterile and of unknown origin.
ALTITUDE Lowland.

Water Lobelia *Lobelia dortmanna*

Very locally common in lakes, mostly in the uplands, growing in clear, nutrient-poor water up to 2m deep, mostly over acid substrates. ✿ Slender, leafless stems grow from a basal rosette on the lake floor, with *3–10 well-spaced, drooping, pale lilac flowers.*

DETAIL Leaves 3–8cm × 0.2–0.4cm, flattened, strap-shaped, blunt, with 2 large hollows in cross-section; bleed white sap. Flowers with a tiny bract at the base of the short stalk, calyx with 5 teeth, corolla 12–20mm long, 2-lipped, upper lip with 2 narrow lobes, lower lip with 3 slightly larger lobes.

GROWTH Perennial.
HEIGHT 20–70cm (120cm).
FLOWERS Jul–Aug (May–Oct).
STATUS Native.
ALTITUDE 0–745m.

Arrowhead *Sagittaria sagittifolia*

GROWTH Perennial.
HEIGHT Flowering spikes
30–80cm (100cm).
FLOWERS Jul–Aug.
STATUS Native.
ALTITUDE Lowland.

Locally common. Shallow water in rivers, ditches, canals, ponds and lakes, usually alkaline and nutrient-rich (relatively tolerant of enrichment by agricultural run-off and phosphates from sewage). ❀ Arrow-shaped leaves, held upright on long stalks, distinctive. Flowers 3-petalled, 20–30mm across, in erect spikes. DETAIL Aerial leaves 5–20cm long, floating leaves pointed-oval, submerged leaves translucent, strap-shaped (in larger or faster-flowing rivers, may only produce submerged leaves). Flowering spikes with several whorls of 3–5 flowers, upper whorls male, with many purple anthers, lower female, smaller and shorter-stalked, with many carpels in a dense head.

Floating Water-plantain *Luronium natans*

Nationally Scarce. Very local. Still or slow-flowing water, usually nutrient-poor, mostly now upland lakes and lightly-used canals. ❀ Rosettes of egg-shaped, floating or terrestrial leaves and flowers are formed in shallow water or on exposed mud, with the *leaf rosettes connected by runners.* In deep water (to 2m) produces rosettes of submerged leaves but does not flower. DETAIL Leaves long-stalked, oval, blunt, 1–3cm long. Submerged leaves green to whitish, flat, very narrow and strap-shaped, 5–20(60)cm long, tapering to a pointed tip. Flowers long-stalked, 12–18mm across, petals 3, each white or pale lilac with a large yellow spot at the base. Stamens 6. SEE ALSO Frogbit (p. 353), which has vaguely similar leaves and flowers.

GROWTH Perennial.
HEIGHT Runners to 75cm.
FLOWERS Jun–Aug.
STATUS Native (Sch 8).
ALTITUDE 0–450m.

Water-plantain *Alisma plantago-aquatica*

GROWTH Perennial.
HEIGHT 20–100cm.
FLOWERS Jun–Aug.
STATUS Native.
ALTITUDE To 460m.

Common. Shallow water or exposed mud on the edge of still or slow-flowing water, preferring sites that are moderately rich in nutrients. ❀ Leaves plantain-like, with tall, very open, well-branched spikes of small flowers, each of the *3 pinkish-white petals with a small yellow blotch at the base*. The *flowers usually only open after midday*. DETAIL Leaves 8–30cm long, oval, rounded to heart-shaped at the base, long-stalked, all arising from the base of the plant. Flowers 7–12mm across with 3 blunt green sepals and 6 stamens. Seeds disc-shaped, in doughnut-shaped clusters. SIMILAR SPECIES **Narrow-leaved Water-plantain** *A. lanceolatum* is widespread but hard to identify. The leaves are narrower, tapering into the stalk at the base. The *flowers open in the morning* and may stay open all day.

Lesser Water-plantain *Baldellia ranunculoides*

GROWTH Perennial.
HEIGHT 5–20cm (50cm).
FLOWERS Jul–Aug
(May–Sep).
STATUS Native.
ALTITUDE 0–320m.

Near Threatened. Very local. Bare mud and shallow water around flushes, streams, ditches, ponds and lakes with alkaline or brackish water and where the vegetation is kept open by low levels of nutrients, grazing, fluctuating water levels or other disturbance. ❀ Flowers as Water-plantain but on a much smaller plant, with *long, narrow leaves* and *seeds in globular heads like a buttercup*. DETAIL Leaves 4–7cm long, tapering into a long stalk, and typically in a basal rosette; submerged leaves strap-shaped. Flowers 12–15(18)mm across, in star-shaped whorls of 6–20 on erect stems. Seeds in tight clusters *c.* 8mm across. The much scarcer subspecies *repens* is found in W Britain and Ireland. More sprawling, with runners and larger flowers (*c.* 15–22mm across), in whorls of 1–5, and smaller seed heads with fewer, warted seeds.

Frogbit *Hydrocharis morsus-ranae*

Listed as Vulnerable, but locally common. Shallow, still, lime-rich water in grazing-marsh ditches, canals, ponds and sheltered bays in larger lakes. Occasionally introduced. ❀ A free-floating aquatic, with rounded leaves and 3-petalled white flowers, 20–30mm across, the petals with a yellow spot at the base and a crumpled, 'tissue

GROWTH Perennial.
FLOWERS Jul–Aug.
STATUS Native.
ALTITUDE Lowland.

paper' texture. Flowers most profusely in hot summers. DETAIL Main stems submerged, to 50(100) cm long, with rosettes of leaves at intervals. Leaves long-stalked, 16–50mm across, rounded, with a deep notch where the stalk is attached. Male and female flowers usually on separate plants, but a small minority of plants have both male and female flowers, but on separate rosettes. Female flowers solitary, on long stalks growing from the rosette, with 6 styles; male flowers stalked, 1–3 together on a common stalk, with 9–12 stamens (some usually represented by sterile staminodes). Ripe seed rarely, if ever, produced in Britain. SEE ALSO Fringed Waterlily (p. 344), which has similar leaves.

Water Soldier *Stratiotes aloides*

Nationally Rare and Near Threatened as a native, confined to the Norfolk Broads and a few sites in Lincolnshire, now mostly grazing-marsh ditches in alkaline water with moderate levels of nutrients. Widely introduced to canals, ponds and lakes elsewhere.

GROWTH Perennial.
FLOWERS Jul–Aug.
STATUS Native and introduced.
ALTITUDE Lowland.

❀ The cabbage-like *rosettes of stiff, sword-like leaves*, which may cover large areas, and the large, 3-petalled flowers, are distinctive. DETAIL Rosettes grow at intervals along submerged runners. Leaves 15–50cm long, stiff, with spines along the edges, brownish-green underwater, becoming green when emergent. Flowers 30–40mm across. All British plants are female (with 6 styles), and reproduction is entirely vegetative.

A free-floating aquatic, normally suspended just below the water's surface, but rising up to ensure the flowers are above the surface; conversely, the rosettes sink to the bottom to avoid harsh temperatures during the winter, rising up again in the spring as gas-filled new leaves develop.

Branched Bur-reed
Sparganium erectum

GROWTH Perennial.
HEIGHT 50–150cm.
FLOWERS Jun–Sep.
STATUS Native.
ALTITUDE 0–510m.

Common. A narrow band along the edge of still or slow-flowing water, typically where 10–20cm deep (sometimes to 100cm), or forming extensive stands in swamps; favours moderate to high levels of nutrients. ❋ Leaves rather iris-like, with erect stems carrying globular heads of tiny flowers, the lower all-female and opening first, the upper flower heads all male. Flower spikes branched. Named after its spiny fruits. **DETAIL** Leaves 10–25mm wide, keeled (narrower submerged leaves may develop in deep or fast-flowing water). Tepals 6, black-tipped, male flowers with 3–8 long stamens. Fruits variably top-shaped with a narrow beak, *c.*13mm long.

Unbranched Bur-reed *Sparganium emersum*

GROWTH Perennial.
HEIGHT 20–60cm (80cm).
FLOWERS Jun–Sep.
STATUS Native.
ALTITUDE 0–500m.

Local. Similar waterside habitats to Branched Bur-reed but typically further from the shore in water 20–100cm deep. Tolerant of disturbance and may persist in the face of dredging and boat traffic. ❋ A smaller version of the above, flowering stems not branched, with 3–10 male heads all together at the tip of the stem. **DETAIL** Aerial leaves 4–10mm wide, sometimes up to 12mm. May have ribbon-like floating or submerged leaves, to 2m × 4–5(10)mm. Fruits egg-shaped, *c.* 13mm long, tapered to a long, fine beak. **SIMILAR SPECIES Floating Bur-reed** *S. angustifolium* is widespread in the N and W, especially in clear, nutrient-poor mountain lakes. Its flower spikes have 2 (1–3) male flower heads, crowded together, with the leaf-like bract of the lowest female head much longer than the spike. Stem and leaves floating, leaves ribbon-like, without a midrib or keel.

Least Bur-reed
Sparganium natans

Rather local. Sheltered, shallow water on the edge of still or slow-flowing water, usually alkaline, with low-moderate levels of nutrients. ❀ Small, with *narrow leaves*. DETAIL Stems and leaves usually floating, leaves to 30(50) cm long × 2–7mm wide, flat, curved. Flower heads stalked or stalkless; lowest female flower head with its stalk not fused to the stem and its leaf-like bract less than 10cm long, *barely longer than flower spike*. Fruits egg-shaped, tapered to a beak, *c.* 5mm long.

GROWTH Perennial.
HEIGHT Stems 6–80cm.
FLOWERS Jun–Jul.
STATUS Native.
ALTITUDE 0–670m.

PONDWEEDS: FAMILY POTAMOGETONACEAE

Broad-leaved Pondweed *Potamogeton natans*

GROWTH Perennial.
FLOWERS May–Sep.
STATUS Native.
ALTITUDE 0–760m.

Common in a range of still or slow-flowing waters, especially where 1–2m deep. ❀ Leaves floating, *egg-shaped*, to 12cm long, with numerous *faintly translucent parallel veins*. DETAIL Leaf stalk often with a *paler flexible joint* just below the leaf blade. Flowers tiny and greenish, with 4 tepals and 4 stamens, grouped together into pencil-thick flower spikes that are held erect above the water. SIMILAR SPECIES **Bog** Pondweed *P. polygonifolius* and **Fen Pondweed** *P. coloratus* both have similar floating leaves, but the veins are *opaque* and they *lack* the flexible joint below the leaf. Bog Pondweed is common in acid waters and may also colonise *Sphagnum* 'lawns' and bare, wet peat. Fen Pondweed has thinner, less leathery leaves and is scarce in shallow, alkaline waters. 18 other pondweeds of the genus *Potamogeton* occur in the British Isles, but most are fully submerged aquatics without floating leaves.

Bulrush
Typha latifolia

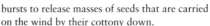

Common in shallow water (to 50cm deep) and the muddy margins of ponds, lakes and ditches, less often by running water; favours sites high in nutrients. Very vigorous: often found in pure stands and can dominate ponds and ditches. ❁ The tiny flowers are grouped into separate male and female spikes, with the female spike being the familiar *fat brown 'cigar'* that persists for months and then bursts to release masses of seeds that are carried on the wind by their cottony down.
DETAIL Stems thick and spongy, especially towards base. Leaves flat, 8–24mm wide. Female flower heads 18–30mm across, *plain-chocolate coloured*. Usually *no gap* between the male and female flower spikes (even when the male flowers have fallen, their scars are obvious); rarely a gap of up to 25mm.

GROWTH Perennial.
HEIGHT 150–300cm.
FLOWERS Jun–Aug.
STATUS Native.
ALTITUDE 0–500m.

Lesser Bulrush
Typha angustifolia

Very similar to Bulrush, but more local, growing in deeper water (at least 50cm deep) and often forming a band along the water's edge, best seen from a boat. Tolerates lower levels of nutrients, and more frequent near the coast. **DETAIL** As Bulrush but leaves *narrower* (3–8mm wide, rarely to 10mm), female spikes *paler, milk-chocolate brown* and *slimmer* (13–25mm across), with a 30–80mm gap between male and female spikes (extremes 5–120mm).

GROWTH Perennial.
HEIGHT 150–300cm.
FLOWERS Jun–Aug.
STATUS Native.
ALTITUDE Lowland.

TREES & SHRUBS

Trees and shrubs are all woody perennials. With trees especially, the various stages of growth from a sapling to full maturity involve huge changes in size and shape, and the conditions in which they grow will also have an impact on their form – an oak tree growing in a crowded forest will assume a very different shape to a specimen tree in parkland with plenty of space around it. Thus, while it is possible with experience to identify many trees from a distance based on their silhouette, for most people this will only ever be a good guess that needs to be confirmed by an examination of leaves, flowers, barks, twigs and buds at close range. The identification of trees and shrubs suffers from another very significant complication: hundreds of non-native species have been planted over the years, sometimes in large numbers. Even a supposedly 'native' hedge may be full of species and varieties that originate from continental Europe. It is a huge subject to treat comprehensively, and only the native species are covered here, plus a few of the commonest introductions.

Scots Pine *Pinus sylvestris*

Abundant as a native in the 'Caledonian Pinewoods' of the Scottish Highlands, in both pure stands and mixed woodland. Commonly planted elsewhere, becoming naturalised on heaths and bogs. Favours stony or sandy, acid ground. ❁ Mature trees have relatively *short, paired needles,* a *reddish-orange upper trunk* and an open, flat-topped canopy.

HEIGHT To 30m (36m).
FLOWERS May–Jun.
STATUS Native & introduced.
ALTITUDE 0–1160m as a native.

Younger trees are pyramidal in outline. **DETAIL** Leaves evergreen, in *pairs* on short side shoots, dark bluish-green, often twisted, 2–8cm long (10–14cm on saplings). Male cones dull yellow, in dense clusters, shedding pollen in late spring; female cones solitary, dark purplish-red in late spring, becoming green in 2nd summer and grey-brown and woody in 2nd autumn, 2.5–7.5cm long, the cone scales *lack prickles on the outer face*; seeds shed in the cone's 3rd spring. **SIMILAR SPECIES Corsican** Pine and **Austrian Pine** *Pinus nigra* (sspp. *laricio* and *nigra*) are very similar but have longer leaves, 8–12cm in Austrian Pine and 10–18cm in Corsican Pine, and as mature trees they have much *duller greyish-brown bark,* including the upper trunk. Both are widely planted, and Corsican Pine is favoured over Scots Pine for forestry in S Britain; also found self-sown, especially on heaths and dunes.

mature female cone & male flowers

female flower

male cones

Juniper
Juniperus communis

Locally common in the N, in woodland and on moorland and rocky slopes on acid soils. Rare in S England on chalk downland and wind-pruned coastal heaths. ✿ Size very variable. A prostrate shrub (especially in higher or more exposed sites), an erect, often conical bush, or a small tree; often appears stunted, old and gnarled. *Leaves in clusters of 3,* needle-like, green, with the *edges rolled down and under to cup the whitish underside*, berries green, ripening to *blue-black* with a whitish bloom. DETAIL Evergreen, leaves 4–20mm long, prickly. Male and female cones usually on different plants, tiny, greenish or pale brown. Male cones 2–3mm long, producing crowded yellow anthers. Female cones 2mm, central scales enlarging to produce the 'berry', initially green but turning blue-black at start of 2nd year and ripening Sep-Oct of 2nd/3rd year.

HEIGHT To 7m (16m).
FLOWERS May–Jun.
STATUS Native.
ALTITUDE 0–1200m.

Yew *Taxus baccata*

Native to mixed woodland, mostly as scattered trees but may form pure stands, usually on well-drained soils on chalk or limestone. Also extensively planted in churchyards, parks and gardens and often bird-sown into local woods. ✿ A large, evergreen bush or tree. Mature trees are well branched with short, massive trunks (sometimes several) and dense foliage; they cast heavy shade. Leaves soft to the touch, in 2 rows, with *no white stripes*. Cones tiny; on female trees they develop into *fleshy red berries*. DETAIL. Leaves abruptly fine-pointed, 10–30mm long, glossy dark green above, paler below; arranged spirally, obviously so on erect leaders, but on many shoots the stalks twist to align the leaves into 2 rows. Cones solitary at the base of a leaf, male and female on separate trees. Male cones 2–3mm across, golden-brown, opening to reveal a multi-headed stamen. Female cones initially green, 1.5mm across, but inner scale swells to form a red berry up to 10mm across, open at the top and enclosing a single large seed (the 'berry' is an aril – a highly-modified scale). 'Irish Yew', often planted, is cylindrical and much more compact.

HEIGHT To 28m.
FLOWERS Feb-Apr.
STATUS Native.
ALTITUDE To 470m.

male cones

Box *Buxus sempervirens*

Nationally Rare as a native, restricted to about 10 sites in S England (inc. Box Hill, Surrey, and Boxwell, Glos), in scrub and woodland on chalk or limestone. Much more widely planted in woods and often self-sown. ❁ Evergreen shrub or small tree with dense foliage of small, oval-oblong, leathery leaves and sprays of tiny yellowish flowers. **DETAIL** Stem green, 4-angled, hairless or downy. Leaves opposite, short-stalked; 10–25mm long, rounded or notched at tip. Flowers in

HEIGHT To 5m (11m).
FLOWERS Mar–April.
STATUS Native & introduced. Cultivated since Roman times and popular for garden hedging and topiary.
ALTITUDE Lowland.

clusters at base of leaves: a single female flower at the centre (sepal-like bracteoles and 3 stigmas) surrounded by 5–6 male flowers (4 sepals and 4 yellow anthers). Fruit a woody pea-like capsule, 7–11mm long, green ripening to pale brown, the 3 styles persisting as spreading 'horns'.

BOG MYRTLE: FAMILY MYRICACEAE

female catkins, July

female catkins

male catkins

Bog Myrtle
Myrica gale

Locally common. Bogs, wet heath and moorland, lightly shaded wet woodland and, very locally in E Anglia, fens, always where there is some movement of groundwater. ❁ A small deciduous shrub with a lovely

HEIGHT To 1.5m (2m).
FLOWERS Apr–May.
STATUS Native.
ALTITUDE To 520m.

resinous, balsamic scent when the *greyish-green foliage* is rubbed. Flowers in small, stiff catkins. Suckers to form dense thickets. **DETAIL** Current year's twigs variably hairy, purplish-brown. Buds 1.5–2mm, oval, blunt, orange- or reddish-brown. Leaves alternate, short-stalked, 2–6cm long, long-oval, narrower towards the base and finely toothed towards the tip, sparsely hairy above when young, more or less hairy below, with *yellow, stalkless, scented glands both sides.* Catkins clustered towards the tip of the previous season's twigs, opening before the leaves. Male and female flowers usually on separate bushes (sometimes separate male and female flowers on the same plant), no petals or sepals but a reddish-brown catkin scale protects each flower; female catkins 4–5mm long with 2 red styles, male catkins 7–15mm long, with 4 reddish-yellow stamens. Fruit a narrowly winged nut.

HEIGHT 10–50cm.
FLOWERS Apr–May.
STATUS Native in a few areas, but mostly of garden origin.
ALTITUDE To 365m.

Mountain Currant *Ribes alpinum*

Nationally Scarce. Native to limestone woods and rocky places in the hills of N England (especially the Peak District), often forming trailing 'curtains' over rocks and cliffs; bird-sown from cultivation elsewhere. Flower clusters held *erect*. **DETAIL** Leaves smaller and more deeply 3-lobed than Red Currant, not net-veined, flowers slightly smaller (5–6mm wide), with bracts *longer* than flower stalks. Dioecious, with male and female flowers on different plants.

Gooseberry *Ribes uva-crispa*

HEIGHT To 100cm.
FLOWERS Late Mar–May.
STATUS In cultivation from the 13th century but not recorded in the wild until 1763. Now widely naturalised, both as a relict of cultivation and bird sown.
ALTITUDE 0–380m.

Scattered in open, usually fairly dry, woodland, scrub and hedgerows. ❀ Sprawling, well-branched, *spiny* deciduous shrub. Flowers small and inconspicuous, fruit the familiar greenish-yellow berry. Commonly cultivated. **DETAIL** Leaves 20–50mm long, 3–5 lobed, toothed, net-veined, usually hairy, with 1 or 3 spines erupting from the stem at the base of the stalk. Flowers in groups of 1–3, drooping, 6–12mm across, hairy; sepals purplish-brown, often bent backwards, petals tiny, white; berry 10–20mm across (much larger in cultivated varieties), sometimes reddish, often bristly-hairy, ripening by mid July.

Red Currant *Ribes rubrum*

HEIGHT 100–200cm.
FLOWERS Apr–May.
STATUS Small-fruited plants
in wet woodland may be
native; larger-fruited forms
occur around abandoned
habitation or are bird-sown
from gardens.
ALTITUDE To 455m.

Fairly common in
woods and hedgerows,
especially in damp
places. ❀ Small
deciduous shrub with dull,
soft, 3–5 lobed leaves
and drooping clusters
of yellowish-green

flowers (often washed purple). Commonly cultivated for
its edible bright red berries. DETAIL Leaves to 8cm wide,
hairless (or sparsely hairy on veins below), net-veined, the
underside with a few reddish oil-glands that are not scented when crushed.
Flowers hairless, *c.* 8mm across, the saucer-shaped receptacle at the base
extended into 5 broad sepals and 5 minute petals. Berries 6–10mm across.

Black Currant *Ribes nigrum*

HEIGHT To 200cm.
FLOWERS Apr–May.
STATUS Introduced to
cultivation from Europe in the
early 1600s, first recorded
in the wild in 1660 and now
widely naturalised. May be
native in E Anglia.
ALTITUDE 0–390m.

Rather local in wet woodland. ❀ As Red Currant but flowers
more cup-shaped and more colourful, berries ripening to
black and leaves aromatic. Commonly cultivated.
DETAIL As Red Currant but leaves larger, sometimes
shiny above, main lobes more pointed, undersides
with tiny, stalkless, yellowish oil-glands that, when
crushed, smell of black currant. Receptacle at base
of flower bowl-shaped, hairy, petals closer in size to
sepals. Berries 10–15mm across.

Flowering Currant
Ribes sanguineum

Common in gardens and used for ornamental hedging, spreading to woods, hedgerows and waste ground. ❀ Deciduous shrub with *drooping flower clusters.* DETAIL Foliage and flowers glandular-hairy, with a fruity scent when crushed. Leaves 3–5 lobed, dull green, hairy below. Flowers 6–10mm across, with 5 pinkish sepals, *fused into a tube at the base,* and 5 short, erect, whitish petals. Fruit a berry, 6–10mm across, purplish-back with a white bloom.

HEIGHT To 2.5m.
FLOWERS Mar–May.
STATUS Introduced to gardens in 1826 and first recorded in the wild by 1916 (N America).
ALTITUDE To 300m.

FAMILY BERBERIDACEAE: BERBERIS

Barberry
Berberis vulgaris

Uncommon. Old hedgerows, woodland edges and waste ground. ❀ *Spiny, deciduous shrub* with *drooping clusters of yellow flowers* followed by bunches of *elongated* red berries. DETAIL Twigs grooved, with *3-forked spines at base of leaf stalks.* Leaves alternate, short-stalked, 2–6cm long, thin and papery in texture with *finely saw-toothed margins, the teeth terminating in fragile, hair-like spines;* moss-green but may be flushed purple. Flowers 6–8mm across with 3 concentric rings of yellow sepals (the outermost tiny) surrounding 3 yellow petals; stamens 6. Fruit *c.* 10mm long.

HEIGHT To 3m.
FLOWERS May–Jun.
STATUS Ancient introduction? Recorded from a Neolithic site but also cultivated in medieval times for its medicinal uses and once widely used as hedging.
ALTITUDE To 700m.

Oregon-grape
Mahonia aquifolium

Planted in woodland as game cover or as an ornamental in parks, churchyards and gardens; bird-sown more widely in scrub and hedgerows, most frequently near houses. ❀ Low, suckering, evergreen shrub with *spiny leaves* and *dense clusters of yellow flowers* followed by blue-black berries with a whitish bloom. DETAIL Leaves with 5–9 glossy leaflets, 3–8cm long, with 5–15 spiny teeth on each side. Flowers fragrant, small, with concentric rings of yellow sepals (the outermost tiny) surrounding 3 yellow petals.

HEIGHT To 1.5m.
FLOWERS Dec–May.
STATUS Introduced from N America in 1823 and recorded from the wild by 1874.
ALTITUDE Lowland.

Dyer's Greenweed *Genista tinctoria*

Fairly common but local. Rough grassland, verges, grassy heaths and scrub, usually on alkaline to slightly acid, *clay* soils. ❀ *Low growing* deciduous shrubby pea, *spineless*, with long clusters of bright yellow flowers; recalls a miniature Broom. DETAIL More or less hairless. Stems green. Leaves shiny green both sides with hairy margins, stalkless, *c.* 2–4cm long, strap-shaped, pointed. Flowers 10–15mm long. Pods 15–30mm long, flat. Subspecies *littoralis*, a prostrate form, is found on the coast of SW England. Leaves broader, pods sometimes downy. SIMILAR SPECIES **Hairy Greenweed** *G. pilosa* has leaves, flowers and pods all very downy. Rare, confined to W Cornwall and W Wales.

HEIGHT 20–60cm (100cm).
FLOWERS Jun–Jul (Aug).
STATUS Native.
ALTITUDE Lowland.

Petty Whin *Genista anglica*

Uncommon and declining, listed as Near Threatened. Damp, grassy heathland; in the lowlands mostly at the transition between dry heath and wet, valley bogs; in the uplands in damp, heathy, unimproved pastures. ❀ *Small, slender, spiny shrub*, often straggling through other vegetation. *Only obvious when in flower*. DETAIL More or less hairless. Stems green when young, *thin and wiry*, with *long, slightly arched spines*. Leaves 2–9mm long, long-oval. Flowers *pale yellow*, 7–10mm long, in clusters at the shoot-tips. Pods 12–20mm long, swollen. Occasionally lack spines (var. *subinermis*).

HEIGHT 50cm (100cm).
FLOWERS Late Apr–Jun.
STATUS Native.
ALTITUDE Mostly lowland, but to 730m.

Broom *Cytisus scoparius*

Common. Verges, scrub, heathland etc., on dry, acid, usually sandy soils. ❀ An erect, spineless, shrubby pea, with long, slender, green shoots. DETAIL Stems 5-angled, silky-hairy in 1st year. Lower leaves stalked, with 3 leaflets, each 6–20mm long, hairy; upper leaves undivided, stalkless. Flowers 15–20mm long, 1–2 together, sometimes orange-yellow; calyx and flower stalks usually hairless. Pods 25–50mm long, flattened, black, long-hairy on edges. A prostrate form (subspecies *maritimus*) is found on cliffs in SW England and W Wales.

HEIGHT To 2.5m.
FLOWERS Apr–Jul.
STATUS Native.
ALTITUDE To 640m.

Gorse *Ulex europaeus*

Very common. Rough grassland, hedges, scrub, cliffs and dunes, heaths and waste ground, usually on acid soils. ❀ A densely spiny, evergreen shrub, sometimes very invasive, forming impenetrable thickets. Flowers strongly coconut-scented, structure typical of the pea family. Gorse can be found in bloom throughout the year, but the *most plants flower Apr–May; summer and autumn flowers are accompanied by numerous buds.* DETAIL Seedlings have clover-like

HEIGHT To 3m.
FLOWERS Jan–Dec.
STATUS Native.
ALTITUDE 0–640m.

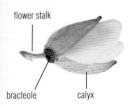

flower stalk

bracteole   calyx

leaves, but older plants are leafless, with green shoots, and *deeply grooved* green spines, 10–30mm long. Calyx split into 2 lobes, 10–20mm long, pale greenish, hairy, with *2 small bracteoles at the base, 1.5–4mm wide, at least twice the width of the flower stalk.* Standard petal 12–18mm long. The pods open with a pop.

Western Gorse
Ulex gallii

Locally abundant. Acid heathland, rough grassland, scrub, waste ground. ❀ Smaller and more delicate than Gorse, but not as petite as Dwarf Gorse. *Only flowers in late summer*, with many dead flowers (but *no flower buds*) through the winter. Virtually no overlap in range with Dwarf Gorse. DETAIL Spines mostly 10–25mm long, stout, *moderately grooved.* Flowers with calyx 8–13.5mm long, *bracteoles tiny, 0.6–0.8mm wide, less than twice width* of flower stalk, standard 11–17mm.

HEIGHT 0.3–1m (2m).
FLOWERS Mid Jul–Oct (Nov).
STATUS Native.
ALTITUDE 0–670m.

bracteole

dried-up in mid winter

Dwarf Gorse *Ulex minor*

Locally abundant. Acid heathland and rough grassland. ❀ Picked out by its weaker, more sprawling stems and shorter, weaker spines. *Flowers late summer.* DETAIL Spines mostly 8–15mm long, *weakly grooved.* Calyx 6–10mm, bracteoles tiny, 0.4–0.6mm wide, standard 7.5–12.5mm. Almost no overlap in range with Western Gorse, but where they meet, to confirm identification, measure 10 flowers and use mean.

HEIGHT 0.1–0.5m (1.5m).
FLOWERS Mid Jul–Oct.
STATUS Native.
ALTITUDE Lowland.

Tree Lupin
Lupinus arboreus

Scattered; may be locally abundant on dunes and other coastal grassland, also inland on verges and waste ground. ❀ Tall semi-evergreen shrub with *elongated clusters of yellow flowers*, often tinged blue (sometimes blue or whitish). DETAIL Stems woody, sparsely hairy. Leaves *palmately* divided into 5–12 narrow leaflets to 6cm long, hairless above, silky-hairy below. Flowers 15–20mm long, lower lip of calyx 7–11mm. SIMILAR SPECIES **Russell Lupin** L. × regalis, with stems not woody, dying down in winter, leaflets over 6cm, lower lip of calyx 5–8mm, and flowers of various colours, is the common garden escape.

HEIGHT To 2m.
FLOWERS Jun–Jul (Sep).
STATUS Introduced to cultivation in 1793 and recorded from the wild by 1945 (California).
ALTITUDE Lowland.

Cherry Laurel
Prunus laurocerasus

ROSES & ALLIES: FAMILY **ROSACEAE**

Commonly planted in parks and gardens, and naturalised or bird-sown in woodland. ❀ Robust *evergreen* shrub with *large, leathery, glossy green leaves*. Flowers in erect clusters, fruit a blackish berry. DETAIL Young twigs and leaf stalks *green*.
Leaves hairless, oval-oblong, 5–20cm long with a *few well-spaced, tiny teeth* along the margins, almond-scented when crushed.
Flowers c. 7–9mm across, petals 5, stamens long, numerous; fruits 10–12mm across.

HEIGHT 2–3m (9m).
FLOWERS Late Mar–May.
STATUS Introduced prior to 1629 and recorded from the wild by 1886 (SE Europe).
ALTITUDE Lowland.

Wall Cotoneaster *Cotoneaster horizontalis*

Widely naturalised, especially in rocky places. ❀ Deciduous shrub, arching to prostrate, but very often supported against a wall. Twigs *all in one plane*, in a distinctive *herring-bone pattern*. Flowers small, pinkish-white, in clusters of 2–3, berries orange-red. DETAIL Leaves 6–12mm long, glossy green above, paler, near hairless below. Anthers white, berries 4–6mm across. SIMILAR SPECIES Over a dozen species of cotoneaster, all originating from the Himalayas and China, are well-naturalised on waste ground, verges, banks and walls, and in scrub (especially on chalk). Some are invasive, and may be a threat to native vegetation. A large and confusing group, many are hard to identify. Some are evergreen, some deciduous, but all have undivided leaves, white flowers, either solitary or in clusters, and the fruits are usually red, but in some species are black, orange or yellow.

HEIGHT 1m (2m).
FLOWERS Apr–Jul.
STATUS Introduced to cultivation around 1879, first recorded from the wild in 1940 and still spreading.
ALTITUDE Lowland.

Bramble
Rubus fruticosus agg.

HEIGHT Stems to 4m.
FLOWERS May–Oct.
STATUS Native.
ALTITUDE 0–490m.

Abundant and familiar thorny shrub. Woodland, scrub, hedgerows, heaths, waste ground, especially on acid soils. ● Leaves deciduous or wintergreen. The white or pale pink flowers are beloved of butterflies, the fruits, Blackberries, beloved of human foragers. DETAIL Stems biannual or perennial, producing flowers in the second year, 5-angled (at least when young), sometimes arching to the ground and rooting at the tip to produce extensive thickets. Leaves with 5 leaflets (sometimes 3 or 7). Stipules fused to the leaf stalk at the base. Flowers 20–32mm across, with 5 petals, numerous carpals and many stamens, receptacle conical. Fruits red, ripening to glossy black, with many 1-seeded segments.

The 'Bramble' is actually a complex of 334 apomictic microspecies. In apomictic plants seed develops directly from the ovule, without the need for fertilisation, and the resulting plants are clones of the seed-parent. (Even in self-pollinated flowers, the ovules (eggs) are fertilised by pollen (sperm) and there is some genetic mixing.) In apomictic plants minor variations that result from occasional chance mutations are passed on unchanged, and this can lead to populations of genetically identical plants that differ in very minor characters – microspecies. As Blackberries are avidly eaten by birds and mammals, the seeds can spread widely and some of the microspecies of Bramble may be found over a wide area (even spreading from Britain to Europe or vice-versa in the stomach of migrating birds). The variations are very subtle, however, and very few botanists can confidently separate them.

Dewberry
Rubus caesius

HEIGHT Stems to 100cm plus.
FLOWERS Jun–Aug.
STATUS Native.
ALTITUDE 0–320m.

Locally common, especially on boulder clays and other alkaline soils: verges, ditches, hedgerows, woodland rides, scrub, rough grassland and dunes. ● Low-growing, deciduous, early-flowering bramble, with relatively large white flowers and *distinctive fruits*. DETAIL Stems slender, low-arching or sprawling to form pillowed tangles; round, hairless, but with a few short (1–2mm), weak, curved prickles and a *whitish bloom*. Leaves divided into 3 leaflets (side leaflets stalkless). Flowers 2–3cm across, with broad white petals. Fruits with a few, large segments, blue-black with a *characteristic strong whitish bloom* (the only sure identification feature); edible but rather insipid.

Cloudberry
Rubus chamaemorus

Locally common on wet, acid, peaty moors and bogs in the Pennines, Cheviots and Scottish Highlands (but rare in the Lakes and Wales). ✿ Low-growing, the distinctive leaves may carpet the ground, but shy to flower in some areas and, when produced, the fruits are often eaten by sheep. **DETAIL** Stems erect, whitish, hairy but *lacking prickles*, growing from extensive creeping rhizomes and dying down in winter. Leaves few, often solitary, 4–7cm across, with 5–7 shallow lobes, wrinkled and sparsely hairy. Flowers solitary, 20–30mm across, with 4–5 white petals; male and female on different plants. Fruit edible, red, turning *orange* when ripe, with 4–20 large segments.

HEIGHT 5–20cm.
FLOWERS Mid May–Jul.
STATUS Native.
ALTITUDE Usually 600–1160m, but down to 90m in N Wales.

Stone Bramble *Rubus saxatilis*

HEIGHT Stems 10–40cm.
FLOWERS Mid May–Aug.
STATUS Native.
ALTITUDE 0–975m.

Locally common in the N and W. Open rocky woodland and shady, N-facing rocky places and scree, usually on alkaline soils, occasionally also rocky heaths and river shingle. ✿ Small, low growing, *small-flowered* bramble, with slender stems and *very weak prickles* (sometimes none). **DETAIL** All stems *annual*, growing from the rootstock in Apr and dying down in Oct. Stems often purplish, hairy; flowering stems short and erect, other stems prostrate, with short, trailing runners that may root at the tip. Leaves divided into 3 oval leaflets, downy beneath; stipules oval to strap-shaped, 5mm long, *arising directly from the main stem* (fused at the base to the leaf stalk in Bramble). Flowers few, 8–15mm across, sepals often purplish, petals white, held erect, narrow, 3–5mm long (no longer than sepals), withering quickly; *receptacle flat* (conical in Bramble). Fruit red, with *1–6 relatively large segments*.

sepals often purplish

Raspberry *Rubus idaeus*

Common everywhere, but most abundant in the N and W. Woods, heaths, scrub, hedgerows, verges and rocky places, often on damp soils, also, usually when bird-sown from gardens, waste and disturbed ground. ❋ Deciduous, perennial shrub, producing the familiar delicious fruit. Stems ('canes') erect, with a whitish bloom and a variable number of straight, weak prickles. Often suckers from the roots to form thickets.

HEIGHT Stems to 1.5m (2.5m).
FLOWERS May–Aug.
STATUS Native.
ALTITUDE 0–855m.

DETAIL Stems little-branched, variably hairy, growing from a tough rootstock; biennial, they produce flowers in their second year. Leaves with 3–5(7) oval leaflets, *white-woolly below;* stipules thread-like. Flowers few, in clusters, often drooping, *c.* 10mm across, petals narrow, *erect;* some plants have only male flowers and produce no fruit. Fruit red, rarely yellow or white, hairy, easily pulling away from the receptacle when ripe.

Field Rose *Rosa arvensis*

Fairly common. Woodland rides, scrub, hedgerows and verges, avoiding very acid soils. ❋ Stems *weak and trailing,* forming low thickets or climbing over other vegetation and hanging down, *wine-red on the sunlit upperside and contrastingly green below;* flowers always creamy-white, with the *projecting styles fused into a slender, hairless, pin-like column.*

HEIGHT Stems to 1m (2m).
FLOWERS Jun–Jul.
STATUS Native.
ALTITUDE 0–410m.

DETAIL Stems with slender, curved prickles. 5–7 leaflets, more or less *hairless.* Flowers 3–5cm across, in groups of 1–6. Sepals often *purplish, not lobed* (outer sepals may have a few small lobes), falling early. Leaf stalks, midribs, and flower stalks with scattered stalked glands. Hips small, with

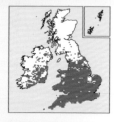

column

disc

a *flat* disc. SIMILAR SPECIES Short-styled Field Rose *R. stylosa* also has the styles fused into a column, but this is *shorter* than the stamens, and the styles soon separate. Disc *conical* (solid when cut through), sepals *all lobed,* prickles stoutly *triangular,* and leaflets dark green, *well spaced, long, sharply pointed.* Well-drained alkaline soils in S England and S Wales.

Dog Rose *Rosa canina*

By far the commonest wild rose
in most of England and Wales.
Rapidly colonises disturbed
ground, but avoids the most
acid soils. ❀ A climber, with long
arching stems and strongly curved
prickles. Flowers white or pale
pink, *sepals lobed, spreading at first
but turning down against the hip
and falling before it ripens.* DETAIL
Leaves glossy, *hairless,* or sparsely
hairy below, with 2–3 pairs of
leaflets. Flowers 3–6cm across, in
groups of 1–6, stalks hairless or

HEIGHT Shoots to 3m (4m).
FLOWERS Jun–Jul.
STATUS Native.
ALTITUDE 0–550m.

with a few gland-tipped hairs. Stigmas in a small conical tuft. Hips with flat
or convex discs, variable in size and shape. Oil glands absent or restricted to
a few on the leaf veins and teeth, stipules and bracts. SIMILAR SPECIES **Hairy
Dog Rose** R. *caesia* Common from the S Midlands northwards (absent S
England and E Anglia). Stems often wine-red on the sunlit side, flowers
short-stalked (the stalks hidden by the large bracts), stigmas woolly, in a
large dome-shaped cluster that conceals the disc, sepals erect, hips large.
Subspecies *vosagiaca* (**Glaucous Dog Rose**) is hairless, leaves contrastingly
blue-grey below. **Round-leaved Dog Rose** R. *obtusifolia*, scarce and local
in S England, has rounded, moderately hairy leaflets, double-toothed, with
small dark red glands on the teeth, hairless flower stalks, and large, deeply-
lobed sepals that fold down, hiding the hips, before falling early.

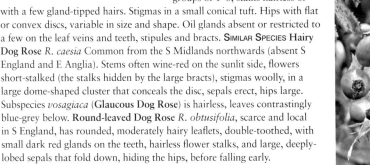

Burnet Rose *Rosa spinosissima*

HEIGHT 10–50cm (120cm).
FLOWERS May–Jul.
STATUS Native.
ALTITUDE 0–730m.

Very locally common. Mainly coastal, on dunes and cliff-top heath
and scrub. Scarce inland, on sandy, mildly acid heathland, also scrub,
hedgerows and mountain ledges on chalk and limestone; increasingly bird-
sown from gardens. ❀ Delicate, low-growing rose, often suckering to form
dense patches. Stems with *abundant slender prickles and bristles. Hips
purplish-black.* DETAIL 3–5 pairs of *small* leaflets, hairless (or sparsely hairy
below), with few or no glands (stipules, leaf stalk and midrib may have
some glands). Flowers solitary, creamy-white (rarely pale pink), 2–4cm
across; no bract at base of flower stalk. *Sepals unlobed, pointed, remaining
on the hip, erect, until it rots.*

Sweetbriar
Rosa rubiginosa

HEIGHT Stems to 2m.
FLOWERS Jun–Jul.
STATUS Native.
ALTITUDE Lowland.

sepals

Fairly common in open scrub and hedgerows, usually on chalky soils. ❀ A free-standing, pink-flowered shrub rose. *The underside of the leaflets and the leaf and flower stalks have many obvious sticky, stalked glands; when rubbed these give off a delicious apples-and-cinnamon scent.* DETAIL Stems erect, straight, with stout, *strongly curved* prickles of various sizes, often mixed with robust bristles, especially towards flowers. Leaflets 5–7, hairy on the veins, specially on undersides, and with gland-tipped teeth. Flowers 2.5–4cm across, in clusters of 1–3, stigmas hairy, sepals slightly lobed, erect to spreading, usually remaining till the fruit ripens. Hips hairless or sparsely glandular-hairy. SIMILAR SPECIES **Small-flowered Sweetbriar** R. *micrantha* has the same sticky, scented glands, but with weaker, arching, often scrambling stems, with the prickles more uniform in size (and *no bristles*), flowers 2–3.5cm across, with hairless stigmas, and sepals mostly bent downwards and falling before the fruit ripens; hips small, urn-shaped. A southern species, on chalky to mildly acidic soils.

Harsh Downy Rose *Rosa tomentosa*

HEIGHT Stems to 3m.
FLOWERS Jun–Jul.
STATUS Native.
ALTITUDE Lowland.

Local. Hedgerows, woodland and scrub on alkaline to mildly acidic soils. ❀ Downy roses have matt, often rather grey- or blue-green leaflets, *hairy* and often velvety to the touch, *especially below*, with dense, *matted hairs* and small, *inconspicuous* stalkless

glands amongst the hairs on the leaf undersides.
Flower stalks glandular-hairy, as are the leaf stalks and midrib, sepals and often hips, while each tooth on the double-toothed leaf margin is tipped by a gland. All these oil glands are very small, at most only slightly sticky, and produce a rather *medicinal scent*. DETAIL *Climbing*, with long, arching stems. Flowers pink (less often white), 3–5cm across. Sepals lobed, *falling before the hip ripens fully*. Hips *long-stalked* (1.5–3.5cm), oval, *opening in disc small*, 20% of total width (cut a hip lengthways with a sharp knife and pull out the styles to see). SIMILAR SPECIES **Sherard's Downy Rose** R. *sherardii* is found in the N and W. An erect *shrub*, with smaller flowers (2.5–4cm), larger hips with a *shorter stalk* (1–1.5cm), a *large, domed mass of stigmas covering much of the disc* and a *larger opening* (33%); sepals more upright, not falling *until the hip is almost ripe*. **Soft Downy Rose** R. *mollis* is found in Wales and N England. Also an erect shrub, but suckering, with slender, *straight* prickles. Leaves *softly hairy both sides*. Flowers mostly *deep pink*. Hips large, *short-stalked* (0.5–1.5cm), the disc *almost covered* by a domed mass of stigmas and with a *large opening* (40–50%). Sepals erect, *more or less unlobed*, attached to the hip by a thick collar and remaining until it rots.

Bird Cherry
Prunus padus

Fairly common in damp woodland and scrub, shady rocky places and sometimes hedgerows. ❀ A deciduous shrub or small tree, spreading by seed and suckers. Flowers sweetly scented, in *horizontal to drooping clusters of 10–40*. DETAIL Bark smooth, grey-brown, releasing a pungent scent when rubbed. Twigs dark reddish-brown, shiny, hairless. Buds hairless. Leaves 5–10cm, very finely toothed, hairless or with tufts of white hairs on the midrib below, with a pair of small yellow-red glands near the tip (as Wild Cherry). Flower stalks 8–15mm, petals toothed, 6–9mm long. Fruits a shiny black sphere, 6–8mm across, bitter tasting.

HEIGHT 3–5m (19m).
FLOWERS May–Jun (after leaves).
STATUS Native, also widely planted.
ALTITUDE 0–650m.

Crab Apple *Malus sylvestris*

Very scattered (often single trees), in old hedgerows, scrub and ancient woodland. ❀ The clusters of pale pink blossom and yellowish-green fruits resemble cultivated Apples, but twigs often *spiny*, and *fruits small and very tart*. DETAIL Twigs may be sparsely hairy when young, but soon *hairless*. Leaves 30–50mm long, oval; leaves and leaf stalk *hairless when mature*. Flower stalk and outside of calyx hairless. Petals 13–30mm, stamens yellow, fruits *c.* 20–30mm across. SIMILAR SPECIES Descendants of the cultivated **Apple** M. *pumila* are *common* in hedgerows and on waste ground – bird-sown or wherever apple cores have been discarded. *Never* spiny, larger, with leaves to 150mm long. *Hairy* on the *underside of the leaves, leaf and flower stalks* and *calyx*, even when mature. Twigs woolly when young, but soon more or less hairless. Fruits to 120mm, but often *much smaller*, yellowish and sour, *inviting confusion* with the real Crab. Separation from Crab Apple may be difficult, as trees that show intermediate characters are frequent. Genetic evidence suggests, however, that intermediates are unlikely to be hybrids, and if these 'intermediates' originate from cultivated Apples, the true Crab Apple may well be rather rare.

HEIGHT To 6m (17m).
FLOWERS Apr–May.
STATUS Native.
ALTITUDE 0–440m.

HEIGHT 1–4m.
FLOWERS Late Mar–May.
STATUS Native.
ALTITUDE 0–500m.

Blackthorn
Prunus spinosa

Abundant in hedgerows, scrub and on rocky slopes. Commonly planted in the past as hedging. ❀ Variably *spiny* shrub or small tree, with *masses of white blossom* in the early spring. Sets abundant fruit (sloes), *blue-black* spheres 8–15mm across with a distinct white bloom that are notoriously *astringent. Suckers strongly* and can form dense thickets.
DETAIL Twigs and branches *blackish*. Short straight side shoots become thorns, young twigs downy. Leaves finely toothed, *small* (2–4cm long), more or less hairy. The flowers usually appear before the leaves, *petals 5–8mm, stamens longer, very prominent*.

Wild Cherry
Prunus avium

Scattered on the margins of deciduous woodland and in hedgerows, also very commonly planted. Suckers to produce groups of clones. ❀ Small to medium-sized tree. The bark (reddish-brown, smooth and shiny on younger specimens) peels to produce characteristic *horizontal lines*. Leaves finely toothed, on long stalks that have 1–2 pairs of *tiny red glands near the tip*. Flowers conspicuous, grouped into *large 'pom-poms'*. Fruits long-stalked, small versions of the familiar cultivated cherry, blackish or dark red, occasionally yellow. DETAIL Twigs hairless. Leaves 6–16cm long, sparsely hairy below. Flowers *long-stalked*, in clusters of 2–6, with a group of large bud scales at the base of the cluster. Petals 8–15mm, arising from a reddish, cup-shaped calyx that is narrowed above, with the sepal-teeth bent downwards; ovary hairless. Fruits 9–12mm across, sweet.

HEIGHT To 29m.
FLOWERS Late Mar–May (just before leaves).
STATUS Native.
ALTITUDE 0–400m.

red glands

Cherry Plum
Prunus cerasifera

A suckering shrub or small tree, commonly used as hedging, also a street tree. ❀ Frequently mistaken for Blackthorn, but generally *lacks spines*. The *first plum to bloom*, making it easy to spot in the early spring; the flowers open with or before the leaves. DETAIL First-year twigs *green* (often purple-brown above), *shiny* and *hairless*. Buds *hairless* except at extreme base. Leaves 3–7cm long, very finely toothed, hairless above, sparsely hairy on midrib below. Flowers 1–2(3) together, *evenly scattered* along the branches, *petals 7–11mm*. Fruit spherical, 20–30mm across, *uniformly* dark red or dark yellow, scarcely bloomed, with a smooth rounded stone, fairly sweet; ripens Jul–early Aug, but often fails to set fruit.

HEIGHT To 8m.
FLOWERS Late Jan–Apr.
STATUS Introduced by the 16th century cultivation by the 16th century (SE Europe and SW Asia)
ALTITUDE Lowland.

Wild Plum *Prunus domestica*
(including **Bullace**, **Damson** and **Gages**)

| Wild Plum | Bullace | Bullace |

HEIGHT To 8m.
FLOWERS Apr–May.
STATUS Thought to originate with a hybrid between Blackthorn and Cherry Plum, plums have been cultivated in Britain since at least 995 and recorded from the wild since 1777.
ALTITUDE Lowland.

Scattered in hedgerows, woodland edges and scrub, usually close to houses. ❀ A large shrub or small tree, originating in cultivation. Very variable. DETAIL First-year twigs *brown to grey*, variably *hairy*; buds often *hairy*. Leaves 3–10cm long, finely toothed; may be hairy below. Flowers 2–3 together, grouped into irregular clusters, *appearing with the leaves*, petals 7–12mm. Three subspecies are recognised, but the differences are not clear-cut: **Wild Plum**, subspecies *domestica*, has relatively large, egg-shaped fruits, up to 45mm long, grooved to one side, purplish or yellowish-red, with a whitish bloom and a rough, very flattened stone; twigs sparsely hairy and spineless. **Bullace**, subspecies *insititia*, has fruits that ripen later (Sep–Nov), are smaller and more rounded (c. 20mm diameter), with a less flattened stone, either greenish-yellow flushed reddish on the upper side (var. *syriaca*, 'White Bullace' – sometimes confused with Cherry Plum), or blue-black with a whitish bloom (var. *nigra*, 'Black Bullace' – sometimes hard to separate from Blackthorn or its hybrids). Fruits tart until frosted, but never as sour as Blackthorn; twigs densely hairy and often spiny. **Damson** (var. *damascena*) is the cultivated form of Bullace, with ovoid, blue-black fruits. **Gages**, subspecies × *italica*, are intermediate but fruit always greenish or yellowish.

HEIGHT To 21m.
FLOWERS May–Jun.
STATUS Native.
ALTITUDE
0–870m.

Rowan *Sorbus aucuparia*

Common in open woodlands, on heaths and moors and in rocky places, favouring acid soils. Often planted as an ornamental, then bird-sown onto waste ground etc. ❀ Small to medium-sized tree, often with multiple trunks. Bark smooth, silvery-grey, becoming fissured with age. Large *clusters of white flowers* are followed by bunches of *orange-red* berries, beloved of birds. Aka Mountain Ash. DETAIL Twigs greyish, smooth, long-hairy when young. Buds conical, dark brown, densely hairy. Leaves alternate, with 5–7 pairs of *coarsely-toothed leaflets*, 3–6cm long, downy below when young. Flowers 8–10mm across with 5 petals, 3–4 styles and *c.* 20 creamy anthers. Berries 8–12mm across.

FAMILY OLEACEAE : ASH, PRIVET ETC.

Ash
Fraxinus excelsior

'keys'

Common in woods, hedgerows, scrub and rocky places, especially on damp or calcareous soils. ❀ A tall tree with smooth, pale grey bark, becoming ridged with age. Lower branches *swept up at tips*, twigs with conspicuous *large black buds*. One of the last trees to come into leaf.

HEIGHT To 18m (38m).
FLOWERS
Late Mar–May.
STATUS Native.
ALTITUDE
0–840m.

Ash pollard

DETAIL Leaves opposite, with 4–6 pairs of *finely-toothed* leaflets, 6–9cm long, hairless or with a hairy midrib below. Flowers in tight clusters, *lacking petals*. Male flowers with 2 purplish-red stamens and yellow pollen, female flowers purple then pale green with a single stigma, hermaphrodite flowers combine both: trees may have either male or female flowers, a mixture of both, switch sexes between seasons or have normal bisexual flowers. The fruits hang well into the winter in conspicuous bunches of 'keys', each seed with a propeller-like wing to one side.

Wild Service Tree *Sorbus torminalis*

Uncommon and scattered in ancient woodland and old hedgerows, often on heavy clays, also rocky places, especially on limestone. Seldom found in large numbers, but suckers freely. Occasionally planted. ❀ Medium-sized tree with *maple-like leaves* that briefly turn bright russet in late autumn. *Clusters of white flowers* are followed by *small, edible, brownish fruits*. DETAIL Bark clean grey, darkening and splitting into small, flaking sections with age. Buds green, rounded. Leaves 5–10cm long, with 3–5 pairs of *pointed, triangular lobes* with finely-toothed margins, the *basal lobes at 90° to the stem*; shiny-green and *hairless above, near-hairless below* when mature. Flowers 10–13mm across with 5 petals, 2–3 styles and 20 creamy anthers; fruits *dark reddish-brown* with fine pale speckling, 8–17mm long, oblong to pear-shaped. SIMILAR SPECIES Maples (p. 391) have opposite leaves. **Swedish Whitebeam** *S. intermedia*, a commonly planted ornamental, is sometimes bird-sown. It has oval, lobed leaves, woolly greenish-white below, and orange-red berries.

HEIGHT To 20m (27m).
FLOWERS Late Apr–Jun.
STATUS Native.
ALTITUDE 0–340m.

Common Whitebeam *Sorbus aria*

Locally common in open woods, scrub and hedgerows on well-drained, often rocky soils, over chalk and limestone and on acid ground. ❀ A conspicuous small to medium-sized tree with *whitish undersides to the leaves*, especially obvious in early summer, and *clusters of white flowers followed by red berries*. DETAIL Bark grey-brown, twigs often reddish-brown. Leaves 6–12cm long, with well-toothed margins, pale green above, with abundant silvery hairs when fresh, *silvery-white and persistently woolly-white below*, with 9–14 pairs of veins. Flowers 10–15mm across, with 5 petals and creamy anthers, fruits 8–15mm across, oval, scarlet.

Over 20 more species of whitebeam are native to Britain. Shrubs or small to medium-sized trees, they are all highly localised and many are very rare. They are apomictic (see p. 366) and produce seed without fertilisation, a strategy that favours the perpetuation of minor local differences and the creation of microspecies.

HEIGHT To 15m (25m).
FLOWERS Late Apr–early Jun.
STATUS Certainly native in S England (north to the N Downs, Chilterns and Cotswolds) and SE Wales, may be native further north, to Derbyshire or even Cumbria; widely-planted as ornamental trees elsewhere, self-sowing back into wild habitats.
ALTITUDE To 455m.

Hawthorn
Crataegus monogyna

Abundant in hedgerows, scrub and woodland. ❁ The 'May-tree' of old, a deciduous shrub or small spreading tree, *very thorny*, with small *lobed leaves* and masses of fragrant white flowers.
DETAIL Sparsely hairy to hairless. Leaves 15–50mm long, *deeply cut into 5 (3–7) more or less pointed lobes*. Flowers in flat-topped clusters, petals 5, 4–6mm long, white or sometimes pink, anthers reddish, *style 1 (occasionally 2)*; fruit (the 'haw') 8–10mm long with 1 (occasionally 2) stony seeds encased in a fleshy red cup.

HEIGHT To 10m (15m).
FLOWERS Late Apr–Jun.
STATUS Native.
ALTITUDE 0–610m.

Hawthorn has been used for hedging for many centuries, with many millions especially grown in the 18th–19th centuries to make the 200,000 miles of new hedges required by the Parliamentary Enclosures. More recent planting sometimes involve non-native stock, which may come into leaf and flower unusually early.

Midland Hawthorn *Crataegus laevigata*

Local and uncommon in ancient woodland, old hedgerows and banks on clay soils, also widely planted. ❁ *Flowers 1–2 weeks earlier* than Hawthorn and blooms more freely in heavy shade.
DETAIL Very like Hawthorn, but twigs not as stiff or as spiny, *leaves shallowly 3-lobed* (even more or less unlobed), with the side lobes more rounded. Flowers slightly larger, petals 5–8mm long, with *2–3 styles*; the fruits have *2–3 seeds*. The two species frequently hybridise, producing fully fertile offspring, making identification tricky; hybrids have both 1- and 2-styled flowers and intermediate leaves.

HEIGHT To 10m.
FLOWERS Apr–May.
STATUS Native.
ALTITUDE Lowland.

Dogwood
Cornus sanguinea

Deciduous shrub, locally common in woods, scrub and hedgerows on chalk, chalky-clay or limestone. Sometimes planted in hedges and common in amenity planting. Suckers vigorously. ❋ Picked out in winter by its purplish-red twigs, in spring by the flat-topped clusters of creamy-white flowers and *prominently* veined leaves; in autumn has blackish berries. DETAIL Twigs, buds and leaves finely hairy, buds dark brown, leaves turning red in autumn, opposite, 4–8cm long, *untoothed*, usually with 3–5 pairs of *curving* veins, sunken on upperside, raised below. Flowers in flat-topped umbels, calyx densely hairy, with 4 tiny lobes, petals 4, 4–7mm long, stamens 4, style 1. Fruits 5–8mm across.

HEIGHT To 4m.
FLOWERS May–Jul
(sporadically in autumn).
STATUS Native.
ALTITUDE Lowland.

Spindle *Euonymus europaeus*

A deciduous shrub or small tree, scattered in woods, scrub and hedgerows, mostly over chalk, limestone and other base-rich soils. Sometimes planted, or bird-sown from gardens. ❋ Shoots green, more or less 4-sided. Inconspicuous greenish-yellow flowers are followed by the unique fruits, *4-angled bright pink capsules* that split open to reveal *bright orange seeds*. DETAIL Hairless. Buds to 5mm, green. Leaves often turning reddish in autumn, short-stalked, opposite, 3–13cm, very finely toothed. Flowers 8–10mm across, in clusters of 3–10, calyx divided into 4 short lobes, 4 greenish-white petals, 4 stamens and 1 style. Fruits 8–15mm across.

HEIGHT 2–5m (8m).
FLOWERS May–Jun.
STATUS Native.
ALTITUDE
0–400m.

Buckthorn *Rhamnus cathartica*

HEIGHT To 8m.
FLOWERS May–Jun.
STATUS Native.
ALTITUDE
0–380m.

Locally common. Woods, scrub and hedgerows, mostly on calcareous soils, also wet carr woodland. ❂ A shrub or small tree, usually with *scattered, sharply-pointed, thorny twigs*. Leaves more or less opposite, with *finely toothed margins*. Flowers small, greenish, in dense clusters, followed by green berries that ripen to glossy black. DETAIL Buds dark brown, *covered by 5 scales*, twigs blackish-brown; the wood turns orange when cut. Leaves 3–9cm long, *hairy*, with 2–5 pairs of conspicuous, *strongly curved* side veins. Flowers *c.* 7mm across, the calyx with 4 widely-spread teeth, 4 tiny petals. *Dioecious*, male and female flowers are on *separate* plants; male flowers with 4 stamens, female with 1 style split into 4 stigmas at the tip and vestigial stamens. Berries 6–10mm across. SEE ALSO Dogwood (p. 377), which has similar leaves, but their margins are untoothed and the twigs are purple.

Alder Buckthorn *Frangula alnus*

HEIGHT To 5m.
FLOWERS May–Jun.
STATUS Native.
ALTITUDE To 450m.

Local in scrub, woodland and sometimes hedges, usually on damp, peaty or sandy soils and especially around bogs and fens; occasionally planted, formerly for charcoal, now often as a food plant for the Brimstone butterfly. ❂ A shrub or small tree, shoots blackish-brown with many *small, slightly raised, silvery scars*. Leaves *alternate, untoothed*. Flowers tiny, in dense clusters, followed by green berries that ripen *through red to black*. DETAIL Buds brown, hairy when young, *lacking protective scales*. Leaves 2–7cm long, hairy below when fresh, with 6–10 pairs of *straight, parallel side veins* that *loop round at the tip to join the next vein*. Flowers bisexual. Calyx 5-toothed, 2–4mm wide, green outside, white inside, containing 5 white petals, 5 stamens and 1 style. Berries 6–10mm across.

ULMUS: ELMS

Tall, deciduous trees with distinctive *asymmetrical leaf bases*. Of the several species, Wych Elm is relatively distinct, spreading via seed and rarely suckering. Separating the other elms is, however, very difficult. They sucker freely, producing groups of clones, and are easily grown from cuttings; all have been extensively planted. Genetic research has shown that several 'species' are, in fact, single clones, while others are hybrids. Until the 1970s elms dominated many landscapes, but Dutch elm disease, a fungal infection spread by bark beetles, decimated most populations. Around 25 million trees were killed, but the roots can survive and produce suckers. As the bark beetles only show an interest in an elm once it has a certain thickness of bark, suckers are not infected until they reach a certain size. The most important identification features are found on adult shoots and leaves, which suckers start to produce after *c.* 10 years, but they are either regularly cut when hedges are flailed, or they are killed by the disease before reaching maturity; either way, suckers are mostly unidentifiable.

Wych Elm *Ulmus glabra*

HEIGHT To 38m.
FLOWERS Feb–Mar.
STATUS Native.
ALTITUDE 0–530m.

Fairly common in woodland and hedgerows, especially in the N and W and on limestone. ❀ Relatively resistant to Dutch elm disease and *likely to be found as a mature tree*; in the open has a broad, rounded outline, the trunk soon dividing into many low, spreading branches. DETAIL Bark brownish, vertically ridged with age. Buds dark brown with *abundant rusty hairs*. Leaves relatively *large*, 7–16cm long, finely-toothed, *roughly hairy on the upperside* ('sandpaper-like'), hairy below, with *12–18 pairs of veins*, Leaf stalk *very short* – less than 3mm, mostly covered by the longer side of the leaf. Flowers *c.* 20mm long, calyx bell-shaped, with 4–5 lobes, the 2 styles surrounded by 4–5 stamens; anthers reddish-purple. The numerous flower clusters give the bare branches a reddish hue in early spring. Fruits 20mm across, the single seed positioned *centrally* in a broad, circular wing, *developing to maturity*, usually in abundance, then shed May–Jun.

asymmetric: the leaf blade extends further down one side of the stalk than the other

English Elm *Ulmus procera*

Formerly common in hedgerows, but mature trees now rare.
DETAIL No *rusty hairs on buds.* Leaves 4–7(9)cm long, often *almost circular*, upperside *rough*, with *8–12 pairs of veins*, leaf stalk 3–8mm *long*. Twigs may develop corky 'wings'. Flowers Jan–Feb, but seed rarely develops to maturity. **Dutch Elm** *U.* × *hollandica* (probably the hybrid with Wych Elm) is intermediate; the leaves can resemble Wych Elm, but with a longer stalk. 'Huntingdon Elm' is a single clone of this hybrid.

Small-leaved Elm *Ulmus minor* agg.

A complex and very variable group, once common in hedgerows, now mostly reduced to suckers. DETAIL No rusty hairs on buds. Leaves *3–7cm long*, upperside *smooth* to *slightly* rough, with 7–12 pairs of veins, leaf stalk *4–10mm*. Flowers as Wych Elm, but seed placed nearer tip of wing; seed frequently fails to develop. Includes 'Plot's Elm' and 'Cornish Elm', distinctive, closely-related populations, perhaps descended from single clones.

Beech *Fagus sylvatica*

Large deciduous tree, casting heavy shade and often in pure stands – few other plants grow in a beechwood. Frequently pollarded. Common in woodland on well-drained soils, and widely planted. ❁ *Bark greyish and smooth*, even in mature trees. Twigs with *long, slender, pointed, reddish-brown buds.* Leaves untoothed; on young trees and hedges the dead leaves remain on the branches through the winter. DETAIL Leaves 4–9cm long with slightly wavy margins and prominent parallel veins; clear green, smooth and shiny, with *silky hairs along the margins, midrib and veins.* Flowers wind-pollinated, male flowers silky-hairy with numerous stamens, in hanging, tassel-like clusters, female flowers in pairs enclosed by a large, long-haired cup that enlarges and becomes woody in fruit, containing the sharply 3-angled nuts (Beech 'mast'), each 12–18mm long.

HEIGHT To 30m (46m).
FLOWERS Apr-May, with the leaves.
STATUS Native to S England and SE Wales, widely planted elsewhere.
ALTITUDE 0–650m.

Family BETULACEAE: BIRCHES

Hornbeam *Carpinus betulus*

Large deciduous tree, casting heavy shade and usually dominant when full-grown, but often coppiced. Favours clay soils. Extensively planted in woods, parks and hedges. ❁ Superficially like Beech but bark smooth and grey with *soft, pale, vertical streaks* (trunk fluted). Leaves rougher to the touch and *double-toothed*, with a pattern of fine teeth overlaid by coarse teeth. DETAIL Twigs brownish-black, *sparsely hairy*. Leaves 3–11cm long, smooth, dull, dark green, sparsely hairy on veins below, with prominent pleated, parallel veins. Flowers wind-pollinated, in catkins, male and female separate. Male catkins 2.5–5cm, each flower with a shell-like bract and numerous stamens, female catkins 2cm, the flowers paired, each with a 3-lobed bract that expands as the nut develops to form 'wings' up to 5cm long, the fruits hanging in bunches.

HEIGHT To 32m.
FLOWERS Apr-May, just before or with leaves.
STATUS Native to SE England and S Wales; widely planted.
ALTITUDE To 380m.

bark   male catkins   female catkins in fruit   nuts with 'wings'

Pedunculate Oak
Quercus robur

Large deciduous tree, common in woodland, parks and hedgerows, and extensively planted, although scarce in the uplands. ✿ Leaves *short-stalked*, but acorns 2–3 together on a *long* stalk. DETAIL Bark rugged. Twigs chestnut-brown, hairless, angled. Buds 4–9mm, hairless, in *clusters* at the tip of the twigs. Leaves 5–12cm long with 3–6 pairs of lobes, with a pair of *small rounded lobes (auricles) at the extreme base*, sometimes partially covering the stalk, *hairless* below or at most with scattered short white hairs (20× lens), *leaf stalk 2–3mm (0–7mm) long*. Flowers pale green, male and female separate, the male in loose, drooping catkins, with 6 hairy green sepals and numerous stamens, the female scattered on slender stalks, each *c.* 2.5mm across (including the enveloping stiff scales that enlarge in fruit to form a 'cup'). Largely wind-pollinated. Fruit the familiar acorn, on a hairless stalk *2–9cm long*.

male catkins

HEIGHT To 30m (40m).
FLOWERS Apr–May, with the leaves.
STATUS Native.
ALTITUDE 0–460m.

lobe (auricle)

tapered into stalk

Sessile Oak
Quercus petraea

Large deciduous tree, common in woodland, parks and hedgerows, especially on poorer, more acid soils and in the uplands of the N and W. ✿ As Pedunculate Oak, but well-grown trees in the open have more upright branches and a narrower crown. Leaves relatively *long-stalked* (over 10mm), acorns almost *stalkless*. DETAIL Twigs grey-brown, hairless. Leaves 5–12cm long, usually with 5–8 pairs of lobes and more or less *tapered* into the stalk at the base, with *basal lobes slight or absent*; underside of leaf minutely but obviously hairy, with *both* star-shaped and simple hairs (20× lens). Leaf stalk 13–25mm (but shorter when young, or on shoots sprouting from the trunk). Acorns on stalks 0–2(4)cm long, the stalks with clusters of hairs. SIMILAR SPECIES *Q.* × *rosacea*, the hybrid with Pedunculate Oak, is *very common*. Intermediate in appearance, it is fertile and back-crosses with the parents, making identification tricky.

HEIGHT To 30m (42m).
FLOWERS Apr–May.
STATUS Native.
ALTITUDE 0–475m.

HEIGHT To 30m.
FLOWERS Apr–May.
STATUS Native.
ALTITUDE To 510m.

Silver Birch
Betula pendula

Common. Woodland and scrub, especially on light, acid soils; often colonises heathland.
❀ Graceful, fast-growing deciduous tree with silvery-white bark and slender, ascending branches. The *outer shoots and twigs often 'weep' downwards* and there are masses of pendant catkins in early spring, followed by a light canopy of small leaves.
DETAIL Bark shiny reddish-brown in young trees, quite soon whitish above with blackish, diamond-shaped scars, and irrupting with age into irregular blackish fissures towards the base. Leaves 2.5–6cm long, triangular-oval, tapering to a point, *distinctly* double-toothed, with *prominent* primary teeth that *curve forward*. Twigs rough, with tiny whitish warts; twigs and leaves *hairless* on mature shoots, but saplings and regrowth can have hairy twigs and leaves. Male catkins pendant, 3–6cm long, female catkins erect, shorter. Seeds tiny, with a pair of large, papery wings, each at least twice the width of the seed and extending well forward of stigmas at tip of seed (total width 3–5mm).

prominent primary teeth

Silver Birch   Downy Birch

Silver Birch   Downy Birch

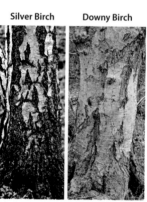

Silver Birch   Downy Birch

Downy Birch Betula pubescens

HEIGHT To 24cm.
FLOWERS Apr–May.
STATUS Native.
ALTITUDE 0–775m.

Common in similar places to Silver Birch, but with a preference for wetter, peatier soils and probably more frequent in the uplands. ❀ Very like Silver Birch, but bark of older trees brown or greyish-white, *rarely* with a fissured, blackish base, and *lacking* blackish diamonds. Twigs usually *held upright*, producing a more rounded crown. DETAIL Leaves *oval*, pointed (but *not* tapering to a *long* point), single- or double-toothed, but without *prominent* primary teeth. Young twigs usually hairy, but *without whitish warts*; in upland areas of N twigs may be near-hairless with sweet-smelling sticky brown glands and sticky buds (subspecies *tortuosa*). Leaves on mature shoots *hairy*, at least on veins below, but becoming hairless when old. Seeds with wings 1–1.5 times width of body, not extending forward of stigmas. SIMILAR SPECIES *B. × aurata*, the hybrid with Silver Birch, is probably common. Usually closest to Downy Birch, but separation of the two species is often difficult and determination of hybrids even harder.

Alder *Alnus glutinosa*

HEIGHT To 12m (29m).
FLOWERS Late Feb–Apr.
STATUS Native.
ALTITUDE 0–470m.

Common on wet ground by rivers, streams, ditches and lakes, and in damp meadows, fens and bogs; may invade open wet habitats, together with willows, to form carr woodland. ❀ A medium-sized tree, with a distinct *purplish* hue in the winter and *small woody cones*. Timber orange when freshly cut. DETAIL Bark purplish-brown, soon becoming dark grey-brown and fissured. Buds *short-stalked*, especially on upper branches, 6–10mm long, blunt, purplish. Twigs hairless, sticky when young, with orange warts. Leaves 30–90mm, near hairless, *blunt*, boldly veined, irregularly double-toothed. Male flowers in 20–60mm-long *catkins*, female catkins *c.* 10mm long and erect, green to purplish, 3–6 together, with crimson stigmas, ripening to persistent, woody cones, 8–28mm long.

Hazel *Corylus avellana*

female flowers

HEIGHT To 6m (12m).
FLOWERS Jan–Mar.
STATUS Native.
ALTITUDE 0–640m.

Common in woodland, hedgerows, scrub and rocky places; frequently planted. ❀ Multi-stemmed shrub with conspicuous yellow catkins ('lambs' tails') in late winter. In late summer produces small clusters of edible nuts, up to 20mm long, green at first but ripening to brown, enclosed in a coarsely-toothed, leafy husk. DETAIL Bark smooth, younger stems rich brown with short horizontal scars, twigs with abundant red-tipped glandular hairs, buds hairy. Leaves 5–12cm long, downy, rough-textured, with double-toothed margins. Male catkins 2–8cm long, appearing in Oct and opening in the New Year; female catkins 5mm long, resembling small buds, from which red styles project.

SALIX: WILLOWS

Willows vary from majestic trees in lowland flood-plains to dwarf shrubs in the high mountains (6 species, not detailed here). All favour damp ground, in marshes, wet woodland and dune slacks. They are commonly planted, for basket-work in the past and more recently for ornament or as biofuel. Willows are dioecious, with male and female flowers on separate trees. Male catkins are white as the buds burst (hence 'pussy willow'), becoming yellow when the anthers are exposed. Female catkins are greenish, and the seeds develop plumes of silky white hairs as they ripen. The leaves are important for identification, and are best examined when mature, Jul–Sep; leaves on suckers, regrowth or heavily shaded plants may not be typical. Identification of willows is made difficult by the numerous hybrids (at least 68 recorded, many being combinations of 3 species), and any willow that does not clearly fit one of the species below is mostly likely a hybrid.

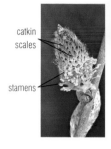

catkin
scales

stamens

Grey Willow
male catkin

below

above

White Willow
Salix alba

HEIGHT 10–25m (33m).
FLOWERS Late Apr–May, with the leaves.
STATUS Ancient introduction.
ALTITUDE 0–350m.

White Willow pollards

A large tree, often pollarded. ❀ *Young leaves silvery-grey, with very fine, dense, silky white hairs above and below*; upperside becoming *dull green* and sparsely hairy with age, underside *remaining silvery-hairy*; at a distance the canopy appears *silvery-white*. DETAIL *Bark deeply fissured*. Twigs densely silky-hairy when young, hairless and satiny with age (bright yellow or orange in some commonly-planted cultivars). Leaves 5–10cm long, very narrow, margins with *numerous tiny teeth*. Catkin scales *yellowish*, male catkins 4–5cm long, stamens 2; female catkins 3–4cm long. **Cricket-bat Willow**, var. *caerulea*, has a bluish tinge to the leaves, which may be almost hairless when mature. SIMILAR SPECIES Weeping Willow *S. × sepulcralis*, the distinctive ornamental, is the hybrid between White Willow and the non-British *S. babylonica*.

Hybrid Crack Willow
Salix × fragilis

The hybrid between Crack and White Willows. Common. ❀ A robust tree, often pollarded.

Twigs and smaller branches *very brittle, snapping easily*. DETAIL *Bark coarsely and deeply fissured*. Twigs thinly hairy when young, soon *hairless and satiny*. Leaves 9–15cm long, *narrow*, *toothed*, silky-hairy when fresh, becoming more or less hairless with age, *glossy green above*, bluish-white below. Male catkins long and slender, with *pale yellow* catkin scales, stamens 2(3). SIMILAR SPECIES Crack Willow *S. euxina* is similar, but twigs shiny yellow-brown, *always hairless*; leaves 5–10cm long, but still broad, *always hairless*. A twiggy bush to 7m or tree to 15m. Introduced, with only males in Britain.

HEIGHT 10–15m (30m).
FLOWERS Apr–May, with the leaves.
STATUS Ancient introduction, most trees are planted, often beside fresh water.
ALTITUDE 0–410m.

female catkins

Osier *Salix viminalis*

Shrub or small tree, often coppiced or pollarded. ❀ Leaves 10–15cm long, *at least 6 times as long as wide*, dull dark green above, *densely silky-hairy below* (hairs flattened), *margins not toothed*. Detail Bark fissured. Twigs often long and straight, densely greyish-hairy when young, becoming satiny. Male catkins short-cylindrical, with *reddish-brown to blackish scales*, stamens 2; the clusters of catkins near the tips of the branches are eye-catching.

HEIGHT 3–6m.
FLOWERS Late Feb–early Apr, before the leaves.
STATUS Ancient introduction, often planted for basketry and more recently for biofuel.
ALTITUDE 0–410m.

Purple Willow *Salix purpurea*

male catkins

A shrub or occasionally a small tree. ❀ Leaves *opposite or almost so*, bluish-green, more or less *hairless*, broadening (and sometimes finely toothed) towards the tip. *Male catkins brightly coloured.* Detail Bark smooth.

Twigs hairless, satiny, yellowish to greyish or purplish-brown. Leaves 3–8cm long, stipules small, soon falling. Male catkins cylindrical, scales *tipped purple-black*, stamen apparently 1 (the separate stamens are fused into 1 filament), *anthers reddish*, becoming yellow as the pollen is shed.

HEIGHT 1.5–5m.
FLOWERS Mar–early May, before the leaves.
STATUS Native.
ALTITUDE 0–440m.

Almond Willow *Salix triandra*

male catkins

Shrub or small tree, sometimes pollarded. ❀ Bark smooth, *peeling in large patches to reveal cinnamon under-bark*. Detail Twigs *hairless*, satiny reddish-brown to yellow-brown above, green below, often conspicuously ridged when young. Leaves 4–11cm long, more or less *hairless*, shiny dark green above, pale green below, *margins finely toothed* (teeth blunt), stipules often relatively large, persistent. Catkin scales uniformly greenish-yellow, male catkins long and slender with *3 stamens per floret*.

HEIGHT To 10m.
FLOWERS Mar–May, with or just before the leaves.
STATUS Ancient introduction, much-planted for basketry, sometimes in osier beds.
ALTITUDE Lowland.

male catkins

Goat Willow
Salix caprea

Shrub or *small tree*, more often on dry ground than other willows.
❀ Leaves 5–12cm long, *broad*, with slightly wavy margins and a few small teeth; dark green above, *densely softly-hairy below*, with very prominent

veins. **DETAIL** Bark irregularly fissured. Twigs sparsely hairy when young, soon becoming hairless; peeling away the bark, the wood of 2nd-year twigs is *smooth* (or at most has short, fine, raised ridges). Leaves with stipules *soon falling*. Male catkins with small *blackish* scales, stamens 2; catkins rather large, egg-shaped, erect, clustered at the branch tips.

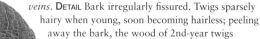

HEIGHT Usually 5–7m.
FLOWERS Mar–Apr, before the leaves.
STATUS Native.
ALTITUDE 0–760m.

HEIGHT To 10m.
FLOWERS Mar–Apr, before the leaves.
STATUS Native.
ALTITUDE Generally lowland, but exceptionally to 665m.

Grey Willow Salix cinerea

The commonest willow in much of Britain. May colonise waste ground and disused railways. ❀ Usually a large, *multi-stemmed shrub*, occasionally a small tree. Leaves 2–9cm long, *narrowly oval*, with a few small teeth, dark green, smooth and often satiny above, sparsely hairy when young, and *hairy below* (usually with some rust-coloured hairs, but hard to see, and only obvious from late summer); veins not *very* prominent. (In E Anglia subspecies *cinerea* has slightly broader, dull grey-green leaves, with wavy, toothed edges.) **DETAIL** Bark fissured with age. Twigs densely hairy when young; *wood of peeled 2nd-year twigs with scattered raised ridges* (often long and very obvious, but may be subtle). Leaves with *large stipules*, usually persistent. Male catkins as Goat Willow but narrower, scales *blackish-brown*, stamens 2.

ssp. cinerea

HEIGHT 5–10m.
FLOWERS Late May–Jun, with the leaves.
STATUS Native.
ALTITUDE 0–410m.

Bay Willow Salix pentandra

A large shrub or small tree.
❀ *Leaves hairless, dark glossy green, with very fine, blunt teeth*; leaf stalk with small oil glands, *young leaves and catkins sticky and pleasantly scented*. **DETAIL** Bark lightly fissured. Shoots and twigs hairless, reddish-brown, *glossy* (as if varnished). Leaves 5–12cm long. Catkin scales pale yellow; male catkins particularly bright golden-yellow as *each floret has 5–8 (4–12) stamens.*

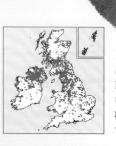

male catkins

Eared Willow *Salix aurita*

male catkins

female catkins

Locally common, mostly in N and W, on heathland, moorland, scrub and along watercourses, on acid soils.
❀ A rounded *shrub*. Leaves *conspicuously wrinkled*, with *wavy margins* and a *twisted tip*, *stipules large and conspicuous*, usually persisting (the 'ears'). Leaves thinly hairy above, greyish and *densely hairy below* (but *no* rusty hairs). **Detail** Twigs sparsely hairy when young; wood of peeled

HEIGHT To 2m (3m). **FLOWERS** Apr–early May, before the leaves. **STATUS** Native. **ALTITUDE** 0–790m.

stipules ('ears')

2nd-year twigs with raised ridges (as Grey Willow). Leaves 2–4cm long. Catkins 1–2cm long, male catkins with dark reddish scales and 2 stamens.

female catkins

male catkins

Creeping Willow *Salix repens*

Locally abundant. Var. *repens* in damp places on heaths and moors; var. *fusca* in E Anglian fens; var. *argentea* in dune slacks. ❀ Very variable small shrub, leaves 1–3.5cm long, generally long-oval and *silky-hairy, at least below*, without large, persistent stipules. **Detail** Var. *repens* has creeping rhizomes, near hairless, low-growing stems and small leaves that are almost hairless above but hairy below; var. *fusca* is more erect, to 1.5m; var. *argentea* has robust, more or less erect silky-hairy stems and large leaves with dense silky white hairs below (and usually also above). Male catkins with reddish-brown scales; stamens 2.

HEIGHT To 1.5m. **FLOWERS** Apr–May, usually before the leaves. **STATUS** Native. **ALTITUDE** 0–855m.

Tea-leaved Willow *Salix phylicifolia*

Locally common. Waterside vegetation and rocky places on alkaline soils. ❀ A shrub or small tree. Leaves 2–6cm long, *leathery, shiny green* above, *hairless* (may be slightly hairy when young), usually with a few shallow, irregular, blunt teeth. Twigs *reddish-brown, satiny*, sometimes downy when young. **Detail** Catkin stalks with *small, leaf-like bracts, catkin scales blackish-brown*, male catkins with 2 stamens. **Similar Species Dark-leaved Willow** *S. myrsinifolia* is very similar, but has duller, hairy twigs, duller, more papery leaves, and more obvious, long-lasting stipules.

HEIGHT To 4m (5m). **FLOWERS** Apr–May, with the leaves. **STATUS** Native. **ALTITUDE** To 720m.

Aspen *Populus tremula*

Locally common (but easily overlooked) in woods and hedgerows, often on damp soils, also heathland, rocky outcrops and riverbanks.

✿ A slender tree, often relatively small and shrubby, *suckering to form thickets*. DETAIL

male catkins

Bark often dark and fissured towards base of trunk, smooth and greyish above, with *horizontal rows of small, dark, diamond-shaped pits*. Twigs grey-brown, downy when young, often with prominent protruding leaf scars and short side spurs. Buds 9–14mm, *pointed* (but bud scales *rounded* at tip), shiny reddish-brown, *sticky* when opening, hairless. Leaves dark green, *hairless* (may be silky-hairy when young), *more or less circular with an irregularly scalloped edge*, stalks hairless or sparsely hairy, long, slender and *flattened*, allowing the leaves to rustle in the slightest breeze. Leaves on suckers can be *much larger, pointed, with a heart-shaped base*. Catkin scales large, dark brown, deeply cut into *long, ragged teeth* and fringed with *silky-white hairs*; male flowers have reddish-purple anthers.

HEIGHT 15–20m (24m).
CATKINS Late Feb–Apr, well before the leaves.
STATUS Native.
ALTITUDE 0–640m.

sucker leaf

Black Poplar *Populus nigra*

HEIGHT To 33m.
FLOWERS Late Mar–Apr, before the leaves.
STATUS Native.
ALTITUDE Lowland.

male catkins

Native Black Poplar (subspecies *betulifolia*) is scarce and local in fields, parks and gardens on major flood plains (*c.* 6,000 mature trees). DETAIL Large tree, mature specimens often *leaning*, branches often heavy and *arching*, upswept at the tips. Bark dark grey-brown, deeply fissured, with distinctive *large swollen bosses* towards the base. No suckers. Twigs yellow-brown, downy when young. Buds *c.* 7–10mm, pointed, shiny dark brown, hairless, sticky; bud scales pointed. Leaves pointed, with small, blunt teeth, shiny dark green above, barely paler below, sparsely hairy when young; base of leaf *without* glands. Catkin scales *deeply toothed* but *not* hairy, male flowers with crimson anthers. Female trees rare: the tiny seeds are embedded in white cotton floss, which is produced in vast quantities and may collect in 'drifts' in the spring. SIMILAR SPECIES Hybrid Black Poplar *P.* × *canadensis* is *much commoner*. Often planted on wet ground and as shelter belts. Tall and straight, *without bosses on the trunk, branches erect*. Many leaves have *1–2 small glands at the base near the stalk*. Mostly male trees are planted, and their long crimson catkins litter the ground in early spring.

Grey Poplar *Populus × canescens*

The hybrid between Aspen and White Poplar. A tall, heavy tree, commonly planted as wind-breaks and for amenity, especially in damp woods and by streams and rivers. Suckers freely and may form thickets. ✿ Rather variable, but note the sex – *female trees are rare.*

DETAIL Bark heavily ridged and blackish towards base, becoming smooth and palegrey above with *horizontal rows of small, black, diamond-shaped pits.* Twigs initially covered with white felted hairs, but *soon* grey-brown and hairless. Buds 6–10mm long, woolly at base, dark brown and hairless towards tip. Leaves of 2 kinds: on short side shoots and at the base of long leading shoots 3.5–6cm across, *rounded, with large, irregular, blunt teeth, recalling Aspen*, dark green above, dull green below with a thin whitish frass that disappears with age; *stalks flattened.* Leaves towards the tips of leaders and on suckers *larger and more distinctly lobed*, uppersides dark green, when young with a layer of woolly white hairs, *undersides with dense felted white hairs*, the hairs thinning (but usually not disappearing) with age. *Catkin-scales deeply and irregularly toothed*, with a fringe of long hairs (as Aspen).

male catkins

short side shoots

leaders & suckers

HEIGHT 20–35m (46m).
CATKINS Mar, before the leaves.
STATUS Introduced *c.* 1700.
ALTITUDE Lowland.

White Poplar *Populus alba*

HEIGHT 15–20m (26m).
CATKINS Late Feb–Mar, before the leaves.
STATUS Reputedly introduced from Holland in the 16th century.
ALTITUDE Lowland.

A medium-sized tree, planted along roadsides, in parks and as windbreaks; salt-tolerant, colonising coastal dunes.
✿ Suckers freely and may form thickets.

Easily confused with Grey Poplar but rather scarcer. Less robust and from a distance can look whitish as the leaves flutter in the wind. Note the sex – *almost all British trees are female.* **DETAIL** Bark as Grey Poplar, but paler and whiter above. Twigs initially *covered with dense white felted hairs*, becoming dark brown and hairless by 2nd year. Buds as Grey Poplar but only 4–5mm long. Leaves on side shoots and the base of leaders oval, with *irregular shallow lobes*, dark green above, with the *underside retaining a fine whitish frass even when mature.* Leaves towards tip of leaders and on suckers *3–5-lobed (like a maple)*, uppersides dark green, covered when young with a dense layer of woolly white hairs, undersides with a dense covering of woolly white hairs that is *retained into maturity.* Stalks white-felted. Catkin scales *oval (not toothed)*, with a fringe of long hairs.

female catkins

short side shoots

leaders & suckers

Sweet Chestnut *Castanea sativa*

A large deciduous tree, formerly much-used for coppice (and old coppice stools have now often 'got away' to become tall trees), also widely planted in hedgerows and parks. ❀ Bark with conspicuous *parallel vertical ridges/fissures*, sometimes gradually corkscrewing around the trunk as they rise upwards in older trees. Leaves long, narrow and *saw-toothed*, fruit the familiar edible chestnut. **DETAIL** Twigs hairless. Leaves 10–30cm long, hairy below when young, short-stalked. Flowers insect-pollinated, in 12–20cm-long spikes, more or less erect, mostly male but with some female flowers at the base; male flowers with numerous stamens, sickly scented, female flowers in groups of 3 in a scaly cup that expands to become a densely spiny husk, 5–7cm across, containing 1–3 shiny brown nuts.

HEIGHT To 35m.
FLOWERS Jul.
STATUS Introduced, probably in the Roman period.
ALTITUDE 0–460m.

FAMILY SAPINDACEAE: MAPLES & HORSE CHESTNUTS

Horse Chestnut *Aesculus hippocastanum*

A large deciduous tree, widely planted in parks, churchyards and as a street tree, less often in woodland. Occasionally self-sown. ❀ The large leaves are *palmately* divided into 5–7 leaflets, the large flowers are carried in conspicuous upright 'candles', and the fruits contain large seeds, the well-known 'conkers'. **DETAIL** Bark smooth when young, then scaly and flaking. Twigs hairless, buds *large*, dark red-brown, *sticky*. Leaves long-stalked, leaflets 8–30cm long, sparsely hairy below when young. Flower clusters 15–30cm high. Sepals 5, petals 4–5, white with a pink or yellow blotch at the base, stamens long and curved. Fruits 5–8cm, green, sparsely spiky, splitting open to reveal 1–2 glossy brown seeds.

HEIGHT To 25m (39m).
FLOWERS May–Jun.
STATUS Introduced to cultivation from SE Europe around 1615 and recorded from the wild by 1870.
ALTITUDE To 505m.

Field Maple *Acer campestre*

HEIGHT To 25m.
FLOWERS May–Jun.
STATUS Native.
ALTITUDE 0–380m.

Common in woods, scrub and hedgerows; frequently cut into a hedge or coppiced. Widely planted. ✿ Medium-sized deciduous tree. Leaves rather small, typically with *5 blunt lobes*. Flowers yellowish-green, in more or less *erect spikes*, fruits paired, with propeller-like wings *spreading at 180°*. DETAIL Bark smooth, vertically furrowed, flaking with age. Twigs pale brown. *Leaves opposite*, 3–5cm long, hairy below, especially on veins, hairless above when mature; leaf stalks reddish, *bleeding milky sap* when cut, sparsely hairy. Flower spikes often with male flowers at the sides, female at the tip. Flowers insect-pollinated, *c.* 6mm across, sepals 5, petals 5, inserted into a nectar-bearing disk, stamens 8, much reduced in female flowers, styles 2. Fruits *downy*. Fruiting erratic; may only produce male flowers following a year of abundant fruiting.

Sycamore *Acer pseudoplatanus*

Large deciduous tree, common in woods, parks and hedgerows, self-sowing freely into almost any habitat in the lowlands, but associated with farmsteads and villages in the uplands. ✿ Leaves *large*, with *5 pointed, coarsely-toothed lobes*. Flowers in *hanging clusters*, fruits hairless, with 2 propeller-like wings *at an angle of* c. *90° to each other*. DETAIL Bark smooth, becoming flaky when old. Twigs hairless. *Leaves opposite*, 7–16cm long, almost hairless (sometimes hairs on veins below), often blotched black by tar-spot fungus. Leaf stalk red, *without* milky sap. Flower spikes 5–20cm long with up to 100 flowers, much as Field Maple, the earliest to open being male.

HEIGHT To 35m.
FLOWERS Late Apr–Jun, with the leaves.
STATUS Introduced in the late 15th century and recorded in the wild from 1632 (Europe).
ALTITUDE 0–580m.

Small-leaved Lime
Tilia cordata

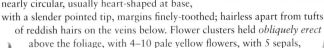

Locally common. Mixed deciduous woodland, wooded limestone cliffs and occasionally hedgerows; sometimes coppiced. ❂ Tall deciduous tree with alternate, *heart-shaped leaves* and clusters of *sweetly-scented flowers appearing mid summer*; the *stalk of each flower cluster is fused halfway to a tongue-shaped papery bract* that forms a parachute for the small, globular fruits. DETAIL Young twigs soon hairless, buds to 5mm, reddish, shiny, hairless. Leaves *mostly 3–6cm long*, nearly circular, usually heart-shaped at base, with a slender pointed tip, margins finely-toothed; hairless apart from tufts of reddish hairs on the veins below. Flower clusters held *obliquely erect* above the foliage, with 4–10 pale yellow flowers, with 5 sepals, 5 petals, numerous stamens and 1 style. Fruit 6–10mm across. SIMILAR SPECIES Lime *T. × europaea*, the hybrid between Small-leaved Lime and Large-leaved Lime *T. platyphyllos* (a rare native), is commonly planted in towns and cities and on rural estates. Trunk with *large bosses, often smothered at the base by a mass of dense twiggy shoots.* Leaves *mostly 6–9cm long*, with tufts of buff hairs on undersides; flower clusters *hanging*.

HEIGHT To 30m (38m).
FLOWERS Late Jun–Jul.
STATUS Native.
ALTITUDE To 600m.

Lime

Small-leaved Lime

FAMILY SANTALACEAE: BASTARD TOADFLAX & MISTLETOE

Mistletoe *Viscum album*

female flowers

male

Abundant in some areas. A partial parasite on a variety of trees, especially apples, limes, hawthorns, poplars, maples and willows, in orchards, hedgerows, parks and gardens (but seldom dense woodland). ❂ Green stems branch repeatedly to form *loose balls of shoots up to 2m across, obvious in the winter and early spring.* Flowers inconspicuous, followed by white berries 6–10mm across in Nov–Dec. DETAIL Evergreen. Leaves in opposite pairs, 2–8cm long, stalkless, oval-oblong with a rounded tip, yellowish-green and leathery. Flowers 3–5 together in tight, bud-like clusters, tiny, stalkless, with 4 rigid, pale green petal-like structures joined at the base. Male and female on separate plants, male with stalkless anthers *on* the petals, female with a stalkless stigma on the ovary.

GROWTH Perennial.
HEIGHT 10–50cm.
FLOWERS Feb–Apr.
STATUS Native.
ALTITUDE Lowland.

Mezereon
Daphne mezereum

Nationally Scarce, listed as Vulnerable. Deciduous woodland, often on relatively bare, steep, rocky slopes, usually on chalk or limestone. Also old chalk pits and Alder carr woodland. Sometimes bird-sown from gardens (conversely, in the past, plants were dug up and taken into gardens from the wild). ❁ A spindly shrub, with fragrant bright pink flowers coming into bloom before the leaves, followed by bright red berries. DETAIL Stem and leaves more or less hairless. Leaves alternate, 3–10cm long, lens-shaped, pointed at the tip and tapering to a short stalk, pale green above, paler below. Flowers in clusters of 2–4 in the axils of the previous season's leaves, buds to 8mm long, with dark-tipped green scales. Flowers *c.* 7mm long, tubular, split at the mouth into 4 sepals, hairy on outside; petals absent. Stamens 8, ovary topped by a single stigma. Berries egg-shaped, 8–12mm long.

HEIGHT To 100cm (200cm).
FLOWERS (Jan) Feb–Apr.
STATUS Native.
ALTITUDE 0–335m.

Spurge Laurel
Daphne laureola

Fairly common. Woods, hedgebanks and plantations, often on heavy, chalky, soils and in heavy shade; sometimes planted. ❁ A small evergreen shrub with glossy green, leathery leaves clustered towards the tips of little-branched, contrastingly pale stems. Flowers in tight clusters, fragrant, small and yellowish-green, produced early in the year. Fruit a black berry. DETAIL Hairless. Leaves alternate, drop-shaped, with a narrow base tapering to a broader, oval tip. Flowers 8–12mm long, tubular, splitting at the mouth into 4 sepals; petals absent. Stamens 8, stigma 1. Berries egg-shaped, 12mm long.

HEIGHT To 100cm (150cm).
FLOWERS Feb–Apr.
STATUS Native.
ALTITUDE Lowland.

HEIGHT 1–3m (9m)
FLOWERS Mar–Apr.
STATUS Native and introduced.
ALTITUDE Lowland.

Sea Buckthorn
Hippophae rhamnoides

Nationally Scarce as a native in dunes and coastal scrub on the E coast of England. Widely planted elsewhere, both inland and by the sea. ❀ Deciduous, spiny shrub or small tree, spreading via suckers and layering to form impenetrable thickets. *Narrow silvery leaves and orange berries distinctive.*
DETAIL Leaves alternate, strap-shaped, 2–8cm, very short-stalked; upperside dull green, dotted with glossy translucent scales, underside covered with silvery scales that turn brown with age. Flowers tiny, appearing just before the leaves in the leaf axils, male and female on separate plants; male with 2 green sepals 3–4mm long, fused at the base and covered with pale brown scales, and 4 stamens, female with 1 yellow stigma emerging from a tiny cup, no petals. Berries 6–10mm across.

female flowers

FAMILY AMARANTHACEAE: GOOSEFOOTS, ORACHES ETC.

HEIGHT To 150cm.
FLOWERS Jul–Sep.
STATUS Native.
ALTITUDE Lowland.

Shrubby Seablite *Suaeda vera*

Nationally Scarce, but very locally abundant on the coast between Lincolnshire and Dorset. Shingle banks and beaches, sea walls and the uppermost margins of saltmarshes. ❀ A salt-tolerant, dense, evergreen shrub that may form thickets, with *tiny flowers and short, succulent leaves.*
DETAIL Stems buried deeply under shingle by winter storms 'layer' and produce new shoots. Leaves cylindrical in cross-section, 5–18mm long, rounded at base and tip. Flowers 1–2 mm across, with 5 succulent green tepals, 5 stamens and 3 minute crimson stigmas. **SEE ALSO** Annual Seablite (p. 177). Small, young plants can be similar, but Shrubby Seablite has leaves that are rounded in cross-section and usually rather shorter.

Crowberry *Empetrum nigrum*

HEIGHT Stems to 120cm.
FLOWERS Mar–May.
STATUS Native.
ALTITUDE 0–1250m.

Locally common on acid moorland. ❀ Heather-like evergreen shrub, low growing, creeping through the vegetation or forming dense mats. Flowers only 1–2mm across, but fruit a *shiny black berry* (although often hard to find). Detail Hairless. Stems reddish, often rooting along their length. Leaves glossy green, needle-like, 3–7mm long, with the margins curled under, largely hiding the white underside. Flowers at the base of the leaves, with 3 sepals and 3 petals, all minute and wine red; stamens 3, with the reddish anthers at the end of *long* filaments, stigma *purplish-black, disc-like*, split into 6–9 segments. Male and female flowers usually on different plants, but mostly bisexual above 650m in Scotland (subspecies *hermaphroditum*). Fruits start green, turning red then black, 4–8mm across.

Bog Rosemary *Andromeda polifolia*

Very local, restricted to shrubby bogs, especially lowland raised bogs, growing amongst *Sphagnum* mosses. ❀ A low-growing, straggly, evergreen shrub, producing short, erect stems from a creeping, woody rhizome. The drooping, heather-like flowers grow in *loose clusters*. Detail Hairless. Leaves alternate, short-stalked, 10–40mm long, oval, but with *strongly turned-down margins* giving a narrower outline; dark green above, with a whitish bloom below

HEIGHT To 35cm.
FLOWERS Late Apr–Jun, sometimes also Sep–Oct in the lowlands.
STATUS Native.
ALTITUDE 0–735m, but mostly lowland.

and *prominently net-veined*. Flowers pale pink fading to whitish, 5–8mm long, with tiny triangular sepals and a globular corolla with 5 short teeth at the mouth. Fruit a dry capsule, but seldom sets seed.

HEIGHT 20–60cm.
FLOWERS Apr–Jun.
STATUS Native.
ALTITUDE 0–1300m.

Bilberry

Bog Bilberry

Bilberry
Vaccinium myrtillus

Abundant on dry, acid soils on heaths, moors, the drier parts of bogs, and in pine, birch and oak woods. ✿ Low-growing, untidy, *deciduous shrub* with *bright green twigs* and small, wine-red, globular flowers followed by *blue-black berries* with a *grape-like bloom*. DETAIL Hairless. *Stems ridged.* Leaves alternate, 10–30mm long, finely-toothed. Flowers 1–2 together in the leaf axils, 4–6mm long, calyx cup-shaped, *untoothed*, corolla with 5 small teeth at the mouth; ovary inferior. Berries 6–10mm across. SIMILAR SPECIES Bog Bilberry *V. uliginosum* is locally common on moorland and bogs in N Scotland, mostly in the uplands, with outliers in N England and Exmoor. Similarly deciduous, but with twigs *brownish, rounded* in cross-section, *leaves blue-green*, conspicuously *net-veined, untoothed*, flowers pale pink, urn-shaped, in clusters of 1–4, the calyx with 5 short, blunt, reddish teeth; berries as bilberry.

Cranberry *Vaccinium oxycoccos*

The berries, although tart, were formerly gathered for jams and sauces. 'Cranberry Sauce' as sold in supermarkets is made from an American species, *V. macrocarpon*, which is cultivated commercially on huge beds of wet sand.

HEIGHT Stems to 30cm (80cm).
FLOWERS Jun–Aug.
STATUS Native.
ALTITUDE 0–760m.

Locally common on bogs and wet heathland, usually amongst *Sphagnum* mosses. ✿ A small creeping shrub with thread-like stems and *tiny, long-stalked, cyclamen-like flowers*. DETAIL Leaves 5–10mm long, alternate, long-oval with down-rolled margins, dark green above, whitish below; evergreen. Flowers 1–5 together, 6–10mm across. Corolla cut into 4 bright pink petals that are pressed backwards to reveal the 8 stamens bundled around the style; filaments dull purple, anthers dull yellow. Berries 8–10mm across, red, sometimes spotted white or brown.

Cowberry
Vaccinium vitis-idaea

Locally abundant on
acid soils on moorland,
the drier parts of bogs,
and in open pine, birch
and oak woodland.
❀ Straggling *evergreen*
shrub with small to
medium-sized, oval,
leathery, dark green
leaves that are pale
green below with *fine
blackish dots*; veins
on underside *obscure*.
Flowers in clusters,
bell-shaped, *tapering*
into the
stalk,

HEIGHT To 30cm.
FLOWERS Late Apr–Jul.
STATUS Native.
ALTITUDE To 1095m.

fruit a bright red berry. DETAIL Young twigs finely hairy.
Leaves alternate, 10–30mm long, stalked, hairless, the
margins slightly inrolled. Flowers pinkish-white, 5–8mm
long, calyx split into 4 short, triangular, reddish lobes,
corolla split *halfway to base* into 4 *spreading petals*; ovary
inferior. Berries 6–10mm across, initially green.

underside

Bearberry *Arctostaphylos uva-ursi*

HEIGHT Stems to 150cm.
FLOWERS Late Apr–Jun.
STATUS Native.
ALTITUDE 0–760m (915m).

Locally common on well-drained moorland, especially in upland
areas. ❀ More or less *prostrate, evergreen* shrub, sometimes mat-
forming, with small, drop-shaped, leathery, dark-green leaves
marked with a *conspicuous network of veins* (most obvious
on underside). Flowers in clusters, urn-shaped, contracting
abruptly into the stalk, fruit a bright red berry. DETAIL Stems
often reddish, minutely hairy. Leaves alternate, 10–20mm long,
stalked, hairless. Flowers white, flushed pink, 4–6mm long,
calyx with 5 small, rounded teeth, corolla with 5 *shallow
spreading teeth*; ovary *superior*. Berries 8–10mm across.

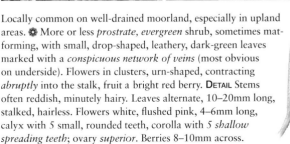

underside

Heather
Calluna vulgaris
HEIGHT To 60cm (150cm).
FLOWERS Mid Jul–Sep.
STATUS Native.
ALTITUDE 0–1095m.

Dominates large
areas of heath, moor
and bog on poor,
acid soils, mostly
either sandy or peaty,
also open woodland
and forestry rides.
✿ A low-growing,
bushy evergreen

with *tiny leaves closely packed along the stems*. The sprays of *very small*
purplish-pink flowers appear in late summer. DETAIL Leaves variably hairy,
1–2mm long, stalkless, in opposite pairs (each pair at 90° to the one
below and partly overlapping, like roof-tiles); the sides curl down and
meet underneath (thus triangular in cross-section) and there are 2 short
downward-pointing spurs at the base. Base of each flower enclosed by tiny
purplish-green bracts, *calyx split into 4 purplish-pink sepals, all longer
than the corolla*, which is 3–4.5mm long and split into 4 petals; stamens 8,
fruit a small capsule.

Bell Heather   *Erica cinerea*

Locally common. Dry heathland, forestry rides and moorland, the last
especially in rocky areas and where protected from heavy grazing.
✿ A low-growing, bushy evergreen with dark green foliage. *Purplish-red,
urn-shaped flowers larger and brighter* than Heather, coming into bloom a
little earlier in the summer. DETAIL Leaves in whorls of 3, very short-stalked,
4–7mm long, the edges curled down and meeting
underneath, hiding the white underside. Corolla
4–6mm long, split into 4 curved teeth at tip;
calyx much shorter, dark green. Stamens 8, the
anthers not projecting from the corolla, style 1.

HEIGHT To 60cm.
FLOWERS (May) Jun–Oct.
STATUS Native.
ALTITUDE 0–1210m.

Cross-leaved Heath *Erica tetralix*

Common. Damp heathland, especially where it grades into valley bogs, and wet, blanket bog in the uplands. ❀ A low-growing, bushy evergreen with leaves in whorls of 4. *Foliage greenish-grey, with dusty-pink, urn-shaped flowers*. DETAIL Young shoots hairy (often glandular-hairy). Leaves short-stalked, 2–5mm long, with abundant short hairs as well as sparse, long, gland-tipped hairs; the edges curl down and under but do not meet towards the base, revealing the pale grey-green underside. Sepals 2mm long, hairy (as leaves), corolla 5–9mm long; stamens 8, the anthers not projecting from the corolla, style 1.

HEIGHT To 70cm.
FLOWERS Jun–Sep.
STATUS Native.
ALTITUDE To 880m.

Dorset Heath *Erica ciliaris*

Nationally Rare but very locally common in its restricted range on wet heaths and wet, acid verges. ❀ Rather like Cross-leaved Heath but foliage not as grey and flowers larger and typically in *elongated spikes* rather than umbel-like clusters. DETAIL Leaves in whorls of 3(4), not downy but with long, usually glandular, hairs, the sides only weakly turned down with the underside clearly visible (whitish with a green midrib, hairless). Corolla 8–12mm long, *curved at the tip with the style projecting*, bright reddish-pink. The anthers *lack* the 2 thread-like downward-pointing basal appendages present in Cross-leaved Heath. SIMILAR SPECIES The hybrid with Cross-leaved Heath, *E. × watsonii*, is frequent. Leaves close to Cross-leaved Heath, flowers nearer Dorset Heath, although the anthers have basal appendages as in Cross-leaved, but rather shorter.

HEIGHT To 60cm.
FLOWERS Mid Jul–Oct.
STATUS Native.
ALTITUDE To 400m.

Cornish Heath *Erica vagans*

Nationally Rare. Restricted as a native to W Cornwall, mostly the Lizard, where locally abundant on heathland, and Co. Fermanagh; occasionally introduced elsewhere. ❀ Leafy spikes of beautiful, *long-stalked, open, pink or lilac, bell-shaped flowers*, appearing in late summer, distinctive. DETAIL Hairless. Leaves 5–10mm long, in whorls of 4–5, the edges rolled under and meeting underneath. Calyx very short, corolla 2.5–3.5mm long, *stamens projecting, with contrasting purplish-brown anthers*.

HEIGHT To 80cm.
FLOWERS Jul–Aug.
STATUS Native.
ALTITUDE Lowland.

Rhododendron *Rhododendron ponticum*

Locally abundant. Woodland, especially on acid soils, also colonises open hillsides. Regenerates freely from seed and can form dense thickets, excluding almost all other plants. The control and removal of Rhododendron is a major headache for conservation managers. ✿ Evergreen shrub with large, leathery, oval to oblong leaves and rounded heads of showy, bell-shaped, mauve-purple flowers. DETAIL Hairless. Leaves 8–20cm long, stalked, sticky-hairy when young. Flowers 4–6cm across. Calyx with 5 very short lobes, petals 5, fused at base. Stamens 10.

HEIGHT To 5m.
FLOWERS May–Jun.
STATUS Introduced to cultivation in 1763 and recorded from the wild by 1894 (Spain).
ALTITUDE 0–600m.

Shallon
Gaultheria shallon

Planted as game cover and naturalised in woods and heathland on acid soils. ✿ Evergreen, suckering shrub with *sprays of small, pinkish-white, heather-like flowers.* DETAIL Twigs zigzag, bristly glandular-hairy when young. Leaves alternate, hairless, leathery, 5–12cm long, finely toothed (the teeth with hair-like bristle tips), oval, rounded to heart-shaped at base and tapering to a point. Flower stalks glandular-hairy. Corolla 7–10mm long, split at the tip into 5 lobes; 10 stamens. Fruit a glandular-hairy, purplish-black berry.

HEIGHT To 1.5m.
FLOWERS May–Jun.
STATUS Introduced to cultivation in 1826 and recorded in the wild by 1914 (N America).
ALTITUDE 0–365m.

FAMILY TAMARICACEAE: TAMARISK

Tamarisk *Tamarix gallica*

Wind-tolerant and commonly planted near the sea, sometimes as hedging; may spread by suckering but rarely self-sown. ✿ *Feathery-leaved* shrub or small tree with sprays of *tiny pink flowers.* DETAIL Twigs reddish. Leaves evergreen to deciduous, reduced to tiny scales 1.5–3mm long that overlap one another on the twigs. Flowers with 5 sepals, 5 petals, 1.5–2mm long, 5 stamens and 3 styles.

HEIGHT To 3m.
FLOWERS Jul–Sep.
STATUS Introduced to cultivation by the late 16th century (W Mediterranean).
ALTITUDE Lowland.

Snowberry
Symphoricarpos albus

Planted as game cover, and commonly naturalised in woods and hedgerows; spreads by suckering, rarely via seed. ❋ Deciduous shrub with clusters of *tiny pink flowers* and distinctive *milky-white berries*. DETAIL Twigs hairless, reddish-brown. Leaves hairless to sparsely hairy, opposite, 2–6cm long, oval, often lobed on non-flowering shoots. Flowers *c.* 8mm long, corolla split at tip into 5 petals, woolly inside; 5 stamens. Berries 8–15mm across.

HEIGHT To 2m.
FLOWERS Jun–Sep.
STATUS Introduced to cultivation in 1817 and recorded from the wild by 1863 (N America).
ALTITUDE 0–385m.

Teaplants *Lycium* spp.

Naturalised in hedgerows and on walls, shingle banks and waste ground, often close to the sea. ❋ *Deciduous, suckering shrubs with purple, star-shaped flowers.* DETAIL Stems slender, arching, may be spiny. Leaves alternate, greyish-green. Corolla cut into 5 petals. Fruit an oval red berry, 10–20mm long. Two species occur. **Duke of Argyll's Teaplant** *L. barbarum*, with leaves 2–10cm long, strap-shaped, widest *around middle*. Flowers 10–17mm across, petals with 3 *little-branched* dark veins. **Chinese Teaplant** *L. chinense* (illustrated) is scarcer, with leaves widest *below middle*, flowers larger (mostly 17mm or more), corolla cut *over halfway* to base, and veins on petals *well branched*.

NIGHTSHADES: FAMILY SOLANACEAE

HEIGHT To 3m.
FLOWERS May–Sep.
STATUS Introduced in the 17th century and recorded from the wild since 1848 (China).
ALTITUDE 0–350m.

Buddleia *Buddleja davidii*

FIGWORTS, MULLEIN: FAMILY SCROPHULARIACEAE

Common in dry, disturbed places, mostly in urban and industrial areas: waste ground, railways, quarries, cliffs, walls and roofs. ❋ Deciduous shrub, a popular and familiar garden plant, with long sprays of fragrant, purple, lilac or white flowers, beloved of butterflies. DETAIL Twigs pale brown, brittle, older branches with peeling bark. Leaves opposite, short-stalked, 2.5–7cm long, long-oval, toothed, white-woolly, dark green above, whitish below. Calyx with 4 lobes, corolla split into 4 petals; 4 stamens.

HEIGHT To 5m (8m).
FLOWERS Jun–Oct.
STATUS Introduced to cultivation in the 1890s. Recorded in the wild by 1922 (China) and still increasing.
ALTITUDE Lowland.

Holly *Ilex aquifolium*

HEIGHT 3–15m (23m).
FLOWERS May–Aug.
STATUS Native.
ALTITUDE 0–600m.

Evergreen shrub or tree, increasingly common in deciduous woodland, also scrub and hedgerows; often planted. ✿ The glossy, dark-green, spiny leaves and red berries are familiar and distinctive. **DETAIL** Bark smooth, greenish. Leaves alternate, 3–12cm long, the lower always with wavy, spiny margins, but leaves out of the reach of grazing animals often spineless. Flowers in small clusters, male and female on separate bushes, 5–7mm across, sepals 4, tiny, petals 4, white, stamens 4 (reduced in female flowers, which have a disc-like, 4-lobed stigma on top of the ovary). Insect-pollinated. Fruits 6–10mm across.

FAMILY OLEACEAE: ASH, PRIVET ETC.

Wild Privet

Ligustrum vulgare

Small shrub. Fairly common in woods, scrub and hedgerows, usually on well-drained alkaline soils. Often planted, and may be bird-sown from gardens. ✿ Mostly deciduous but sometimes retaining its leaves through the winter. Privets are familiar as garden hedging, and the native Wild Privet has the same clusters of fragrant, creamy, waxen flowers and black berries as Garden Privet, but with *smaller, narrower, leaves*. **DETAIL** Buds 1.5–3mm long, dark purplish or greenish, the scales *fringed by hairs* in winter.

HEIGHT To 3m (5m).
FLOWERS Jun–Jul.
STATUS Native.
ALTITUDE 0–490m.

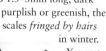

Young twigs pale brown, *densely but minutely hairy when fresh* (as are flower stalks). Leaves opposite, 3–6cm long. Flowers 4–5mm across, calyx a tiny greenish-cream cup, corolla tube 3mm long, splitting into 4 petals, stamens 2, style 1. Fruit 6–8mm across. **SIMILAR SPECIES Garden Privet** *L. ovalifolium* has larger, more broadly oval leaves, hairless bud scales, young twigs and flower stalks, and the corolla tube 5–6mm long, *longer than the lobes*.

Garden Privet   **Wild Privet**

Elder *Sambucus nigra*

HEIGHT To 10cm.
FLOWERS Jun–Jul.
STATUS Native.
ALTITUDE 0–470m.

Common in woodland, scrub, hedgerows and waste ground, especially on rich soils. ✿ A deciduous shrub or small tree with *large clusters*, 10–20cm across, of tiny, *sickly sweet-scented, whitish flowers*, followed by heavy, *drooping* bunches of purple-black berries. DETAIL Bark pale brown with prominent vertical furrows that deepen as the bark becomes more corky with age; older twigs have a white pith inside and give off an unpleasant odour when crushed. Leaves slightly hairy, cut into 2–7 pairs of finely-toothed leaflets, 3–9cm long; stipules *very small or absent*. Flowers 5mm across, corolla split into 5 petals, stamens 5, with *creamy-white anthers*, stigmas 3–5. Berries 6–8mm in diameter, rarely greenish-yellow, edible.

Dwarf Elder
Sambucus ebulus

HEIGHT 60–120cm.
FLOWERS Jul–Aug.
STATUS Ancient introduction.
ALTITUDE Lowland.

Very local in hedgerows and on waste ground, creeping via rhizomes to form dense stands. ✿ *Easily mistaken for an umbellifer* (pp. 296-311). Flower clusters as Elder, but smaller (7–10cm across), and produced in *midsummer*, followed by *erect* bunches of green berries that ripen to black (but rapidly stripped by birds); flowers often washed purplish-pink on outside, with *reddish-purple* anthers. DETAIL Stems robust, grooved and unbranched, but dying down in winter. Leaves slightly hairy, rather foul-scented when bruised, cut into (2)3–8 pairs of strap-shaped, finely toothed, pointed leaflets, 6–8cm long. Conspicuous *oval, leaf-like stipules* at the junction of the leaf-stalk and stem. Berries mildly toxic.

Guelder Rose *Viburnum opulus*

HEIGHT 2–4m.
FLOWERS May–Jul.
STATUS Native.
ALTITUDE 0–400m.

Deciduous shrub, locally common in woods, scrub, hedgerows, willow and alder carr and the banks of rivers and streams; favours moist, neutral to alkaline soils. Sometimes planted as an ornamental or in hedgerows, and also bird-sown from gardens. ❀ Large, showy, *hydrangea-like flower clusters* are followed by *startlingly red berries*, while the palmately-lobed leaves turn red in autumn. **DETAIL** Twigs greyish or reddish, hairless. *Leaves opposite*, 5–8cm long, deeply cut into 3 lobes with coarsely and irregularly toothed margins; hairless above, very finely hairy below. Flower stalks with a pair of tiny disc-like glands near the tip. Outer flowers large (15–20mm across) and sterile, lacking stigmas and stamens; central flowers fertile but smaller, 5–6mm across. Calyx with 5 tiny teeth, corolla divided into 5 petals, 5 stamens with creamy-yellow anthers. Fruit a shiny red berry 8–11mm across; the fruit is yellow in some cultivars.

Wayfaring Tree *Viburnum lantana*

HEIGHT 2–6m.
FLOWERS Late Apr–Jun.
STATUS Native.
ALTITUDE Lowland.

Deciduous shrub, locally common in woods, scrub, hedgerows, mostly on chalk and limestone. Sometimes planted as an ornamental or in 'native' hedges well away from the native range. ❀ Leaves oval, finely toothed, usually pointed and *conspicuously veined*. Flowers in dense clusters, sickly fragrant, creamy, followed by the berries, which turn red then purplish-black – there is often a *characteristic mixture of colours*. **DETAIL** Twigs pale brown, downy. Leaves opposite, 5–10cm long, oval, regularly toothed, sparsely hairy above, *silvery below with sparse woolly hairs*. Flowers all fertile, 5–6mm across, calyx hairless to sparsely hairy, with 5 triangular teeth, corolla divided into 5 petals; 5 stamens, with creamy-yellow anthers. Fruit a *flattened berry*, c. 8mm long.

underside

upperside

Traveller's Joy *Clematis vitalba*

Locally common. Scrub, hedgerows, woodland margins, railway banks, dunes, mostly on chalk, limestone or other alkaline soils. ❀ A deciduous woody climber, sprawling over the vegetation and *twining around supports with its leaf stalks*. The *creamy flowers, c.* 20mm across, are delicately scented. Easily picked out in autumn and winter by the *fluffy seed heads*, which give it the alternative name, 'Old Man's Beard'. DETAIL Leaves opposite, silky-hairy when young but soon sparsely hairy, cut into 3(5) leaflets, 3–10cm long, oval, tapering to a pointed tip and rounded to heart-shaped at the base, variably lobed or toothed. Flowers in dense clusters, with *4 hairy, creamy to greenish-white sepals*, c. 10mm long, and numerous stamens and carpels; petals absent. In fruit the style elongates into a feathery white plume.

GROWTH Perennial.
HEIGHT To 30m.
FLOWERS Jul–Sep.
STATUS Native.
ALTITUDE 0–305m.

Hop *Humulus lupulus*

HOP: FAMILY CANNABACEAE

'cones'

female

male

GROWTH Perennial.
HEIGHT Shoots to 8m.
FLOWERS Jul–Aug (Sep).
STATUS Native.
ALTITUDE Lowland.

Locally common. Wet woodland, hedgerows and scrub on moist soils. Also a frequent escape, both in the native range and northwards into Scotland. ❀ A vigorous climber, scrambling up through other vegetation in a clockwise direction. Stem and leaves *roughly hairy*, with the leaves deeply cut into 3–5 coarsely-toothed lobes. Flowers tiny, the female flower heads *enlarging in fruit to form 'cones' 3–5cm long* (to 10cm in cultivated varieties); the cones are used to flavour beer. DETAIL Leaves opposite, dark green, 10–15cm long, with yellow, disc-shaped, stalkless glands below. Male and female flowers on separate plants. Male with 5 pale green sepals and 5 stamens in loose, branched clusters. Female flowers with 5 minute scales and an ovary with 2 styles enclosed by a bract, grouped in globular heads 15–20mm long. The bracts enlarge greatly and eventually turn pale brown to form the 'cone'.

White Bryony
Bryonia dioica

male

GROWTH Perennial.
HEIGHT Stems to 5m.
FLOWERS May–Sep.
STATUS Native.
ALTITUDE Lowland.

Fairly common. Hedgerows, woodland borders and scrub, on well-drained, often alkaline soils. ❀ Climbs via unbranched tendrils that grow from the leaf axils and *coil into spring-like clockwise spirals*. Clusters of *greenish-white flowers (delicately veined darker)* are followed by berries that ripen from green through yellow to red. DETAIL Roughly hairy. Grows from a large, tuberous rootstock. Leaves alternate, 5–8cm long, deeply lobed, roughly hairy. Calyx green, with 5 teeth, corolla cut into 5 petals. Male and female flowers on separate plants, male 12–18mm across with 3 stamens, female 10–12mm across with a single style, the disc-like tip split into 3 stigmas. Berries 5–9mm across.

Black Bryony *Tamus communis*

Common. Hedgerows, woodland edges and waste ground, on a variety of soils. ❀ A twining climber (twines clockwise), without tendrils, with *glossy, 'ace of spades' leaves* and sprays of *very small greenish-white flowers* followed by *trailing strings of red berries.* DETAIL Hairless. Grows from a large, tuberous rootstock. Leaves alternate, 5–15cm long, stalked. Flowers 3–6mm across, bell-shaped, split into 6 yellowish-green sepals. Male and female flowers on separate plants, male with 6 stamens, female with a single stout style that is split at the tip into 3 down-curved stigmas, these in turn with 2-lobed tips. Berries 10–13mm across.

GROWTH Perennial.
HEIGHT Stems to 5m.
FLOWERS Late May–Jul.
STATUS Native.
ALTITUDE Lowland.

male

male

Ivy *Hedera helix*

Common. Woodland, hedgerows, walls and rocky places. ❀ Familiar evergreen climber (may also sprawl on the ground), supporting itself with *numerous short roots along the stems* and capable of covering the tallest tree. *Flowers in the autumn, followed by purplish-black berries in the spring.* **Detail** Young twigs downy. Leaves 4–10cm long, shiny dark green, sometimes with pale marbling, those on creeping and climbing stems 3–5 lobed, often cut more than half way to the base, those on flowering stems more diamond-shaped and not lobed.

GROWTH Perennial.
HEIGHT Shoots to 30m plus.
FLOWERS Sep–Nov.
STATUS Native.
ALTITUDE To 610m.

non-flowering stem

flowering stem

Flowers in umbels, with 5 tiny sepals, 5 yellowish-green petals, 3–4mm long, 5 stamens and 1 style. **Similar Species** Atlantic Ivy *H. hibernica* is native to W Britain and Ireland but introduced almost everywhere. Leaves typically larger and less deeply lobed, rarely with paler marbling, but identification confirmed by the star-shaped hairs on the young leaves of creeping or climbing stems (10× lens), which all *lie flat to the surface* and are often dirty yellow; in Ivy the hairs are *whitish*, with the *rays of the star spreading in 3 dimensions*.

HONEYSUCKLE, ELDERS ETC: **Family CAPRIFOLIACEAE**

Honeysuckle *Lonicera periclymenum*

GROWTH Perennial.
HEIGHT Stems to 6m (10m).
FLOWERS Jun–Sep.
STATUS Native.
ALTITUDE 0–610m.

Common in hedgerows, scrub and woodland on a variety of soils. ❀ A deciduous climber, the older stems woody, with peeling bark, the younger shoots purplish. Flowers unmistakable and very fragrant, especially in the evening, followed by tight clusters of sticky red berries. **Detail** Leaves hairy, opposite, 3–7cm long, oval, short stalked. Flowers stalkless, in whorls at the tip of the stems, variably glandular-hairy. Calyx very small, with 5 teeth, corolla a long tube (40–50mm long), split at the tip into a 4-lobed upper lip and 1-lobed lower lip, creamy, variably washed red. Stamens 5, projecting from the corolla tube, as does the single style, tipped with a lobed stigma.

INDEX OF ENGLISH AND SCIENTIFIC NAMES

gentians 196–197	periwinkles 198	buglosses etc. 199–201	comfreys 202–203	Hound's Tongues 204	forget-me-nots 205–207
Dodder 208	bindweeds 208–209	nightshades 210–212	foxgloves 213	speedwells 214–218	fluellens 219
toadflaxes 220–221	plantains 222–223	figworts 224–225	mulleins 226	dead-nettles and mints 227–239	
dead-nettles and mints 227–239		monkeyflowers 240	eyebrights 241	Yellow Rattle, bartsias 242	louseworts 243
broomrapes 244–246	butterworts 247	Bogbean 248	Vervain 248	bellflowers 249–252	thistles etc. 253–260
blue daisies 261	yellow daisies 262–291	cudweeds 272–273	Daisy, fleabanes 276–277	Mugworts 279	Hemp Agrimony 281